# Polymeric Drugs & Drug Delivery Systems

# Polymeric Drugs & Drug Delivery Systems

EDITED BY

## *Raphael M. Ottenbrite, Ph.D.*

Department of Chemistry
Virginia Commonwealth University

## *Sung Wan Kim, Ph.D.*

Department of Pharmaceutics and Pharmaceutical Chemistry
University of Utah

Routledge
Taylor & Francis Group

LONDON AND NEW YORK

# Polymeric Drugs and Drug Delivery Systems

First published 2001 by Technomic Publishing Company, Inc.

Published 2018 by Routledge
2 Park Square, Milton Park, Abingdon, Oxon, OX14 4RN
52 Vanderbilt Avenue, New York, NY 10017

*Routledge is an imprint of the Taylor & Francis Group, an informa business*

First issued in paperback 2019

Main entry under title:
   Polymeric Drugs and Drug Delivery Systems

Bibliography: p.
Includes index p. 311

Library of Congress Catalog Card No. 00-109967

ISBN 13: 978-0-367-45542-2 (pbk)
ISBN 13: 978-1-56676-956-3 (hbk)

# Table of Contents

## 22. BIOACTIVE POLYMERIC COMPOSITES BASED ON HYBRID AMORPHOUS CALCIUM PHOSPHATES ........ 301

JOSEPH M. ANTONUCCI, DRAGO SKRTIC,
ARTHUR W. HAILER, and EDWARD D. EANES

# Preface

OVER the last quarter century we have seen the field of biorelated polymers for medical applications undergo a dramatic transition from the pragmatic and/or serendipic approach to applying basic research principles. Specifically, we have seen the development of many new polymeric materials for intended applications and solutions to problems related to those applications. The development during this time has been dynamic with the consistent emergence of new findings. Consequently, one can anticipate a literal explosion of new clinical products and applications that will be derived from this multidisciplinary field in the next millenium.

The basis of this book is to expose the reader to the important areas of synthetic biorelated polymers systems and the potential impact they will have in the 21st Century. Consequently, we deliberated over the appropriate areas to be covered in this book, what value these would provide, and who could benefit. The chapters are written to emphasize the chemical and physical properties of several unique polymer systems and the many stages involved in their physiological adaptations to achieve an intended utilization. The importance of multidisciplinary knowledge and skills are unprecedented since the field encompasses chemistry, materials science, engineering, biochemistry, biophysics, pharmacology, physiology, and clinical studies.

There are 22 chapters in the book and they cover the most important aspects of polymers as drugs, prodrugs, drug delivery systems, and in situ prostheses. The major features promulgated are synthesis, derivatization, degradation, characterization, application, and evaluation techniques as well as new biodegradable materials, assemblies, hydrogels, telechelic polymers, derivatized polysaccharides, micro- and nanoparticles, mimetic protein networks, and interpenetrating polymers. Polymer drug design for enhanced physiological drug distribution, drug targeting, time-controlled release, and sensor-responsive release are also presented. In addition, accounts are given on in situ probes, microparticle diagnostic agents, and sensor devises.

We wish to thank the contributors to this publication who are outstanding representatives of the multidisciplinary sciences necessary to so fruitfully accomplish the work that has been so elegantly described.

# CHAPTER 1

# Semitelechelic Poly[N-(2-hydroxypropyl)-methacrylamide] for Biomedical Applications

ZHENG-RONG LU[1]
PAVLA KOPEČKOVÁ[1]
JINDRICH KOPEČEK[1]

## INTRODUCTION

SEMITELECHELIC (ST) polymers are low molecular weight ($M_n < 20,000$) linear macromolecules having a reactive functional group at one end of the polymer chain and one terminal end. The single functional group is able to conjugate or graft the macromolecules to other molecular species or surfaces. The modification of therapeutic proteins or biomedical surfaces with the synthetic polymers improves their properties. For example, the covalent attachment of macromolecules to therapeutic proteins results in an increase of their resistance to proteolysis, reduction of their antigenicity, and prolongation of intravascular half-life [1–3]. The modification of enzymes with synthetic polymers permits their use as catalysts of organic reactions in organic solvents due to increased stability [4,5]. Modification of biomedical surfaces may reduce their biorecognizability and/or prevent protein adsorption [6,7]. Due to the multifunctionality of proteins or surfaces, the polymers containing multifunctional groups may lead to cross-linking or loop formation; semitelechelic polymers seem to be the macromolecules of choice. The ST polymers permit one point attachment and avoid the cross-linking of the substrates caused by multipoint attachment. The water-soluble ST polymers can also be conjugated to hydrophobic anticancer drugs to form more soluble, more easily formulated, and deliverable drugs [8].

[1]Departments of Pharmaceutics and Pharmaceutical Chemistry/CCCD, and of Bioengineering, University of Utah, Salt Lake City, UT, 84112, USA.

Methoxy poly(ethyleneglycol) (mPEG) was the most frequently used semitelechelic polymer for over 2 decades. It has been successfully used for the modification of various proteins, biomedical surfaces and hydrophobic anticancer drugs (for reviews see References [3,9,10]. Recently, a number of new semitelechelic (ST) polymers, such as ST-poly(N-isopropylacrylamide) (ST-PNIPAAM) [11–15], ST-poly(4-acryloylmorpholine) (ST-PAcM) [16], ST-poly(N-vinylpyrrolidone) (ST-PVP) [17], and ST-poly[N-(2-hydroxypropyl)methacrylamide] (ST-PHPMA) [18–21] have been prepared and shown to be effective in the modification of proteins or biomedical surfaces.

R = reactive functional group

ST-PVP          ST-PAcM          ST-PNIPAAM          ST-PHPMA

HPMA copolymers are water-soluble biocompatible polymers, widely used in anticancer drug delivery (reviewed in Reference [22]). HPMA copolymers containing reactive groups at side-chain termini were previously used for the modification of trypsin [23], chymotrypsin [23,24], and acetylcholinesterase [25]. The modification dramatically increased the acetylcholinesterase survival in the blood stream of mice and the thermostability of modified enzymes when compared to the native proteins. However, the modification involved multipoint attachment of the copolymers to the substrates, which may cause crosslinking. To modify proteins or biomedical surfaces by one point attachment, semitelechelic polymers should be used.

## SYNTHESIS OF SEMITELECHELIC HPMA POLYMERS

Semitelechelic HPMA polymers were synthesized by free radical polymerization of HPMA using 2,2′-azobis(isobutyronitrile) (AIBN) as the initiator and alkyl mercaptans as chain transfer agents. Alkyl mercaptans with different functional groups, namely, 2-mercaptoethylamine, 3-mercaptopropionic acid, 3-mercaptopropionic hydrazide, and methyl 3-mercaptopropionate, were used as the chain transfer agents; ST HPMA polymers

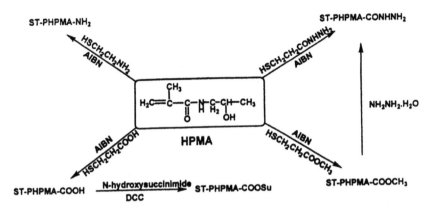

**Scheme 1** Synthesis of ST-PHPMA polymers [20].

with amino, carboxy, hydrazo, and methyl ester end groups, respectively, were prepared (Scheme 1) [20]. The functional end groups of the ST HPMA polymers can be transformed to other active functional groups. For example, ST-PHPMA-COOCH₃ was transformed to ST-PHPMA-CONHNH₂ by the reaction with an excess of hydrazine [20]. Semitelechelic polymers with *N*-hydroxysuccinimide (HOSu) ester end groups ST-PHPMA-COOSu were synthesized by esterification of ST-PHPMA-COOH with a large excess of *N*-hydroxysuccinimide with dicyclohexyl carbodiimide (DCC) as coupling agent. The ST-PHPMA-COOSu will react with amino groups on proteins and on biomedical surfaces. The molecular weight of the ST-PHPMA after polymer analogous esterification did not change, confirming the assumption that the secondary OH groups of the HPMA monomer were not reactive under the experimental conditions used [20].

The free radical polymerization of HPMA in the presence of mercaptans involves two different initiation mechanisms (Scheme 2) [26]. One is the initiation by RS• radicals from chain transfer agent; the other appears to be the direct initiation by the primary isobutyronitrile (IBN) radicals formed by the decomposition of AIBN [27]. The RS• are formed by either the free radical transfer reaction of alkyl mercaptans with the IBN radicals or the chain transfer reaction of an active polymer chain with the mercaptans. The initiation by the RS• radicals produces the ST polymers with a functional group at one end of the polymer chain. The initiation by IBN radicals leads to nonfunctional polymer chains with an IBN end group. The presence of the polymers with IBN end groups effects the purity and the functionality of ST polymers. As expected, the production of nonfunctionalized polymer chains is affected by reaction conditions. The polymerization is mainly terminated by chain transfer reaction with the mercaptans, but other termination mechanisms, such as disproportionation and recombination, take place depending on the reaction conditions [26].

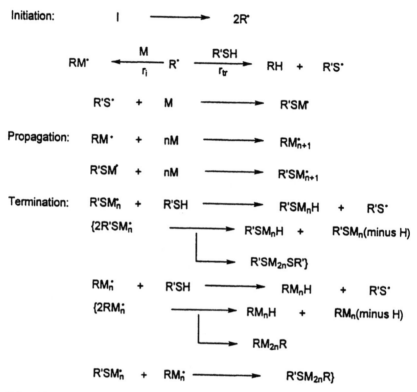

**Scheme 2** The mechanism of the chain transfer free radical polymerization of HPMA in the presence of alkyl mercaptans [26].

The concentrations of both chain transfer agent and initiator are important for the polymerization when the concentration of HPMA is constant. The molecular weight of the ST HPMA polymers was regulated by the concentration ratio of mercaptan (S) to HPMA (M), according to the Mayo equation, $1DP_{n(end)} = 1/DP_{n,o} + C_s[S]/[M]$ [27]. $DP_{n(end)}$ and $DP_{n,o}$ are the number average degrees of the polymerization in the presence and in the absence of the chain transfer agents S; $C_s$ is the chain transfer constant; $C_s = k_u/k_p$, where $k_u$ is the rate constant for chain transfer and $k_p$ is the rate constant for propagation. Figure 1 shows the dependence of the molecular weight of the ST PHPMA polymers on the concentration ratio of mercaptans to HPMA. High [S]/[M] produces ST HPMA polymers with low molecular weight and higher purity, and vice versa. The concentration ratio of AIBN to HPMA does not have a significant effect on the chain length of the polymers in the presence of chain transfer agents but has an impact on the purity of the ST HPMA polymers. A low [AIBN]/[HPMA] ratio produces ST HPMA polymers with high purity, i.e., a smaller content of polymers with IBN end

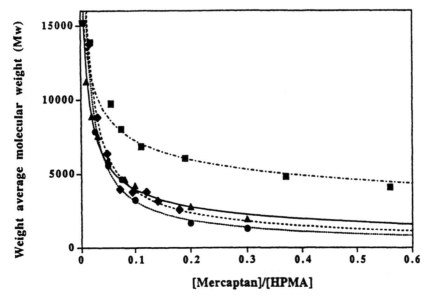

**Figure 1** The effect of concentration ratio of mercaptans (■, 2-mercaptoethylamine; ♦, methyl 3-mercaptopropionate; ▲, 3-mercaptopropionic acid; ●, 3-mercaptopropionic hydrazide) to HPMA on the weight average molecular weight (SEC) of ST HPMA polymers (data from Reference [20]).

groups, and a narrower molecular weight distribution [26]. The dependence of the relative content of nonfunctional macromolecules on the concentrations of both mercaptan and initiator suggested that there might be a competition between the initiation by IBN radicals and the free radical transfer reaction of the radicals with mercaptans.

The efficiency of the alkyl mercaptans also has an influence on the chain transfer polymerization. The chain transfer constants of 2-mercaptoethylamine, 3-mercaptopropionic acid, methyl 3-mercaptopropionate, and 3-mercaptopropionic hydrazide are 0.08, 0.34, 0.38, and 0.8, respectively [20]. This is reflected in the molecular weights obtained, the lower the $C_S$ the higher the molecular weight of the ST-PHPMA at a constant [S]/[HPMA] ratio. 2-Mercaptoethylamine produced higher molecular weight polymers at the same [S]/[HPMA] ratio compared with the other mercaptans (Figure 1). The efficiency of chain transfer agent also affects the production of polymer chains with IBN end groups. At identical [S]/[HPMA] ratios, the most efficient chain transfer agent, 3-mercaptopropionic hydrazide, produces fewer polymer chains with IBN end groups compared to the other mercaptans [19].

By analysis of the ST HPMA polymers by MALDI-TOF MS in the (more sensitive) refectron mode it was possible to recognize macromolecules with small mass differences formed by different termination mechanisms. For example, ST HPMA polymer chains with a mass difference of 2 Da, formed by disproportionation termination were observed in the mass spectra. As expected, the polymerization was mainly terminated by chain transfer reaction, and only a very small amount of active polymer chains terminated by disproportionation. The polymerization can be terminated by recombination when the concentration of mercaptan was very low; in fact, macromolecules terminated by recombination were found in the MALDI-TOF mass spectrum of a ST-polymer produced at a very low molar ratio of mercaptan to HPMA [26].

Because some of the macromolecules were initiated by the primary radicals to form macromolecules with initiator end groups, the functionality of ST HPMA polymers could be improved by using an initiator containing the same functional group as the chain transfer agent.

## CHARACTERIZATION OF SEMITELECHELIC HPMA POLYMERS

Several different methods were used to characterize the ST HPMA polymers. The average molecular weights of the ST polymers were determined by size exclusion chromatography (SEC). The functional end groups of ST HPMA polymers were determined by different chemical assays based on the properties of the end groups. The amino groups in ST-PHPMA-NH$_2$ were determined by ninhydrin and trinitrobenzene sulfonic acid (TNBS) assays. The carboxyl groups of ST-PHPMA-COOH were determined by titration. The methyl ester groups in ST-PHPMA-COOCH$_3$ were determined by proton NMR and by hydrolysis with excess KOH followed by titration of the remaining KOH with HCl. The hydrazo end groups in ST-PHPMA-COONHNH$_2$ were determined by the TNBS assay [20].

The polymers were also characterized by MALDI-TOF MS [28]. This new technique can accurately determine the molecular weight of a protein and a polymer chain and provide important information on the structure of repeating units and end-group compositions of synthetic polymers [29,30]. In contrast, the conventional SEC can only provide the molecular weight distribution; in addition, calibration standards with similar structures are required for the calculation of the average molecular weights. Currently, mass spectrometry is able to measure the mass of a macromolecule up to $10^6$ Da. For synthetic polymers, however, the new technique could not provide accurate molecular weight distribution data for the polymers with a broad polydispersity (*PD*). The method underestimated peaks corresponding to higher masses, resulting in lower values of the average molecular weights compared with SEC. Consequently, the average molecular weights of the ST HPMA polymers (*PD* > 1.1) calculated from the mass spectra were lower than those determined by SEC. However, the average molecular weights determined by MALDI-TOF MS for the narrow dispersed polymers showed good agreement with those obtained by SEC. For example, there was a good agreement between the average molecular weights of fractionated ST-PHPMA (*PD* < 1.1) determined by the MALDI-TOF MS and the SEC, respectively. The average molecular weights of ST-PHPMA fractions (*PD* < 1.1) obtained from the MS were only slightly lower than those from SEC, and the difference between two methods was less than 7% [20]. The mass spectrometry is a fast, precise method to determine the molecular weight of protein–polymer conjugates; it is difficult to obtain suitable standards for the calibration of SEC to characterize such conjugates.

Although the MALDI-TOF MS has a limitation to determine the average molecular weight of synthetic polymers with high polydispersity, the accurate determination of the molecular weight of individual macromolecules provides important information on the chemical structure of the end groups and the composition of individual polymer chains of semitelechelic polymers. Shown in Figure 2 is a MALDI-TOF mass spectrum of a ST-PHPMA with COOCH$_3$ end groups. Two main peak series corresponding to macromolecules with different end groups in the polymer sample can be identified. The mass increment between peaks with identical end groups was 143.2, the mass of a HPMA monomer unit. Peak series a) at $m/z = n \times 143.2 + 23(Na^+) + 68.1 + 1.0$ [$n$ is the number of monomer units;

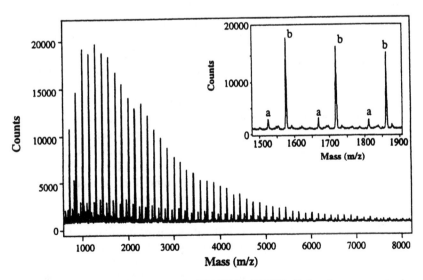

**Figure 2** MALDI-TOF mass spectrum of a ST-PHPMA-COOCH$_3$. Peak series a represent polymer chains with initiator (IBN) end groups; peak series b represent polymer chains with methyl ester end groups.

mass of the initiator residue $(CH_3)_2[CN]C$ (IBN) is 68.1] corresponds to polymer chains initiated by IBN radicals and terminated by proton transfer from methyl 3-mercaptopropionate (MMP). The peak series b) at $m/z = n \times 143.2 + 23(Na^+) + 120.2$ (molar mass of methyl mercaptopropionate $= 120.2$ Da) represents the semitelechelic polymer chains H-(HPMA)$_n$-SCH$_2$CH$_2$COOCH$_3$ initiated by the radical formed from MMP (R'S') and terminated by proton transfer from MMP. The MALDI-TOF mass spectra of ST HPMA polymers clearly showed the presence of polymer chains formed by the direct initiation with primary IBN radicals. Consequently, the polymer chains with IBN end groups were found in the mass spectra of almost all the ST HPMA polymers. As discussed above, the relative peak intensity (reflecting the relative content) of the polymer chains with IBN end groups in the mass spectra of the ST HPMA polymers varied with the reaction conditions and the chain transfer constant of particular mercaptan.

## MODIFICATION OF PROTEINS WITH SEMITELECHELIC HPMA POLYMERS

The modification of therapeutic proteins with synthetic polymers is one of the methods to enhance their pharmacological activity. Synthetic polymers can be conjugated to proteins by reacting the active functional groups of the polymers with the functional groups of the protein, such as amino, carboxyl, hydroxyl, and sulfhydryl groups. The most frequently used method is the modification of the amino groups of the proteins containing synthetic polymers with active ester groups. There are also reports on the modification of the hydroxyl [10] and sulfhydryl [31] groups of the proteins. Carboxyl group modification is not frequently used [10,32], although the carboxyl group is a common group in proteins. Here, we focus the discussion on the modification of the amino and carboxyl groups of the proteins.

$\alpha$-Chymotrypsin (CT) was used as the model protein to evaluate the consequences of its modification with the ST HPMA polymers. There are 17 carboxyl groups and 17 amino groups in $\alpha$-chymotrypsin. The amino and carboxyl groups of the protein were modified with ST-PHPMA-COOSu and ST-PHPMA-CONHNH$_2$, respectively. The amino-directed modification of CT was performed by directly reacting the protein with excess ST-PHPMA-COOSu at neutral pH (7.0–7.5) and 4°C in 20 mM in CaCl$_2$ aqueous solution. The carboxyl group-directed modification with polymers containing hydrazo was performed by reacting the protein with an excess of ST-PHPMA-CONHNH$_2$ at pH 4.5–5.0 in the presence of 1-ethyl-3-(3-dimethylaminopropyl)carbodiimide hydrochloride (EDC). At the acidic pH used, the amino groups (p$K_a$ = 6.8–8.0 for $\alpha$-amino, 10.4–11.1 for $\epsilon$-amino of lysine) on proteins are deactivated due to protonation. However, the hydrazo groups ($pK_a \approx 3.0$) remain active to react with the carboxyl groups of the same proteins in the presence of a coupling agent (EDC). The pH of the reaction mixture during both modifications may change, and dilute NaOH or HCl should be added to maintain suitable pH [20].

The conjugates were characterized by MALDI-TOF mass spectrometry, and the mass spectrum of a carboxyl-modified CT conjugate is shown as an example (Figure 3). The mass spectrum showed a broad peak for the conjugate because of the random conjugation of different molecular weight macromolecules to the enzyme. Nevertheless, the molecular weights of the conjugates can be calculated from the molecular weight distribution in the corresponding mass spectrum. The modification of CT with narrow fractions of ST HPMA polymers produced conjugates possessing a uniform structure compared with those prepared from unfractionated semitelechelic macromolecules [20].

The conjugation degree or the number of polymer chains on each protein molecule depends on the molecular weight of the polymers and on the conjugation conditions. Lower molecular weight polymers produced conjugates with a higher conjugation degree. This is probably due to a smaller hydrodynamic volume resulting in a lower steric exclusion effect. A higher concentration ratio of ST-PHPMA to CT gave conjugates with a higher conjugation degree when same molecular weight polymers were used for the conjugation. When the polymer enzyme molar ratio was low in the carboxyl-directed modification, a much larger excess of coupling agent was necessary to achieve a relatively high conjugation degree [20].

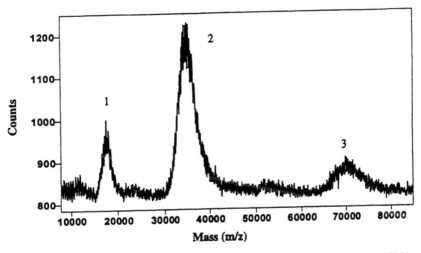

**Figure 3** The MALDI-TOF mass spectrum of the chymotrypsin conjugate with a ST-PHPMA-NHNH$_2$ fraction ($M_w = 1,400$). The peaks 1, 2, and 3 are the double-charged, single-charged conjugate, and single-charged double conjugate aggregate, respectively.

It seems that a maximum of about 10 ST-PHPMA chains can be attached to 1 CT molecule. The conjugation degree was lower than that of the chymotrypsin conjugates with PEG-SC (succinimidyl carbonate of mPEG; up to 14 chains per one CT molecule) [20,33]. This might be attributed to the solution structure difference between the two polymers. PHPMA has a random coil structure in aqueous solution, whereas PEG possesses a more extended one. The coiled PHPMA may cover more of the enzyme surface, and the steric effect may prevent more PHPMA macromolecules from attaching to the enzyme.

The chemistry of the modification was important for the biological activity of the protein–polymer conjugate. Listed in Table 1 are the Michealis-Menten kinetic constants of the enzymatically catalyzed cleavage of Z-Gly-Leu-Phe-NAp (Z, benzyloxycarbonyl; NAp, p-nitroanilide) by the CT conjugates and the native enzyme. The carboxyl group modified conjugates I and II showed a lower activity than native CT [20]. The amino group modified conjugates III, IV, V, and VI showed higher reactivities than native CT, similarly to literature data [21,32]. This indicates that the modification mode of the protein affects the activity of the conjugates. Similar results have been also reported by other research groups. Sakane and Pardridge [33] have shown that carboxyl-directed pegylation of brain-derived neurotrophic factor preserved the biological activity of the conjugate, whereas the amino group-directed modification did not. Pettit et al. [34] have shown that amino group-directed pegylation of interleukin-15 alters the biological activity of the conjugate. How the modfcation mode affects the biological activity of the conjugates is not clear; nevertheless, modification of proteins with synthetic polymers definitely alters the conformation of the protein, and the

TABLE 1. Enzymatic Activity of the ST HPMA Polymer Modified Chymotrypsin in the Cleavage of Z-Gly-Leu-Phe-NAp.

| Conjugate | Bond Mode[a] | Conjugate $M_n$[b] | Conjugate $M_w$[b] | Np[c] | $k_{cat}/K_M(M^{-1}S^{-1})$ |
|---|---|---|---|---|---|
| Native Chymotrypsin | | | | | 2,140 |
| I | P-NHNHCO-E | 32,300 | 32,500 | 3 | 1,070 |
| II | P-NHNHCO-E | 46,600 | 47,800 | 8 | 1,100 |
| III | P-CONH-E | 40,900 | 41,700 | 6 (5.8) | 4,200 |
| IV | P-CONH-E | 44,900 | 46,600 | 3.6 (3.7) | 4,400 |
| V | P-CONH-E | 49,400 | 51,000 | 6.5 (7.3) | 4,200 |
| VI | P-CONH-E | 36,100 | 37,800 | 4.3 (4.3) | 4,000 |

[a]P represents polymer, and E the enzyme.
[b]Determined by MALDI-TOF MS.
[c]The number of polymer chains attached to each enzyme molecule; the data in the parentheses were obtained from TNBS assay (data from Reference [20]).

charge character of protein may be also changed due to the modification. In the case of CT, it is known that a carboxyl group of an aspartic acid is involved in the active site of chymotrypsin [35]. The conjugation of some of the carboxyl groups might be the cause for the decrease in the activity of the carboxyl group-modified chymotrypsin conjugates. It appears that the conjugation degree and the molecular weight of the polymers did not have a pronounced effect on the activity of chymotrypsin-ST HPMA conjugates toward Z-Gly-Leu-Phe-NAp.

For high molecular weight substrates, for example, PHPMA-Gly-Leu-Phe-NAp, there was not much difference in the activity of both carboxyl group modified conjugates and the amino group modified conjugates. The activity of all the conjugates mentioned above exhibited lower activity than CT. In this case, the steric hindrance of the polymer chains of the polymer substrate and the conjugates dominates and makes the formation of enzyme substrate more difficult, resulting lower activity of the conjugates [20].

## MODIFICATION OF BIOMEDICAL SURFACES
## WITH SEMITELECHELIC HPMA POLYMERS

Semitelechelic HPMA polymers also effectively modify biomedical surfaces. ST-PHPMA-NH$_2$ of different molecular weights was used to modify the surface of nanospheres based on a copolymer of methyl methacrylate, maleic anhydride, and methacrylic acid [18]. The polymer chains were covalently attached to the surface; the efficiency of the polymer binding to the surface and the thickness of the coating layer depended on the molecular weight of the polymers. High molecular weight polymers gave thicker coatings but relatively low efficiency of binding. However, the thickness of the ST PHPMA layer was less compared with commonly used PEG based on the molecular weight. This was attributed to the fact that PEG has an extended conformation in water. However, the random coiled PHPMA chain may cover more surface space similar to modified proteins; the occupied area per PHPMA molecule is around 150 A$^2$ [18].

The modification of biomedical surfaces with ST HPMA polymers decreased the biorecognizability of surfaces. The surface modification reduced the protein adsorption to the nanospheres compared with the unmodified nanospheres. The thickness of the coating layer and/or the molecular weight of polymers also affected the protein adsorption; consequently, a thicker coating resulted in lower protein adsorption [18]. The protein repulsion of the modified surface may be caused by the change of the surface energy [36] and surface charge after the modification. The modified nanospheres had an increased intravascular half-life after intravenous administration to rats, and the intravascular half-life increased with the increase of the molecular weight of ST PHPMA. The accumulation of the modified nanospheres in the liver decreased in a molecular weight–dependent manner. The higher the molecular weight of the polymers, the lower the accumulation of the modified nanospheres in the liver. The molecular weight dependence of the biorecognition seems to indicate the influence of the hydrodynamic thickness of the coating layer on the process of opsonization and capture by Kupffer cells of the liver and macrophages of the spleen [18].

The modification of the nanospheres with ST HPMA polymers significantly changed the surface structure and property of the nanospheres, which resulted in substantial changes in the biorecognizability and biodistribution of the nanospheres. The biocompatibility of HPMA polymers bodes well for the future application of ST PHPMA in the modification of biomedical surfaces.

## CONJUGATION WITH HYDROPHOBIC ANTICANCER DRUGS

HPMA copolymers are well-known carriers for anticancer drugs; two HPMA copolymer-adriamycin conjugates are now in clinical trials [37]. The ST HPMA polymers are also promising in the modification of anticancer drugs. Usually, the anticancer drugs are hydrophobic organic compounds; their poor aqueous solubility diminishes their bioavailability and therapeutic efficacy. Their conjugation to biocompatible water-soluble polymers can increase their water solubility and thereby improve the therapeutic index. For example, taxol, a poorly soluble anticancer drug, has been conjugated to one end of PEG via an ester bond to increase its water solubility [8,39]. The ester bond between the polymer and the drug can be hydrolyzed to release the drug *in vivo*. The water-soluble ST HPMA polymers can also be used for the conjugation of anticancer drugs via the functional end groups. The conjugation of anticancer drugs to ST HPMA polymer not only provides good solubility and a controlled drug release mode but has the potential to overcome multidrug resistance [39]. Compared to PEG, the ST HPMA polymers have the advantage that different functional groups may be introduced to the polymers during the synthesis.

## CONCLUSIONS

The functional semitelechelic HPMA polymers can be readily prepared by free radical polymerization in the presence of functional mercaptans. The functional groups and chain length of the ST polymers can be controlled by the choice of a particular mercaptan and the reaction conditions. ST HPMA polymers can be used for the modification of proteins and biomedical surfaces by one-point attachment. The activity of the modified α-chymotrypsin was changed based on the chemistry of the modification. The modification of the surface of nanospheres increased their intravascular half-life in rats and reduced their biorecognizability.

## REFERENCES

1. Abuchowski, A., Kazo, G. M., Verhoest, C. R., van Es, T., Kafkewitz, D., Nucci, M. L., Viau, A. T., and Davis, F. F. *Cancer Biochem. Biophys.* 1984, 7, 175–186.
2. Gaertner, H. F., and Offord, R. E. *Bioconjugate Chem.* 1996, 7, 38–44.
3. Zalipsky, S. *Bioconjugate Chem.* 1995, 6, 150–165.
4. Takahashi, K., Ajima, A., Yoshimoto, T., Okada, M., Matsushima, A., Tamaura, Y. and Inada, Y. *J. Org. Chem.* 1985, 50, 3412–3415.
5. Wang, P., Sergeeva, M. V., Lim, L. and Dordick, S. J. *Nature Biotechnology* 1997, 15, 789–793.
6. Gaertner, H. F. and Offord, R. E. *Bioconjugate Chem.* 1996, 7, 38–44.
7. Lee, J. H., Kopečková, P., Kopecek, J., and Andrade, J. D. *Biomaterials* 1990, 11, 455–464.
8. Greenwald, R. B., Gilbert, C. W., Pendri, A., Conover, C. D., Xia, J., and Martinez, A. *J. Med. Chem.* 1996, 39, 424–431.
9. Harris, J. M., Ed. *Poly(ethylene glycol) Chemistry, Biotechnical and Biomedical Applications*, Plenum Press, New York, NY, 1992.
10. Harris J. M. and Zalipsky S., Eds. *Poly(ethylene glycol) Chemistry and Biological Applications*, ACS, Washington, DC, 1997.
11. Gewehr, M., Nakamura, K., Ise, N., and Kitano, H. *Makromol. Chem.* 1992, 193, 249–256.
12. Takei, Y. G., Aoki, T., Sanui, K., Ogata, N., Okano, T., and Sakurai, Y. *Bioconjugate Chem.* 1993, 4, 42–46.
13. Takei, Y. G., Aoki, T., Sanui, K., Ogata, N., Sakurai, Y. and Okano, T. *Biomaterials* 1995, 16, 667–673.
14. Takei, Y. G., Matsukata, M., Aoki, T., Sanui, K., Ogata, N., Kikuchi, A., Sakurai, Y., and Okano, T. *Bioconjugate Chem.* 1994, 5, 577–582.
15. Chen, G. and Hoffman, A. S., *J. Biomater. Sci. Polymer Edn.* 1994, 5, 371–382.
16. Ranucci, E., Spagnoli, G., Sartore, L. and Ferruti, P. *Macromol. Chem. Phys.* 1994, 195, 3469–3479.
17. Caliceti, P., Schiavon, O., Morpurgo, M. and Veronese, F. M., *J. Bioact. Compat. Polym.* 1995, 10, 103–120.
18. Kamei, S. and Kopeček, J. *Pharmaceutical Res.* 1995, 12, 663–668.
19. Lu, Z.-R., Kopečková, P., Wu, Z., and Kopeček, J. *Polymer Preprints*, 1998, 39(2), 218–219.
20. Lu, Z.-R., Kopečková, P., Wu, Z., and Kopeček, J. *Bioconjugate Chem.* 1998, 9, 793–804.
21. Ulbrich, K. and Oupický, D. *Eighth International Symposium on Recent Advances in Drug Delivery Systems*, 1997, pp. 215–218, Salt lake City, UT, February 24–27.
22. Putnam, D. and Kopeček, J. *Adv. Polym. Sci.* 1995, 122, 55–123.

23. Chytrý, V., Kopeček, J., Sikk, P., Sinijärv, R., and Aaviksaar, A. *Makromol. Chem. Rapid Commun.* 1982, 3, 11–15.
24. Kopeček, J., Rejmanová, P., and Chytrý, V. *Makromol. Chem.* 1981, 182, 799–809.
25. Lääne, A., Aaviksaar, A., Haga, M., Chytrý, V., and Kopeček, J. *Makromol. Chem. Suppl.* 1985, 9, 35–42.
26. Lu, Z.-R., Kopečková, P., Wu, Z., and Kopeček, J. submitted.
27. Heitz, W. In *Telechelic Polymers: Synthesis and Applications;* Goethals, E. J., Ed.; CRC Press: Boca Raton, Florida; 1989, 61–94.
28. Russell, D. H. and Edmondson, R. D. *J. Mass Spectrom.* 1997, 32, 263–276.
29. Schadler, V., Spickermann, J., Rader, H. J., and Wiesner U. *Macromolecules* 1996, 29, 4865–4870.
30. Rader, H. J., Spickermann, J., and Mullen K. *Macromol. Chem. Phys.* 1995, 196, 3967–3978.
31. Chilkoti, A., Chen, G., Stayton, P. S., and Hoffman, A. S. *Bioconjugate Chem.* 1994, 5, 504–507.
32. Chiu, H. C., Zalipsky, S., Kopečková, P., and Kopeček, J. *Bioconjugate Chem.* 1993, 4, 290–295.
33. Sakane, T. and Pardridge, W. M. *Pharmaceutical Res.* 1997, 14, 1085–1091.
34. Pettit, D. K., Bonnert, T. P., Eisenman, J., Srinivasan, S., Paxton, R., Beers, C., Lynch, D., Miller, B., Grabstein, K. H., and Gombotz, W. *J. Biol Chem.* 1997, 272, 2312–2318.
35. Voet, D. and Voet J. G. *Biochemistry* (2nd ed.), John Wiley & Sons, New York, 1995, 371–410.
36. Gombotz, W. R., Guanghui, W., Horbett T. A. and Hoffman, A. S. In *Poly(ethylene glycol) Chemistry, Biotechnical and Biomedical Applications,* Harris, J. M., Ed. Plenum Press, New York, 1992, 247–261.
37. Cassidy, J., Vasey, P., Kaye, S. B. and Duncan, R. In *Proceedings 2nd Int. Symp. Polymer Therapeutics* Kumamoto, Japan, 1997, 18.
38. Greenwald, R. B., Pendri, A., Bolikal, D. and Gilbert, C. W. *Bioorg. Med. Chem. Letters* 1994, 4, 2465–2470.
39. Minko, T., Kopečkova, P., Pozharov, V. and Kopeček J. *J. Control. Release,* 1998, 54, 223–233.

# CHAPTER 2

# Thermoresponsive Polymeric Micelles for Double Targeted Drug Delivery

J. E. CHUNG[1]
TERUO OKANO[1]

## INTRODUCTION

PIPAAm in aqueous solution is well known to exhibit a reversible thermoresponsive phase transition at 32°C [1]. This transition temperature is a lower critical solution temperature (LCST). PIPAAm is water soluble and hydrophilic, showing an extended chain conformation below its LCST, and undergoes a phase transition to an insoluble and hydrophobic aggregate above the LCST. This phase transition occurs with in a narrow temperature window and is reversible with reversing temperature changes. When the phase transition occurs by heating through its LCST, negative entropy dominates the otherwise exothermic enthalpy of the hydrogen bonds formed between the polymer polar groups and water molecules— the initial driving force for dissolution. Increased entropy results from release of water molecules oriented around nonpolar polymer regions forming a clathrate-like structure [2,3]. Thermal destruction of the specific water orientations around hydrophobic polymer regions facilitates polymer–polymer association by hydrophobic interactions, resulting in polymer precipitation. Increasing temperature promotes the release of water molecules from the water clusters surrounding the hydrophobic isopropyl groups. Polymer mobility must be the greatest at polymer chain ends. Dehydration will be initiated at these ends, rapidly triggering a polymer phase transition by their hydrophobic contribution acting as nuclei to accelerate further polymer dehydration. Indeed, single-point grafted

[1]Institute of Biomedical Engineering, Tokyo Women's Medical University, Kawada-cho 8-1, Shinjuku-ku, Tokyo 162-8666 Japan.

linear PIPAAm has shown faster phase transition phenomena due to its high polymer end mobility over relatively immobile PIPAAm chains attached to a surface with random multipoint crosslinks along the chain [4,5]. Also for PIPAAm hydrogels, freely mobile ends of combtype grafted PIPAAm gels lead to dramatically increased de-swelling rates above the LCST compared to conventional, random crosslinked gels in which both ends of PIPAAm chains were relatively immobile [6,7]. This observation strongly suggests that the PIPAAm phase transition is initiated at the ends of a polymer chain due to their high mobility. Exploiting the high chain end mobility will lead to more effective control of the thermoresponsive properties of modified PIPAAm compared to statistically modified PIPAAm.

We have constructed thermoresponsive systems utillizing semitelechelic PIPAAm chains with freely mobile ends, synthesized by telomerization using telogens as a following reaction [6]: $nM + HS\text{-}X \rightarrow H\text{-}(M)n\text{-}S\text{-}X$.

Telomerization was effective to regulate quantitative incorporation of functional groups to one end of PIPAAm chains [7,8]. Molecular weight of the semitelechilic PIPAAm determined from GPC data was in good agreement with that determined by the end-group assay. This indicates that each macromolecule carries one amino or carboxyl end group.

We have utilized thermoresponsive properties of PIPAAm and its gels as on–off switches for drug release [6,7], chromatography systems [9–11], and attachment/detachment of cells [12–14] (Scheme 1). Hydrophobic chains of collapsed PIPAAm above its LCST interact with cells and proteins. Although below the LCST, PIPAAms are highly hydrated flexible chains and

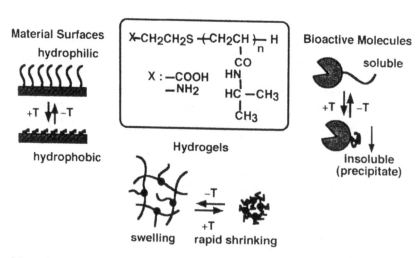

**Scheme 1** Thermoresponsive systems constructed using semitelechelic poly(*N*-isopropylacrylamide).

do not readily interact. We have already reported thermoresponsive surfaces grafted with PIPAAm chains for novel hydrophobic liquid chromatography matrices modulating separation and solute–surface partitioning by temperature control [9–11]. We have also reported that poly(styrene) surfaces grafted with PIPAAm chains become a hydrophilic/hydrophobic switchable surface. This surface can control cell attachment and detachment by thermal modulation without any cell damage [12–14]. Cells attach and proliferate normally when cultured on hydrophobic PIPAAm surfaces above the LCST. Upon cooling these cultures below the LCST, viable cells spontaneously detach from hydrating PIPAAm grafted surfaces as a result of decreased interactions between cells and the PIPAAm surface. Utilization of PIPAAm thermoresponse allows construction of new materials systems that reversibly modulate interactions with cells, including aspects of cellular morphology and cellular metabolic functions.

Several types of drug carriers such as microspheres, liposomes, and polymer have been investigated to achieve targetable drug delivery, especially for anticancer drugs. However, nonselective scavenging of such carriers by the reticuloendothelial system (RES) is a serious problem even when monoclonal antibodies are used to carry the drug [15,16].

A-B–type block copolymers of PIPAAm containing a hydrophobic segment exhibit thermoresponsive soluble/insoluble changes and can form coreshell structured polymeric micelles. Polymeric micellar structures comprising hydrophilic outer shell of soluble PIPAAm segments as annuli surrounding hydrophobic aggregated inner core microdomains hydrophobic segments in aqueous solution below the LCST (Scheme 2). The hydrophobic inner core of the micelle can contain hydrophobic drugs, whereas the PIPAAm outer shell plays an important role in the aqueous solubilization and temperature response. The hydrophilic outer shell that prevents interaction of the inner core with biocomponents and other micelles can be suddenly switched to a dehydrated, hydrophobic state at a specific tissue site by local heating above the LCST. Therefore, utilizing preferable characteristics of thermoresponsive polymeric micelles as a drug carrier system improves targeting efficacy according to both effect of passive targeting and temperature modulation by local heating.

We have shown that polymeric micelles constructed of block copolymers of poly(ethylene oxide) (PEG) and poly(L-asparate) containing the anticancer drug (adriamycin, ADR) selectively accumulate at solid tumor sites by a passive targeting mechanism. This is likely due to the hydrophilicity of the outer PEG chains and micellar size (<100 nm) that allow selective tissue interactions [17,18]. Polymeric micelle size ranges are tailored during polymer synthesis steps. Carefully selection of block polymer chemistry and block lengths can produce micelles that inhibit nonselective scavenging by the reticuloendothelial system (RES) and can be utilized as targetable drug

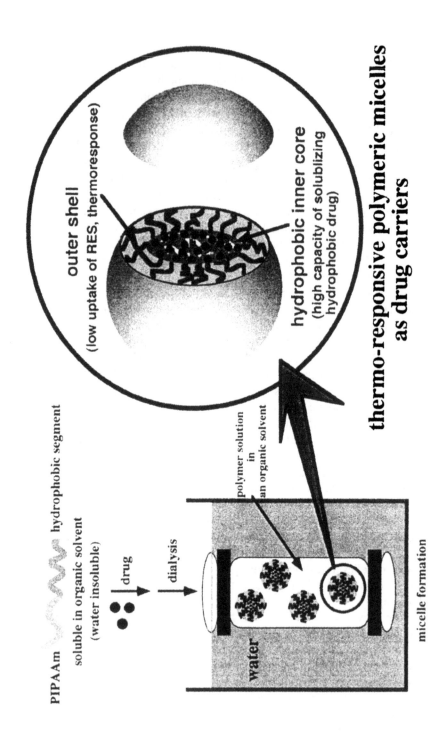

**Scheme 2** Thermoresponsive polymeric micelle structures and functions.

18

carriers to enhance permeability and retention (EPR) effects at tumor tissues [19,20]. From this perspective, thermoresponsive micelles comprising block copolymers of PIPAAm and a hydrophobic segment are able not only to utilize spatial specificity (site targeting) in a passive manner due to the highly hydrated outer shell surrounding the hydrophobic core loaded with drugs and their size but also to increase their spatial specificity in combination with a physical targeting mechanism achieved by introducing a thermoresponsive polymer segment. Weinstein and coworkers studied thermosensitive liposomes to achieve temperature modulated, targeted drug delivery [21]. However, the conventional liposome formulations have only limited value *in vivo* due to nonselective scavenging by the RES and slow response to the temperature changes [21,22]. In addition, drug release from liposome carriers was not clearly defined only when liposome carriers were heated due to liposome unstability, even though the drug release was enhanced upon heating. Clear and sensitive thermoresponse of PIPAAm opens up opportunities to construct effective drug delivery system in conjunction with localized hyperthermia. Delivery to a specific site can be enhanced by local heating/cooling procedures, inducing changes in the physical properties of polymeric micelles. Therefore, thermoresponse of PIPAAm is expected to accomplish multiple functions for a double-targeting system in both passive and stimuli-responsive manners, enhancing vascular transport and drug release, and/or embolization induced by local tissue temperature changes. Selective accumulation of micelles at malignant tissue sites could be increased by micellar adsorption to cells mediated by hydrophobic interactions between polymeric micelles and cells. Simultaneously, this strategy can also achieve temporal drug delivery control: drug is released and expresses its bioactivity only for a period defined by local heating and cooling.

The thermoresponsive character of micellar structures constructed by self-assembly of the modified PIPAAm chains is not always consistent with that of PIPAAm, especially with regard to its LCST and the thermal response transition rates. The thermoresponsive properties and structures of the assembled supramolecular micellar architecture depend upon the molecular structure of a single modified PIPAAm chain that is the building block of the supramolecular assembly. We have reported that thermoresponse and structures of molecular assemblies formed from alkyl-terminated PIPAAms depended on the alkyl group hydrophobicity [23]. Moreover, high chain end mobility led to more effective control of the thermoresponsive properties of hydrophobically modified PIPAAm as well as subsequent induced phase separation into hydrophilic and hydrophobic microdomains compared to statistically modified PIPAAm [24], that is, a design of the molecular architecture of modified PIPAAm directly results in desired supramolecular architecture of thermoresponsive micellar structures formed by self-assembly of the modified PIPAAms. In order to design and facilitate a reversibly thermoresponsive micelle for a drug delivery system, we have exploited a polymer micelle formation mechanism with structural stability and temperature response based on intra- or intermolecular hydrophilic/hydrophobic interactions and a block copolymer molecular architecture [23–27].

## PREPARATION OF THERMORESPONSIVE POLYMERIC MICELLES (SCHEME 3)

### SYNTHESIS OF A-B BLOCK COPOLYMERS OF PIPAAm WITH VARIOUS HYDROPHOBIC SEGMENTS

Amino-, hydroxyl-, and carboxyl-semitelechilic PIPAAm (PIPAAm-NH$_2$, PIPAAm-OH, PIPAAm-COOH) were synthesized by telomerization using 2-aminoethanethiol hydrochloride (AESH·HCl), 2-mercaptoethanol (ME), and 3-mercaptopropionic acid (MPA) as telogens, respectively [6–8]. *N*-Isopropylacrylamide (IPAAm), a telogen and benzoyl peroxide were dissolved in DMF. This solution was repeatedly degassed under reduced pressure in freeze–thaw cycles and sealed in an ampule. Polymerization was carried out at 70°C and stopped by freezing after 4 h. After evaporating most of the DMF, polymers were precipitated into an excess of diethyl ether; the polymer was reprecitated twice more and dried *in vacuo*. An excess of triethylamine (TEA) was added dropwise to the polymer solution of PIPAAm-NH$_2$·HCl in THF at room temperature in order to convert the hydrochloride end group to the free PIPAAm-NH$_2$. The polymer was precipitated in an excess of diethyl ether, reprecipitation was carried out twice more and dried *in vacuo*. The dried polymers were dissolved in MeOH and dialyzed against MeOH through a dialysis membrane (Spectra/Por® CE, MWCO = 500) at 4°C for 3 days. After evaporation of MeOH, the product was dissolved in water and lyophilized. Semitelechilic PIPAAm molecular weight was determined by gel permeation chromatography [GPC, TOSOH, SC-8020, poly(styrene) standard] in DMF containing LiBr (20 mM; elution rate: 1 mL/min) at 40°C. Terminal groups were titrated by nonaqueous potentiometric titration [28].

Carboxyl terminated PSt (PSt-COOH) and carboxyl-terminated PBMA (PBMA-COOH) were also prepared by radical polymerization using 3-mercaptopropionic acid (MPA) as a chain transfer agent and purified by precipitation in a large excess of MeOH [24,25]. The polymer product was obtained as a white powder by lyophilization.

Stearoyl-terminated PIPAAms (PIPAAm-C$_{18}$H$_{35}$) was obtained by the reaction of the primary amino end group of PIPAAm-NH$_2$ with a large excess of acyl chlorides [23]. A block copolymer of PIPAAm and poly(styrene) (PIPAAm-PSt) was obtained by a condensation reaction

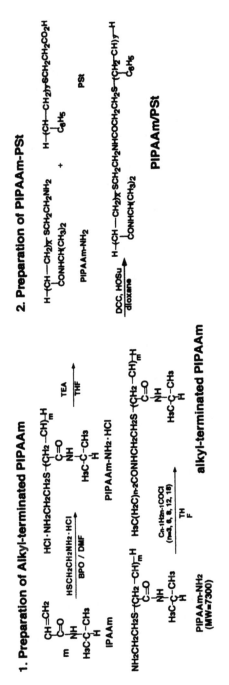

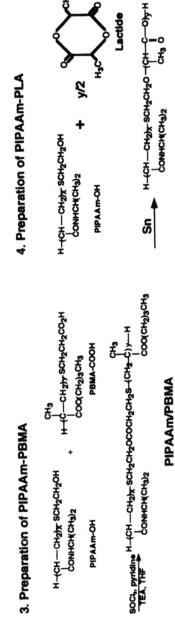

**1. Preparation of Alkyl-terminated PIPAAm**

**2. Preparation of PIPAAm-PSt**

**3. Preparation of PIPAAm-PBMA**

**4. Preparation of PIPAAm-PLA**

Scheme 3 Preparation of thermoresponsive polymeric micelles.

21

between the terminal carboxylic end group of semitelechilic PSt-COOH ($M_w$ 4,700) and the primary amino group of semitelechilic PIPAAm-NH$_2$ (42,000) [25]. Block copolymers of PIPAAm-PBMA were obtained by reactions of PIPAAm hydroxyl groups with activated terminal groups of PBMA [27]. The PIPAAm-PSt and PIPAAm-PBMA showed different solubility properties from either PIPAAm or the hydrophobic homopolymers. Water at 30°C diethyl is a good solvent for PIPAAm and diethylether for the hydrophobic homopolymers, respectively. However, the block copolymer product was insoluble in both of these solvents. Titrations confirmed that almost 100% of the terminal groups of PIPAAm chains reacted with either stearoyl chloride, PSt-COOH, or PBMA-COOH. PIPAAm-PLA block copolymer was obtained by ring opening polymerization of DL-lactide using the terminal hydroxy group of PIPAAm with tin(II)2-ethyl-hexanoate as a catalyst. Control of the PLA block chain length was achieved by changing the amount of DL-lactide monomer in the reaction mixture.

## MICELLE FORMATION FROM PIPAAm BLOCK COPOLYMERS

Micelle solutions of PIPAAm-C$_{18}$H$_{35}$ was prepared by direct dissolution of the polymer in cold water (4°C) due to its good water solubility [23]. Each solution of PIPAAm-PSt, PIPAAm-PBMA, and PIPAAm-PLA was prepared by dissolving each copolymer in DMF, N-ethylacetamide, and DMAc, respectively. The solutions were put into a dialysis bag (MWCO = 13,000) and dialyzed against distilled water at 10°C, 20°C, and 4°C, respectively, for 24 hours. The micelles were purified with ultrafiltration membrane of 200,000 molecular weight cut off at 4°C. The aqueous solution was lyophilized to leave a white powder of micelles.

## THERMORESPONSIVE STRUCTURAL CHANGES OF POLYMERIC MICELLES

### Core-Shell Micellar Structure Formation

In general, incorporation of hydrophobic groups into PIPAAm chains decreases the LCST [29–31]. Hydrophobic groups alter the hydrophilic/hydrophobic balance in PIPAAm, promoting a PIPAAm phase transition at the LCST, water clusters around the hydrophobic segments are excluded from the hydrophobically aggregated inner core. The resulting isolated hydrophobic micellar core does not directly interfere with outer shell PIPAAm chain dynamics in aqueous media. The PIPAAm chains of the micellar outer shell therefore remain as mobile linear chains in this core-shell micellar structure. As a result, the thermoresponsive properties of PIPAAm in the outer PIPAAm chains of this structure are unaltered [23–27,32].

Micellar solutions of PIPAAm-$C_{18}H_{35}$ with a stearoyl group at a terminal position as well as other terminally modified PIPAAm show nearly the same LCST and the same phase transition rate as for freely mobile, linear PIPAAm chains irrespective of the attached highly hydrophobic end groups. In contrast, random copolymers of IPAAm and stearyl methacrylate (P(IPAAm/SMA)) exhibited an LCST shifted to lower temperature proportionally to hydrophobic SMA mole fraction, even above their CMC values. Moreover, phase transition behavior of P(IPAAm/SMA) was considerably less sensitive to temperature changes compared to terminally modified PIPAAm. The LCST shifted toward lower temperatures demonstrates incomplete phase-separated microdomains, that is, parts of the hydrophobic comonomer segments are exposed to the aqueous media and, simultaneously, parts of the outer shell PIPAAm chains are mixed in the inner core. Hydrophobic aggregation of the comonomers entangles parts of the PIPAAm main chain in the inner core. Simultaneously, hydrophobic aggregates remain partially exposed to water. Moreover, PIPAAm chains surrounding hydrophobic aggregates are immobile, hydrated loops without a freely mobile end since PIPAAm chains incorporate random hydrophobic sequences that aggregate. Such a structure slows down the phase transition rate of PIPAAm [4,5]. Therefore, hydrophobic group location on a PIPAAm copolymer influences the thermoresponsive properties of a polymeric micelle [24]. Terminal modification of PIPAAm appears to be essential in the design strategy to fabricate polymeric micelles as stable carriers that maintain similar thermoresponses as linear PIPAAm in their outer shells.

A choice of the terminal hydrophobic segment length of the linearly modified PIPAAm is also important to facilitate both formation of clearly phase-separated, core-shell micellar structures and preservation of the core-shell structures during thermoresponsive structural changes by heating through the LCST. The influence of intramolecular hydrophobic/hydrophilic balance for the core-shell micellar structure formation and structural changes upon heating have been researched using alkyl-terminated PIPAAms with various alkyl chain length ($-C_3 \sim -C_{18}$) [23]. Core-shell micellar structures of PIPAAm-$C_{12}H_{23}$ and PIPAAm-$C_{18}H_{35}$, with a considerably long hydrophobic alkyl chain at a terminal position, isolated the hydrophobic inner core from the aqueous media and does not influenced the LCST of the PIPAAm outer shell. For $C_3$–$C_8$ terminated PIPAAm samples incorporated with relatively short hydrophobic alkyl chains, the LCST observed was reduced with increasing terminal alkyl chain length. It is expected that polymer chains aggregate to form a more stable structure by isolating the hydrophobic segments from the aqueous media as much as possible even though the hydrophobic affinity is weak. However, in the case of $C_3$–$C_8$ terminated PIPAAm, the hydrophobic alkyl chain association is loose but remains in contact with water. When LCSTs for single unimers, determined by DLS measurements at a fixed angle (90°), were compared with the LCSTs for each micelle measured by absorbance above the CMC the LCSTs of $C_3$–$C_8$ terminated PIPAAms were consistent with the LCSTs above the CMC, which were reduced with increasing alkyl chain length [23]. By contrast, the LCSTs of $C_{12}$ and $C_{18}$-terminated PIPAAm unimers were much lower than their micellar LCSTs, similar to PIPAAm.

Polymeric micelles with selected chemistries and molecular architecture of block copolymers, such as PIPAAm-$C_{18}H_{35}$, PIPAAm-PSt, PIPAAm-PBMA, and PIPAAm-PLA micelles, showed the same LCST and the same thermoreponsive phase transition kinetics as those for PIPAAm irrespective of the hydrophobic segment incorporation. This confirms two points: (a) that hydroxyl groups or amino goups of PIPAAm termini completely react with the hydrophobic segment end groups and (b) that the block copolymers form core-shell micellar structures with hydrophobic inner cores completely isolated from the aqueous phase.

The hydrophobic inner core formation of the polymeric micelle aqueous solutions was also characterized by fluorescence spectroscopy using pyrene and 1,3-bis(1-pyrenyl) propane ($PC_3P$) as fluorescence probes. The fluorescence spectrum of pyrene at the low concentration possesses a vibrational band structure that exhibits a strong sensitivity to the polarity of the pyrene environment [33]. The ratio ($I_1/I_3$) of the intensity of the first band ($I_1$) to that of the third band ($I_3$) was monitored as a function of each polymeric micelle concentration [34]. As the concentrations of polymeric micelles, such as PIPAAm-$C_{18}H_{35}$, PIPAAm-PSt, PIPAAm-PBMA, and PIPAAm-PLA, increase a large decrease in $I_1/I_3$ was observed. This indicates partitioning of the hydrophobic probe into a hydrophobic environment. From these plots, it is possible to estimate a concentration corresponding to the onset of hydrophobic aggregation of the hydrophobic segments. This concentration determined from the midpoints of the plots for $I_1/I_3$ changes was rather low, providing evidence for the facile formation of stable micelles. The values were 80, 10, 20, and 10 mg/L for PIPAAm-$C_{18}H_{35}$, PIPAAm-PSt, PIPAAm-PBMA, and PIPAAm-PLA, respectively. This property is necessary for their use in aqueous milieu such as body fluids. Short alkyl ($C_3$–$C_8$)-terminated PIPAAm solutions showed only small decreases in polarity. These small polarity changes indicate that pyrene is partitioning into the polymer-rich phase and not into a clearly separated alkyl chain phase distinct from PIPAAm. Hydrophobic driving forces for the intermolecular aggregation of shorter alkyl terminal groups appears too weak when balanced with the hydrophilicity of their highly hydrated PIPAAm main chain.

$PC_3P$ is a sensitive probe for local viscosity measurement by forming an intramolecular excimer [35,36]. The extent of excimer emission depends upon the rate of conformational change of the chain linking the two pyrenyl groups, leading to a stable "sandwich" conformation between excited and unexcited aromatic moieties [37]. This motion is impeded by local viscosity. As a consequence, the excimer to monomer emission intensity ratio ($I_E/I_M$) provides a measure of the microviscosity of the $PC_3P$ local environment. Shown in Figure 1 are representative emission spectra for

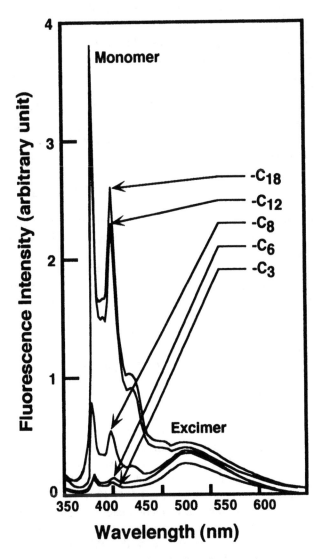

**Figure 1** Representative emission spectra for $PC_3P$ in alkyl-terminated PIPAAm aqueous solutions above the CMC (20,000 mg/L). $\lambda_{ex} = 333$ nm, $[PC_3P] = 2.2 \times 10^{-7}$ M, 20°C (Reference [23], p. 37).

PC$_3$P in alkyl-terminated PIPAAm solutions above their CMC. The monomer emission for alkyl-terminated PIPAAm solutions showed a significant dependence on the incorporated alkyl chain length because the PC$_3$P probes were more soluble in the hydrophobic microdomain with increasing alkyl chain length. Excimer emission was essentially unaffected by alkyl chain length above C$_6$. This shows that the motion of PC$_3$P is suppressed by the microviscosity created by hydrophobic alkyl chain aggregation. According to the ratios ($I_E/I_M$) of PC$_3$P dissolved in solutions of alkyl-terminated PIPAAms, values for C$_{12}$ and C$_{18}$ polymer solutions were markedly lower than the other samples. The values were 0.128 and 0.127 for PIPAAm-C$_{12}$H$_{23}$ and PIPAAm-C$_{18}$H$_{35}$, respectively. These results indicate that C$_{12}$ and C$_{18}$ polymers formed hydrophobic microdomains that were relatively rigid, allowing for poor intramolecular PC$_3$P motions. The emission ratios ($I_E/I_M$) for PC$_3$P in the PIPAAm-PSt and PIPAAm-PBMA polymeric micelle solutions showed a high viscosity ($I_E/I_M$ = 0.068 and 0.044, respectively) for the micelle inner core, implying stability for a concentrated drug payload.

### Reversible Structural Change Responding to Temperature Change

Polymeric micelles with a stable core-shell micellar structure such as PIPAAm-C$_{18}$H$_{35}$, PIPAAm-PSt, PIPAAm-PBMA, and PIPAAm-PLA micelles solutions showed nearly constant sizes and unimodal distributions below their LCST due to the hydrophilic PIPAAm outer shell. Micelles had an almost identical average diameter with unimodal distribution before and after a heating/cooling cycle through their LCST even though these solutions showed a polydispersed and increased average size near the LCST as a sign of intermicellar aggregation. This indicates that intermicellar aggregates formed upon heating redisperse to the initial micellar structures upon cooling below the LCST. However, random copolymer solutions of P(IPAAm/SMA) with increased micellar sizes did not show reversibility through the same heating/recooling cycles. Short alkyl chain terminated PIPAAm solutions such as PIPAAm-C$_6$H$_{11}$ and PIPAAm-C$_8$H$_{15}$ also showed similar phenomena, perhaps because the hydrophobic segments exposed to the aqueous media due to weak hydrophobic aggregation forces in shorter alkyl-terminated PIPAAms enhanced hydrophobic aggregation and entanglement above the LCST, which would disturb rehydration of PIPAAm chains below the LCST [23]. As described above, P(IPAAm/SMA) forms relatively incompletely phase separated hydrophobic microdomains remaining exposed to aqueous media. The architectural difference between this and the PIPAAm-C$_{18}$H$_{35}$ micelle would explain the thermal irreversibility in the same way as for the short alkyl chain terminated PIPAAms.

PIPAAm-$C_{18}H_{35}$ and PIPAAm-PBMA micelle solutions (Figure 2) show increasing polarity of the pyrene environment with increasing temperature through the LCST; however, the solutions showed a reduced but constant micropolarity below the LCST. PIPAAm solutions [Figure 2(a)] showed an abrupt decrease in polarity when temperature was raised through its LCST, indicating transfer of pyrene into the precipitated polymer-rich phase [23]. On the other hand, PIPAAm-$C_{18}H_{35}$ and PIPAAm-PBMA micelle solutions showed lower polarity than that for PIPAAm solutions over the entire temperature region due to the presence of hydrophobic micellar cores. Aggregation of collapsed PIPAAm outer shells could induce micelle structural deformation, which would increase the pyrene microenvironment polarity observed by the increase in pyrene polarity above the LCST. If so, then structural deformations that allows this change in pyrene partitioning turns to the initial micelle structure with increasing rehydration of the PIPAAm chains below the LCST. A small hysteresis around the LCST has been observed due to the delayed hydration of the PIPAAm chains upon cooling [23,27]. The clear reversibility of the structural changes supports the contention that the micelles do not undergo a serious structural change in their inner core regions such as destruction or fusion in the thermal cycle through the LCST, although the polymer core-shell micellar structure can sensitively undergo conformational changes upon heating through the LCST.

Shown in Figure 3 are representative emission spectra for $PC_3P$ in PIPAAm, PIPAAm-$C_{18}H_{35}$ and PIPAAm-PBMA solutions above their CMC as a function of temperature. PIPAAm solutions show a continuous reduction in $I_E/I_M$ with increasing temperature still below the LCST, since hydrophobic polymer-rich phases solubilizing $PC_3P$ probes begin to stiffen as the polymer chains dehydrate [Figure 3(a)]. However, the $I_E/I_M$ values discontinuously decrease with temperature increase through the LCST, implying a phase transition in PIPAAm chains. Above the LCST, it remains essentially unaffected by any further temperature increase. This suggests that the motion of $PC_3P$ is suppressed by the microviscosity created by the contracted hydrophobic polymer chain aggregation. On the other hand, the $I_E/I_M$ ratios for $PC_3P$ dissolved in PIPAAm-$C_{18}H_{35}$ and PIPAAm-PBMA micellar solutions are markedly lower than those for PIPAAm solutions over the entire temperature region due to the highly compact cores of aggregated PBMA chains [Figure 3(b)]. Interestingly, the micelle solutions showed increases in $I_E/I_M$ as the temperature increase through the LCST, irrespective of the PIPAAm phase transition. This is evidence that a decrease in rigidity of the inner micellar cores occurs above the LCST. Therefore, conformational changes by aggregated and collapsed outer shell PIPAAm chains might induce some deformation of the inner core structure, resulting in both a micropolarity increase and microrigidity decrease. These results

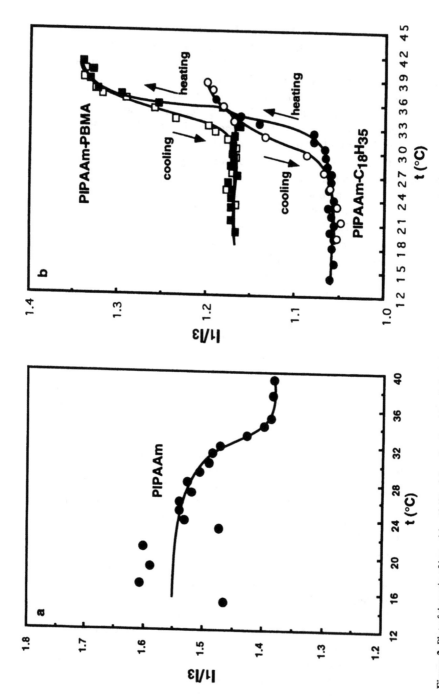

**Figure 2** Plot of the ratio of intensities ($I_1/I_3$) of the vibrational bands in the pyrene fluorescence spectrum as a function of temperature for PIPAAm (a). PIPAAm-C$_{18}$H$_{35}$ and PIPAAm-PBMA (b). $\lambda_{ex} = 340$ nm, [pyrene] = $1.6 \times 10^{-7}$ M, 1°C/min, [polymer] = 5,000 mg/L.

28

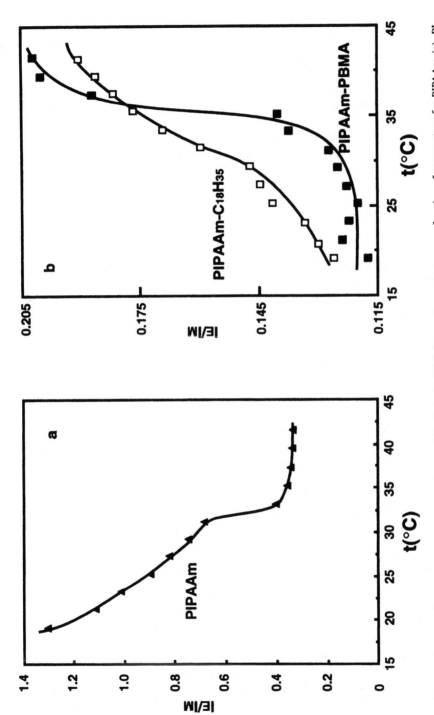

**Figure 3** Plot of the $(I_E/I_M)$ ratio intensities of the vibrational bands in the $PC_3P$ fluorescence spectrum as a function of temperature for PIPAAm (a). PIPAAm-$C_{18}H_{35}$ and PIPAAm-PBMA (b). $\lambda_{ex} = 333$ nm, $[PC_3P] = 2.2 \times 10^{-7}$ M, 1°C/min, [polymer] = 20,000 mg/L.

indicate the potential for these polymeric micelle structural changes to selectively modulate drug release from the inner cores upon heating in a manner distinct from that of PIPAAm gels.

To activate drug delivery using a thermoresponsive polymeric micelle system for both temporal and spatial selectivity, the hydrophobic aggregation forces that form these structures must be strong enough to form stable micelles as well as load hydrophobic drugs. Simultaneously, this structure must allow reversible structural changes upon heating to selectively initiate drug release. In particular, the inner core structural changes induced by the outer shell structural changes above the LCST facilitate drug release enhancement by selection of inner core polymer chains with relatively lower Tg than the outer shell LCST.

## ON–OFF SWITCHABLE DRUG RELEASE FROM THERMORESPONSIVE POLYMERIC MICELLES

### ADRIAMYCIN (ADR) INCORPORATION INTO PIPAAm-PBMA MICELLES

PIPAAm-PBMA block copolymers form a micellar structures by self-association of the hydrophobic PBMA segments in water, a good solvent for PIPAAm chains below the LCST but a nonsolvent for the PBMA chains. This amphiphilic system produces stable and monodispersed micelles from polymer/N-ethylacetamide (good solvent for the both polymer blocks) solutions dialyzed against water. Hydrophobic drugs can be physically incorporated into the inner micelle cores with PBMA chains by hydrophobic interactions between the hydrophobic segments and drugs.

PIPAAm-PBMA micelle formation and drug loading resulting from solvent exchange during dialysis was significantly affected by interaction of the solvents with both polymers and drugs, solvent exchange speed, and solution temperature. With the gradual decrease in organic solvent composition during dialysis, a spontaneous hydrophobic association of PBMA segments both with themselves and with hydrophobic drugs is a driving force for micelle formation and drug loading into micelle cores. Therefore, drug choice and degree of drug loading are defined by the interactions of drugs with both the polymer hydrophobic segments and solvent. Polymeric micelles loaded with the highest drug concentration possible were formed simultaneously without precipitation by optimizing the hydrophobic interactions among the polymer, the drug, and the solvent. Control of the optimum loading/aggregation conditions should provide information critical to generalize the optimum drug loading conditions in polymeric micelles.

Optimum hydrophobic interactions to form PIPAAm-PBMA micelles containing ADR was regulated by modulating hydrophilic/hydrophobic block lengths of the polymer, concentration of both the polymer and ADR in the dialysis bag, triethylamine (TEA), and the dialysis temperature. PIPAAm ($M_W$ = 6,100)-PBMA ($M_W$ = 8,900) was most successful for most micelle formation and drug (ADR) loading (9.6 wt.%) with *N*-ethylacetamide as the solvent for both the polymers and ADR and other selected conditions. The polymer-ADR solution of *N,N*-dimethylformaldehyde or dimethylsulfoxide precipitated during dialysis against water. A Spectra/Por dialysis membrane with a pore size of $M_w CO$ = 12,000–14,000 provided optimum solvent exchange rates for micelle formation of this PIPAAm-PBMA/ADR combination. The micelle-ADR product was a transparent red solution at ca. 20°C (dialysis temperature) with highly hydrated PIPAAm segments in water. Other higher or lower dialysis temperatures than 20°C led to precipitation. Optimum dialysis temperature was defined by PIPAAm hydration degree balanced with hydrophobic aggregation forces for micelle formation as well as solvent diffusion rates of unloaded drugs.

To remove the hydrochloride from ADR.HCl 1.3 a molar equivalent of TEA was added to the ADR solution dropwise prior to mixing with a polymer solution. A larger amount of TEA addition resulted in precipitation during dialysis, whereas adding less TEA failed to improve ADR loading due to weak hydrophobic interactions. In addition, higher polymer concentrations and higher ratios of ADR/polymer in the dialysis solution increased ADR loading. However, a polymer concentration higher than 1.3 wt.% and a ratio of ADR/polymer higher than 1.1 led to precipitation. Supernatants of the solutions including precipitates showed large aggregate size distributions (diameter > 500 nm) and also showed high drug loading. ADR loaded into PIPAAm-PBMA micelles under selected optimum conditions exhibited monodispersed size distribution.

## THERMORESPONSIVE ADR RELEASE FORM POLYMERIC MICELLES

ADR release profiles from polymeric micelles in water showed drastic changes with temperature alterations through the LCST (Figure 4). ADR release was accelerated upon heating through the LCST, whereas ADR release was suppressed below the LCST. Temperature-accelerated ADR release was consistent with a temperature-induced structural change of the polymeric micelle [Figures 2 and 3(b)]. ADR release from polymer micelles is switched thermally on/off in response to reversible structural changes of micelles modulated by temperature changes through the LCST (Figure 5). ADR release begins upon heating above the LCST and halted simply by cooling below the LCST; ADR is released once again upon another heating cycle.

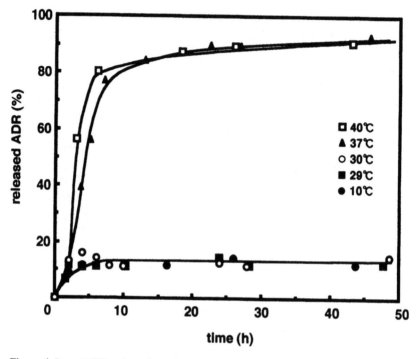

**Figure 4** Drug (ADR) release from thermoresponsive PIPAAm-PBMA micelles containing ADR [27].

## CYTOTOXICITY OF POLYMERIC MICELLES MODULATED BY TEMPERATURE CHANGES

Shown in Figure 6 is *in vitro* cytotoxic activity of PIPAAm-PBMA micelles loaded with ADR or micelles without ADR at 29°C (below the LCST) and at 37°C (above the LCST) compared with that of free ADR. *In vitro* cytotoxic activity was measured using bovine aorta endothelial cells. Bovine aortic endothelial cells were obtained as previously reported using dispase for cell dissociation from freshly harvested bovine aorta [13]. The cells plated at a density of $3 \times 10^4$ cells/well, were exposed with free ADR or micelles loaded with ADR at below and above the LCST for 5 days. In order to assay cytotoxicity of the free ADR or micelles loaded with ADR, culture medium was replaced with 10% FBS-supplemented phenol red-free DMEM containing 10% alamar Blue, a dye that is subject to reduction by cytochrome c activity and changes the color from blue to red [38]. After 4-hour incubation, reduction of the dye was estimated by absorbance at 560 and 600 nm. PIPAAm-PBMA polymeric micelles loaded with ADR showed higher cytotoxic activity than that of free ADR at 37°C (above the LCST)

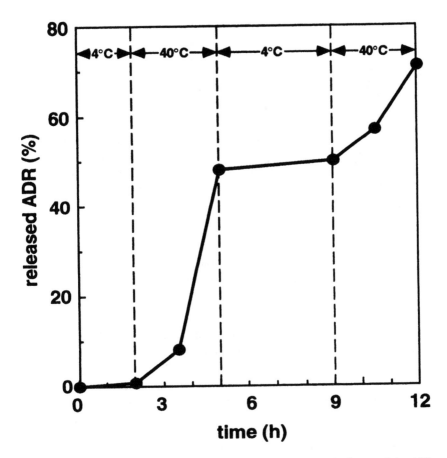

**Figure 5** On–off switched drug (ADR) release from PIPAAm-PBMA micelles containing ADR responding temperature changes.

while exhibiting lower cytotoxic activity than that of free ADR at 29°C (below the LCST). The blank micelles showed no cytotoxicity both below and above the LCST. Cytotoxicity is well correlated with micelle structural changes (Figure 2) and ADR release (Figure 4) selectively initiated by heating through the LCST. These results show that ADR cytotoxicity against endothelial cells can be reversibly switched using ADR-loaded PIPAAm-PBMA micelle carriers combined with culture temperature modulation. Higher cell cytotoxicity by polymeric micelles-ADR over that seen for the same amount of free ADR in culture suggests different routes for drug uptake action by cells caused by the carrier properties. Cells treated by micelles-ADR are observed to be red and have visible aggregates surrounding the cells. Such a phenomenon was observed only for cells treated by

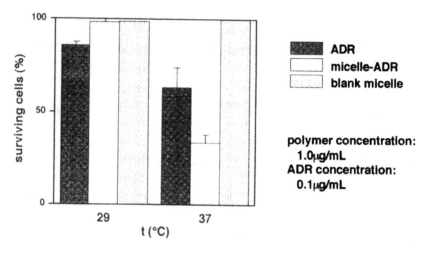

**Figure 6** *In vitro* cytotoxicity of free ADR (0.1 µg/mL) and thermoresponsive PIPAAm-PBMA micelles containing ADR (0.1 µg/mL) against bovine aorta endothelial cells at 29°C (below the LCST) and 37°C (above the LCST). Incubation time: 4 days.

micelles-ADR. We previously reported that hydrophobic PIPAAm chains, which collapse above their LCST, actively interact with cells, whereas hydrated PIPAAm chains below the LCST do not [12–14]. These micelles have been shown to undergo both physical property changes from hydrophilic to hydrophobic as well as structural changes, switching on drug release upon heating through the LCST. Enhanced interaction between the polymeric micelles and cells above the LCST may provide higher drug uptake by the cells through a more effective route. This has been the focus in this current research. Observed results indicate that thermoresponsive polymeric micelles can express specific drug toxicity elicited by local heating. This results from changes in the physical properties of the polymeric micelles, inducing drug release and/or enhanced adsorption to cells mediated by hydrophobic interactions between cells and polymeric micelles (Scheme 4).

Thermoresponsive polymeric micelles with PIPAAm block copolymers can be expected to combine passive spatial targeting specificity with a stimuli-responsive targeting mechanism. We have developed LCSTs of PIPAAm chains with preservation of the thermoresponsive properties such as a phase transition rate by copolymerization with hydrophobic or hydrophilic comonomers into PIPAAm main chains. Micellar outer shell chains with the LCSTs adjusted between body temperature and hyperthermic temperature can play a dual role in micelle stabilization at a body temperature due to their hydrophilicity and initiation of drug release by hyperthermia resulting from outer shell structural deformation. Simultaneously, micelle interactions with cells could be enhanced at heated sites due

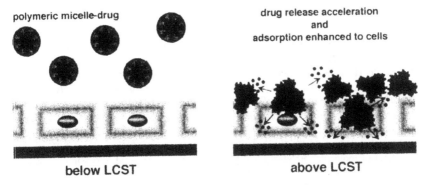

polymeric micelle-drug

drug release acceleration
and
adsorption enhanced to cells

below LCST

above LCST

**Scheme 4** Drug action mechanisms of thermoresponsive polymeric micelles.

to thermal changes in the outer shell structure, namely, the micelle-drug system is expected to deliver drug in a body only when or where heated by hyperthermia. Reversible and sensitive thermoresponsive drug delivery from the polymeric micelles comprises a novel multifucntional drug delivery system achieving both spatial targeting and temporal dosing control in conjunction with localized hyperthermia (Scheme 5).

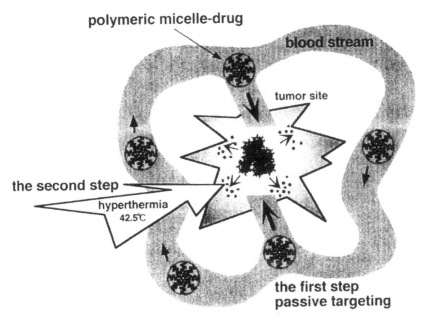

polymeric micelle-drug

blood stream

tumor site

the second step

hyperthermia
42.5℃

the first step
passive targeting

**Scheme 5** Double targeting with both passive and stimuli-responsive manners by thermoresponsive polymeric micelles.

## REFERENCES

1. Heskins, M., and Guillet, J. E. Solution properties of poly(N-isopropylacrylamide), *J. Macromol. Sci. Chem.*, 1968, A2, 1441–1455.
2. Bae, Y. H., Okano, T., and Kim, S. W. Temperature dependence of swelling of crosslingked poly(N, N'-alkyl substituted acrylamides) in water, *J. Polym. Sci., Polym. Phys.*, 1990, 28, 923–936.
3. Schild, H. G., Poly(N-isopropylacrylamide) experiment, theory and application, *Prog. Polym. Sci.*, 1992, 17, 163–249.
4. Matsukata, M., Aoki, T., Sanui, K., Ogata, N., Kikuchi, A., Sakurai, Y., and Okano, T. Effect of molecular architecture of poly(N-isopropylacrylamide)-trypsin conjugates on their solution and enzymatic proprties, *Bioconjugate Chem.*, 1996, 7, 96–101.
5. Takei, Y. G., Aoki, T., Sanui, K., Ogata, N., Sakurai, Y., and Okano, T. Dynamic contact angle measurement of temperature-responsive surfaces properties for poly(N-isopropylacrylamide) grafted surfaces, *Macromolecules*, 1994, 27, 6163–6166.
6. Okano, T., Katayama, M., and Shinohara, I. The influence of hydrophobic and hydrophilic domains on water wettability of 2-hydrooxyethyl methacrylate/styrene copolymers, *J. Appl. Polmer Sci.*, 1978, 22, 367–377.
7. Yoshida, R., Uchida, K., Kaneko, Y., Sakai, K., Kikuchi, A., Sakurai, Y., and Okano, T. Comb-type grafted hydrogels with rapid de-swelling response to temperature changes, *Nature*, 1995, 374, 240–242.
8. Kaneko, Y., Sakai, K., Kikuchi, A., Yoshida, R., Sakurai, Y., and Okano, T. Influence of freely mobile grafted chain length on dynamic properties of comb-type grafted poly(N-isopropylacrylamide) hydrogels, *Macromolecules*, 1995, 28, 7717–7723.
9. Kanazawa, H., Yamamoto, K., Matsushima, Y., Takai, N., Kikuchi, A., Sakurai, Y., and Okano, T. Temperature responsive chromatography using poly(N-isopropylacrylamide)-modified silica, *Anal. Chem.*, 1996, 68, 100–105.
10. Kanazawa, H., Kashiwase, Y., Yamamoto, K., Matsushima, Y., Kikuchi, A., Sakurai, Y., and Okano, T. Temperature responsive liquid chromatography 2. Effect of hydrophobic groups in N-isopropylacrylamide copolymer-modified silica, *Anal. Chem.*, 1997, 69, 823–830.
11. Yakushiji, T., Sakai, K., Kikuchi, A., Aoyagi, T., Sakurai, Y., and Okano, T. Graft architectural effects on thermo-responsive wettability changes of poly(N-isopropylacrylamide)-modified surfaces, *Langmuire*, 1998, 14, 4657–4662.
12. Yamada, N., Okano, T., Sakai, H., Karikusa, F., Sawasaki, Y., and Sakurai, Y. Thermo-responsive polymeric surfaces; control of attachment and detachment of cultured cells, *Makromol. Chem., Rapid Commun.*, 1990, 11, 571–576.
13. Okano, T., Yamada, N., Sakai, H., and Sakurai, Y. A novel recovery system for cultured cells using plasma-treated polystyrene dishes graphted with poly(N-isopropylacrylamide), *J. Biomed. Mater. Res.*, 1993, 27, 1243–1251.
14. Okano, T., Yamada, N., Okuhara, M., Sakai, H., and Sakurai, Y. Mechanism of cell detachment from temperature-modulated, hydrophilic-hydrophobic polymer surfaces, *Biomaterials*, 1995, 16, 297–303.
15. Yang, H. M., and Reisfeld, R. Doxorubicin conjugated with a monoclonal antibody directed to a human melanoma-associated proteoglycan suppresses the growth of established tumor xenografts in nude mice, *Proc. Natl. Acad. Sci. U.S.A.*, 1988, 85, 1189–1193.

16. Thédrez, P., Saccavini, J. C., Nolibé, D., Simoen, J. P., Guerreau, D., Gestin, J. F., Kremer, M., and Chatal, J. F. Biodistribution of indium-111-labeled OC 125 monoclonal antibody after intraperitoneal injection in nude mice intraperitoneally grafted with ovarian carcinoma, *Cancer Res.,* 1989, 49, 3081–3086.

17. Yokoyama, M., Miyauchi, M., Yamada, N., Okano, T., Sakurai, Y., Kataoka, K., and Inoue, S. Characterization and anticancer activity of the micelle-forming polymeric anticancer drug adriamycin-conjugated poly(ethylene glycol)-poly(aspartic acid) block copolymer, *Cancer Res.,* 1990, 50, 1693–1700.

18. Yokoyama, M., Okano, T., Sakurai, Y., Ekimoto, H., Shibazaki, C., and Kataoka, K. Toxicity and antitumor activity against solid tumors of micelle-forming polymeric anticancer drug and its extremely long circulation in blood, *Cancer Res.,* 1991, 51, 3229–3236.

19. Maeda, H., Seymour L. W., and Miyamoto, Y. Conjugates of anticancer agents and polymers: advantages of macromolecular therapeutics *in vivo, Bioconjugate Chem.,* 1992, 3, 351–361.

20. Matsumura, Y., and Maeda, H. A new concept for macromolecular therapeutics in cancer chemotherapy: Mechanism of tumoritropic accumulation of proteins and the antitumor agent smancs. *Cancer Res.,* 1986, 46, 6387–6392.

21. Weinstein, J. N., Magin, R. L., Yatvin M. B., and Zaharko, D. S. Liposomes and local hyperthermia: selective delivery of methotrexate to heated tumors, *Science,* 1979, 204, 188–191.

22. Khoobehi, B., Peyman, G. A., McTurnan, W. G., Niesman, M. R., and Magin, R. L. Externally triggered release of dye and drugs from liposomes into the eye. An *in vitro* and *in vivo* study, *Opthalmology,* 1988, 95, 950–955.

23. Chung, J. E., Yokoyama, M., Suzuki, K., Aoyagi, T., Sakurai, Y., and Okano, T. Reversibly thermo-responsive alkyl-terminated poly(*N*-isopropylacrylamide) core-shell micellar structures, *Colloids Surfaces (B: Biointerfaces),* 1997, 9, 37–48.

24. Chung, J. E., Yokoyama, M., Aoyagi, T., Sakurai, Y., and Okano, T. Effect of molecular architecture of hydrophobically modified poly(*N*-isopropylacrylamide) on the formation of thermo-responsive core-shell micellar drug carriers. *J. Contr. Rel.,* 1997, 53, 119–130.

25. Cammas, S., Suzuki, K., Sone, Y., Sakurai, Y., Kataoka, K., and Okano, T. Thermo-responsive polymer nanoparticles with a core-shell micelle structure as site-specific drug carriers. *J. Contr. Rel.,* 1997, 48, 157–164.

26. Kohori, F., Sakai, K., Aoyagi, T., Yokoyama, M., Sakurai, Y., and Oakano, T. Preparation and characterization of thermally responsive block copolymer micelles comprising poly(*N*-isopropylacrylamide-*b*-DL-lactide). *J. Contr. Rel.,* 1998, 55, 87–98.

27. Chung, J. E., Yokoyama, M., Yamato, M., Aoyagi, T., Sakurai Y., and Okano, T. Thermo-responsive drug delivery from polymeric micelles constructed using block copolymers of poly(*N*-isopropylacrylamide) and poly(butylmethacrylate), *J. Contr. Rel.,* 1999, 62, 115–127.

28. Siggia, S., Hanna, J. G., and Kervenski, J. R. Quantitative analysis of mistures of primary, secondary, and tertiary aromatic amines, *Anal. Chem.,* 1950, 22, 1295–1297.

29. Taylor, L. D., and Cerankowski, L. D. Preparation of films exhibiting a balanced temperature dependence to permeation by aqueous solutions, *J. Polym. Sci. Poly. Chem.,* 1975, 13, 2551–2570.

30. Takei, Y. G., Aoki, T., Sanui, K., Ogata, N., Okano, T., and Sakurai, Y. Temperature-responsive bioconjugates. 2. Molecular design for temperature-modulated bioseparations, *Bioconjugated Chem.*, 1993, 4, 341–346.

31. Yoshida, R., Sakai, K., Okano, T., and Sakurai, Y. Modulating the phase transition temperature and thermo-sensitivity in N-isopropylacrylamide copolymer gels, *J. Biomater. Sci. Polymer Ed.*, 1994, 6, 585–598.

32. Winnik, F. M., Davidson, A. R., Hamer, G. K., and Kitano, H. Amphiphilic poly(N-isopropylacrylamide) prepared by using a lipophilc radical initiator: Synthesis and solution properties in water, *Macromolecules*, 1992, 25, 1876–1880.

33. Dong D. C., and Winnik, M. A. The Py scale of solvent polarities, *Can. J. Chem.*, 1984, 62, 2560–2565.

34. Kalyanasundaram, K., and Thomas, J. K. Environmental effects on vibronic band intensities in pyrene monomer fluorescence and their application in studies of micellar system, *J. Am. Chem. Soc.*, 1977, 99, 2039–2044.

35. Almeida, L. M., Vaz, W. L. C., Zachariasse, K. A., and Madeira, V. M. C. Fluidity of sarcoplasmic reticulum membranes investigated with dipyrenylpropane, an intramolecular excimer probe, *Biochemistry*, 1982, 21, 5972–5977.

36. Zachariasse, K. A., Vaz, W. L. C., Sotomayor C., and Kühnle, W. Investigation of human erythrocyte ghost membrane with intramolecular excimer probes, *Biochim. Biophys. Acta*, 1982, 688, 323–332.

37. Melnick, R. L., Haspel, H. C., Goldenberg, M., Greenbaum, M., and Weinstein, S. Use of fluorescent probes that from intramolecular excimers to monitor structural changes in model and biological membranes, *Biophys. J.*, 1981, 34, 499–515.

38. Alley, M. C., Scudiero, D. A., Monks, A., Hursey, M. L., Czerwinski, M. J., Fine, D. L., Abbott, B. J., Mayo, J. G., Shoemaker, R. H., and Boyd, M. R. Feasibility of drug screening with panels of human tumor cell lines using a microculture tetrazolium assay. *Cancer Res.*, 1988, 48, 589–601.

# pH/Temperature-Sensitive Polymers for Controlled Drug Delivery

SOON HONG YUK,[1] SUN HANG CHO[1]
SANG HOON LEE,[1] JUNG KI SEO[2]
JIN HO LEE[2]

## INTRODUCTION

As the solvent in a polymer solution becomes poorer, e.g., through a temperature change, a phase transition will eventually take place. There have been a number of reports on the phase transition polymers in response to various external stimuli such as pH [1–5], temperature [6–10], light [11–14], and chemical substances [15–20]. These polymer systems have been model systems for understanding the fundamental and classic problems in polymer physics.

With the advance of pharmaceutical science, it has been recognized that constant release is not the only way to maximize drug effectiveness and minimize side effects and that the assumptions used for constant release rate sometimes fail due to physiological conditions. From this perspective, zero-order drug release is not acceptable in all cases and externally modulated or self-regulating drug delivery systems have been developed as novel approaches to deliver drugs as required. To realize such drug delivery systems, it is important to construct a system where the drug itself senses environmental stimuli and responds appropriately to control the drug release. For this purpose, the phase transition polymers have been intensively exploited as a candidate material during last decade [21].

[1]Advanced Materials Division, Korea Research Institute of Chemical Technology PO Box 107, Yusung, Taejeon, Korea 305–600.
[2]Department of Macromolecular Science, Han Nam University, 133 Ojeong Dong, Daedeog Ku, Taejeon, Korea 300–791.

In this study, we demonstrate new pH/temperature-sensitive polymers with transitions resulting from both polymer–polymer and polymer–water interactions and their applications as stimuli-responsive drug carriers [22–23]. For this purpose, copolymers of (N,N-dimethylamino)ethyl methacrylate (DMAEMA) and ethylacrylamide (EAAm) [or acrylamide (AAm)] were prepared and characterized as polymeric drug delivery systems modulated for pulsatile and time release.

## EXPERIMENTAL SECTION

## POLYMER PREPARATION

### Materials

AAm monomer was purchased from Junsei Chemical Co., Japan. DMAEMA monomer, ammonium persulfate (APS), and tetramethylethyldiamine (TEMED) were purchased from Aldrich. Bovine insulin, N,N-azobis(isobutyronitrile) (AIBN) and glucose oxidase (GOD) were purchased from Sigma Chemical Co. DMAEMA monomer was distilled before use. Other reagents were used as received.

### Synthesis

Copolymers of DMAEMA and AAm were prepared by free radical polymerization in water at room temperature using APS as initiator and TEMED as accelerator. The feed compositions for poly(DMAEMA-co-AAm) are shown in Table 1. The initiator and accelerator concentrations were 2 mg/mL

TABLE 1. Feed Composition for Copolymers[a] in the Study.

| Code | DMAEMA | | AAm | | $M_w/10^{5b}$ |
|---|---|---|---|---|---|
| | g | mol% | g | mol% | |
| Poly DMAEMA | 4.2 | 100 | — | — | 2.8 |
| Copolymer I | 11.4 | 80 | 1.46 | 20 | 3.2 |
| Copolymer II | 9.58 | 67 | 2.40 | 33 | 3.8 |
| Copolymer III | 7.15 | 50 | 3.65 | 50 | 3.5 |
| Copolymer IV | 5.72 | 40 | 4.38 | 60 | 3.0 |
| Poly AAm | — | — | 7.2 | 100 | |

[a]Polymers were synthesized in 90 mL of $H_2O$.
[b]Measured by laser scattering.

of APS and 2.4 μL/mL of TEMED, respectively. All polymers were purified by dialysis against distilled-deionized water at room temperature and freeze-dried.

EAAm was synthesized in our laboratory as described previously [24]. Copolymers of DMAEMA and EAAm were prepared by free radical polymerization as follows: 7.8 g of distilled monomers (mixtures of DMAEMA and EAAm) and 0.02 g of AIBN as an initiator were dissolved in 100 mL of a (50/50 by volume) water/ethanol mixture. The feed compositions for poly(DMAEMA-*co*-EAAm) are shown in Table 2. The ampoule containing the solution was sealed by conventional methods and immersed in a water bath held at 75°C for 15 h. After polymerization, all polymers were dialyzed against distilled-deionized water at 4°C and freeze-dried.

## PH/TEMPERATURE-INDUCED PHASE TRANSITION

### Transmittance Measurements

The phase transition was traced by monitoring the transmittance of a 500 nm light beam on a Spectronic 20 spectrophotometer (Baush & Lomb). The concentration of the aqueous polymer solution was 5 wt%, and the temperature was raised from 15 to 70°C in 2° increments every 10 min. To observe their pH/temperature dependence, the phase transitions of polymers in citric-phosphate buffer solution versus temperature at two pH values (4.0 and 7.4) were measured.

### FT-IR Measurement

For Fourier transform infrared (FT-IR) measurement, thin films of polymers were cast from 0.5 wt% distilled-deionized water onto CaF$_2$ plates at room temperature. Most of the water in the films was removed by evaporation at 50°C in a vacuum oven for 24 h. FT-IR spectra of the dried film were

TABLE 2. Feed Composition for Copolymers in the Study.

| Code | DMAEMA | | EAAm | | $M_w/10^{4a}$ |
|---|---|---|---|---|---|
| | g | mol% | g | mol% | |
| Poly DMAEMA | 14.2 | 100 | — | — | 2.8 |
| Copolymer 1 | 11.4 | 80 | 1.9 | 20 | 1.3 |
| Copolymer 2 | 8.5 | 60 | 3.9 | 40 | 2.4 |
| Copolymer 3 | 7.1 | 50 | 4.9 | 50 | 2.9 |

[a]Measured by laser scattering.

measured on a Magna IR spectrophotometer (Nicolet Inc.) using 64 average scans at a resolution of 4 cm$^{-1}$.

### $^{13}$C-NMR Measurement

$^{13}$C-NMR spectra were recorded on a Bruker DRX-300 (300 MHz). The 8–10% (w/v) solution of polymer in CDCl$_3$ was used. The triad tacticity was determined from the peak intensities of the α-methyl carbon in the spectra.

## TEMPERATURE-CONTROLLED SOLUTE PERMEATION

### Preparation of Polymer Membrane

A series of crosslinked copolymer gels composed of DMAEMA and AAm were prepared using methylenebisacrylamide as a crosslinker for the preparation of polymer membrane. The feed compositions for the polymer membranes are listed in Tables 1 and 2. The polymerization was carried out between two Mylar sheets separated by a rubber gasket (1-mm diameter) and backed by glass plates. After polymerization, the gel was immersed in distilled water for 3 days to remove unreacted compound. The thickness of gel membrane was 1 mm in swollen state (20°C).

### Swelling Equilibrium Measurements

After immersion in water at a desired temperature, the gel was removed from the water and tapped with a filter paper to remove excess water on gel surface. The gel was repeatedly weighed and reimmersed in water at a fixed temperature until the hydrated weight reached a constant value. After equilibration at one temperature, the gel was reequilbrated at a higher temperature. The swelling, defined as the weight of water uptake per unit weight of dried gel, was calculated by measuring the weight of swollen gel until weight changes were within 1% of the previous measurement.

### Permeation Experiments

Two-compartment glass permeation cells were used for solute permeation study as a function of temperature. Hydrocortisone was used as a model solute. The volume of each compartment was 6 cm$^3$ and the area for the diffusion was 1.77 cm$^2$. Stirring was maintained at 600 rpm for all experiments via internal bar magnet. The donor compartment was filled with 2 wt% hydrocortisone aqueous solution (the hydrocortisone water solubility at 25°C is 280 μg/cm$^3$) and the receptor compartment was filled with water.

The entire receptor compartment was sampled and the drug concentration was assayed at 248 nm using a UV spectrophotometer (Shimadzu, Japan).

## GLUCOSE-CONTROLLED INSULIN RELEASE

### Preparation of Insulin-Loaded Matrix

Lyophilized copolymer was ground down to colloidal dimensions (<1 μm) using a laboratory planetary mill (Pulverisette, Fritsch GmbH). A 110-mg sample of copolymer powder, 20 mg of bovine insulin, and 20 mg of GOD were mixed, and the mixture was compressed into a disk-shaped matrix of 5-mm thickness and 15-mm diameter.

### Measurement of Weight Loss of Insulin-Loaded Matrix in Response to Glucose

After immersion in phosphate buffer solution (PBS) for a desired time at 37°C, the insulin-loaded matrix was removed and dried in a vacuum oven at room temperature. The percent of weight of the matrix was determined as a function of time.

### Release Experiments

Insulin release from the insulin-loaded matrix was measured in response to alternating changes of glucose concentration when 150 mg of an insulin-loaded matrix was introduced to 100 mL of PBS at 37°C. The amount of insulin released was measured by taking 1 mL of the release medium at a specific time and immersing the matrix in a fresh medium. Insulin was determined by reverse-phase HPLC, using a Resolvex $C_{18}$ (Fisher Scientific) and 0.01 N HCl/acetonitrile (80/20-50/50, v/v%) mobile phase over 30 min at a flow rate of 1 mL/min. The eluate was monitored by optical absorption at 210 nm.

## RESULTS AND DISCUSSION

### PH/TEMPERATURE-INDUCED PHASE TRANSITION

As reported previously [22–23], polyDMAEMA and PolyEAAm have lower critical solution temperatures (LCSTs) at 50° and 80°C, respectively. Interestingly, poly[DMAEMA-*co*-EAAm (or AAm)] exhibits the LCST between 0° and 50°C, depending on the copolymer composition.

The effect of the EAAm (or AAm) content on the LCST in water is shown in Figure 1. When the temperature of a polyDMAEMA aqueous solution

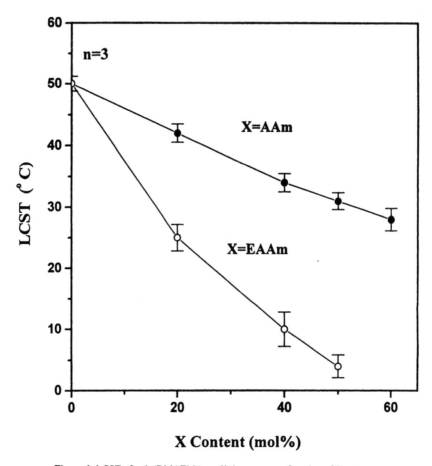

**Figure 1** LCST of poly(DMAEMA-*co-X*) in water as a function of *X* content.

was raised above 50°C, the polymer precipitated from the solution. This is due to the hydrophobic interaction between (*N,N*-dimethylamino)ethyl groups above the LCST. With the incorporation of EAAm (or AAm) in the copolymer, the LCST was shifted to a lower temperature. In general, the LCST should increase with increasing hydrophilicity of the polymer [25]. However, a LCST shift to a lower temperature was observed with the incorporation of the hydrophilic EAAm (or AAm). In addition, the LCST shift of poly(DMAEMA-*co*-EAAm) is larger that in the poly(DMAEMA-*co*-AAm).

FT-IR studies were used to characterize the intra/imtermolecular interaction in the copolymer as shown in Figures 2 and 3. For copolymers of EAAm (or AAm) with DMAEMA, a rather broad band centered at approximately 3,300 cm$^{-1}$ (3,3340 cm$^{-1}$) was observed, which was assigned to

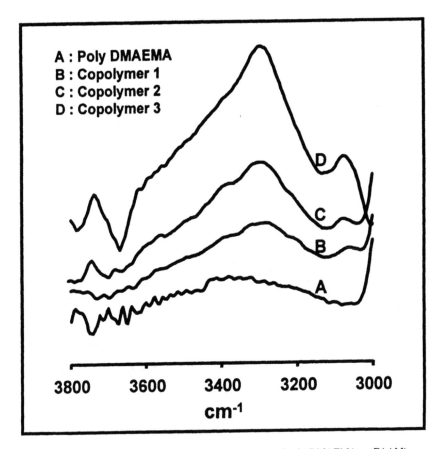

**A : Poly DMAEMA**
**B : Copolymer 1**
**C : Copolymer 2**
**D : Copolymer 3**

D

C

B

A

3800    3600    3400    3200    3000

**cm⁻¹**

**Figure 2** N-H stretching region of the infrared spectrum of poly(DMAEMA-*co*-EAAM).

ethylamide (or amide) groups that are hydrogen bonded to DMAEMA. The absorbance at approximately 3,300 cm⁻¹ (3,340 cm⁻¹) increased with the EAAm (or AAm) content in the copolymer. The disubstituted amide is known to be a powerful hydrogen bond acceptor. As for the unsubstituted amide such as AAm, its role as a hydrogen bond donor or acceptor has also been demonstrated [26]. Therefore, EAAm, a monosubstituted amide can be a hydrogen bond donor or acceptor depending on the environment. Because the dimethylamino group in DMAEMA is known to be a powerful hydrogen acceptor [27], it is reasonable to suggest efficient hydrogen bondings between ethylamide (or AAm) and *N, N*-dimethylamino groups.

From these results, it can be concluded that the LCST shift to the lower temperature is due to the formation of hydrogen bonds, which protect (*N,N*-dimethylamino)ethyl groups from exposure to water and result in a

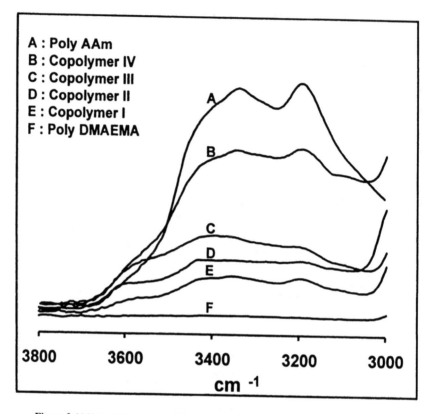

**Figure 3** N-H stretching region of the infrared spectrum of poly(DMAEMA-*co*-AAm).

hydrophobic contribution to the LCST as shown schematically in Figure 4. Because of the more hydrophobic nature of EAAm, poly(DMAEMA-*co*-EAAm) exhibits a more significant temperature responsiveness than poly (DMAEMA-*co*-AAm).

To understand the intra/intermolecular interaction in more detail, we compared the LCST change of the mixture of polyEAAm (or AAm) and polyDMAEMA in appropriate ratio with that of copolymer. With an increase in polyEAAm (or polyAAm) content, the LCST of polymer mixture was shifted to lower temperatures, which was observed in the copolymers of DMAEMA and EAAm (or AAm) as shown Figure 1.

However, the difference was observed in the LCST change of polymer mixture depending on the preparation method of polyDMAEMA. As presented previously, polyDMAEMA can be prepared in water containing APS as an initiator and TEMED as an accelerator or water/ethanol solvent mixture containing AIBN as an initiator. PolyDMAEMA prepared in the

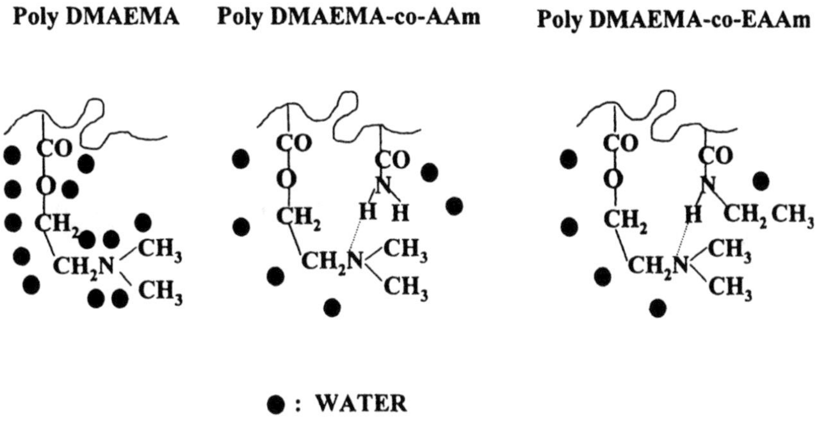

**Poly DMAEMA**  **Poly DMAEMA-co-AAm**  **Poly DMAEMA-co-EAAm**

● : WATER

**Figure 4** Schematic representation of intra/intermolecular interaction via hydrogen bond.

water shows the LCST at 50°C and that prepared in the water/ethanol mixture shows an LCST at 30°C. The LCST of polyDMAEMA prepared in water was shifted to the lower temperature with the addition of polyEAAm (AAm) content, which was observed in the aqueous copolymer solution. However, a polymer aggregate was formed with the addition of a small amount of polyEAAm (AAm) in the case of polyDMAEMA prepared in the water/ethanol mixture. This indicates that there is a difference in the stereochemical configuration of polyDMAEMA, depending on the preparation method. To understand this behavior of polyDMAEMA, the tacticity of poly DMAEMA was investigated. Figure 5 shows the $\alpha$-CH$_3$ portion of the $^{13}$C-NMR spectra of polyDMAEMA prepared at two different conditions. The relative tacticities calculated from the peak areas are given in Table 3. The data show that polyDMAEMA prepared in the water does not contain syndiotactic triads; however, the polymer prepared in water/ethanol mixture contain syndiotactic triads. There have been a number of studies on the tacticity effect on polymer properties. Especially, poly(2-hydroxyethyl methacrylate) (HEMA) was investigated from the viewpoint of stereochemical configuration. In this study, several differences between the isotactic and syndiotactic chains were reported [28]. First, the isotactic chain has no intrachain hydrogen bonds linking its ester side chains. On the other hand, every sequence in the syndiotactic chains bring two pendant hydroxy groups into close proximity where they can form a hydrogen bond that stabilizes the local chain conformation. These differences in stereochemical configuration can lead to the significant change in physical properties. Therefore, polyDMAEMA with syndiotatic triads has stable intramolecular interaction that is more favorable for the temperature-induced phase transition

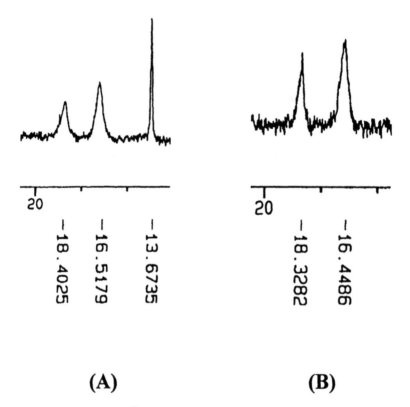

**Figure 5** α-CH₃ portion of the ¹³C-NMR spectra of poly DMAEMA; (a) prepared in water/ethanol, (b) prepared in water.

caused by a hydrophobic interaction between (dimethylamino) ethyl groups above the LCST and has a LCST at 30°C that is lower than that of poly-DMAEMA without syndiotatic triads. This may lead to the aggregation of polyDMAEMA with syndiotactic triads with the addition of small amount of polyEAAm (or polyAAm).

TABLE 3. Chemical Shift for Various α-CH₃.

| Polymer | δα-CH₃ (ppm) | | | Area | | |
|---|---|---|---|---|---|---|
| | i | h | s | i | h | s |
| Poly DMAEMA prepared in water/ethanol solvent mixture | 18.36 | 16.49 | 13.63 | 3.28 | 5.49 | 2.95 |
| Poly DMAEMA prepared in water | 18.32 | 16.45 | — | 3.34 | 5.42 | — |

i: Isotactic triads, h: atactic triads, s: syndiotactic triads.

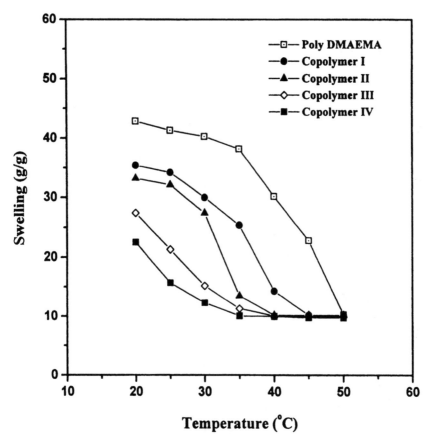

**Figure 6** LCST of poly(DMAEMA-*co*-EAAM) in citric-phosphate buffer solution at pH 4.0 and 7.4.

## THERMOSENSITIVE PERMEATION

Based on the solution property of poly (DMAEMA-*co*-AAm) in response to temperature, the temperature dependence of equilibrium swelling of poly (DMAEMA-*co*-AAm) gel as a function of chemical composition was observed as shown in Figure 6. The transition temperature of copolymer gel between the shrunken and swollen state was shifted to the lower temperature with increases in AAm content in the gel network. This is attributed to the hydrogen bond in the copolymer gel network and its hydrophobic contribution to the LCST. Copolymer II gel was selected as a model polymer network for permeation study because it showed the sharp swelling transition around 34°C.

The permeation of hydrocortisone across the gel membrane composed of copolymer II was observed in response to pulsatile temperature changes between 20 and 40°C. As shown in Figure 7, higher permeation was observed

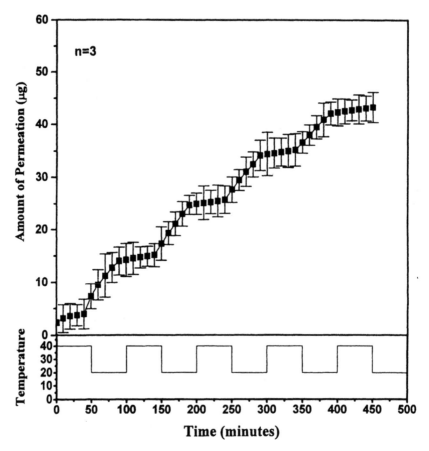

**Figure 7** Equilibrium swelling change of poly(DMAEMA-*co*-AAm) gel in response to temperature change.

at 20°C and reduced permeation was observed at 40°C. This is attributed to the permeation of hydrocortisone through swelled or deswelled gel network depending on the temperature.

### GLUCOSE-SENSITIVE INSULIN RELEASE

The temperature-induced behavior of copolymers at two pHs (4.0 and 7.4) is observed as shown in Figure 8. At pH 4, no LCST was observed with poly-DMAEMA, and the LCST of all copolymers was increased compared to that at pH 7.4. At pH 4.0, (*N,N*-dimethylamino)ethyl groups of DMAEMA are fully ionized. An increasing electrostatic repulsion between charged sites on DMAEMA disrupts the hydrogen bonds between EAAm and DMAEMA. These interfere with the hydrophobic interactions between (*N,N*-dimethyl-

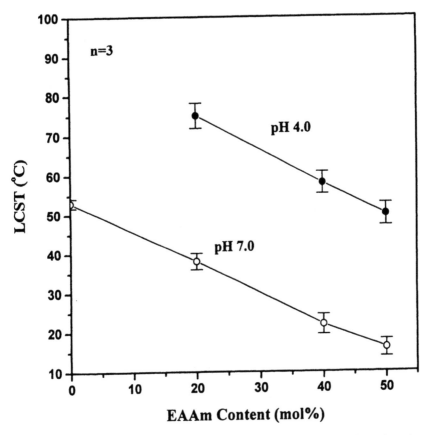

**Figure 8** Permeation of hydrocortisone across the poly(DMAEMA-*co*-AAm) gel membrane in response to pulsatile temperature change.

amino)ethyl groups above the LCST and the hydrophobic contribution to the LCST due to hydrogen bonding.

Based on the pH/temperature-responsiveness of polymer, glucose-controlled insulin release system was designed as shown schematically in Figure 9. In the presence of glucose, gluconic acid generated by the glucose-GOD reaction protonates dimethylamino groups of poly(DMAEMA-*co*-EAAm), inducing the LCST shift to a higher temperature from the surface of the insulin-loaded matrix. This leads to the disintegration of the matrix with polymer dissociation from the surface with the insulin release. Poly(DMAEMA-*co*-EAAm) with 50 mol% of EAAm was selected as a model polymer for the preparation of an insulin-loaded matrix considering its pH/temperature responsiveness (see Figure 8).

Shown in Figure 10 is the weight loss of the matrix in PBS at two different glucose concentrations. We found that 100% of the initial weight had

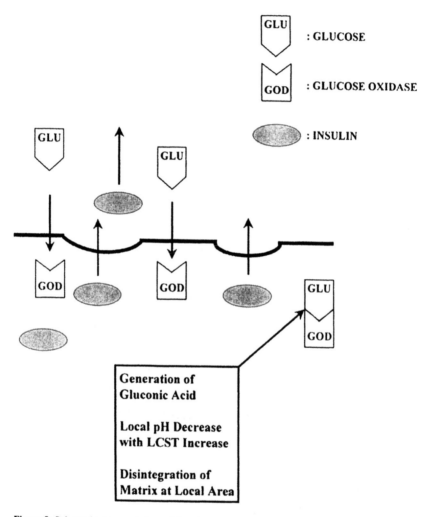

**Figure 9** Schematic representation of the glucose-controlled insulin release using poly(DMA-EMA-*co*-EAAm).

been lost at 5 g/L of glucose concentration during 24 hours, whereas 10% of the weight had been lost at 0.5 g/L of glucose concentration. These results indicate that this polymer system responds to the change of glucose concentration in the presence of GOD.

Shown in Figure 11 is the alternating insulin release rate in response to an alternating exposure of the insulin-loaded matrix to high and low aqueous glucose solutions. Minimal release was observed at the lower concentration of glucose. The large deviation in the release rate was attributed to inhomogeneous mixing of the components in the insulin-loaded matrix causing its irregular dissolution. The release rate was not maintained constantly from

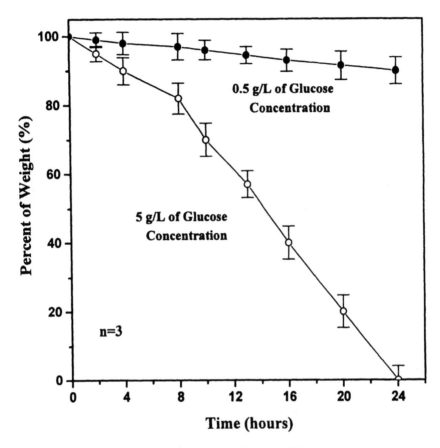

**Figure 10** Disintegration of the polymer matrix in phosphate buffer solution at two glucose concentrations.

cycle to cycle. As the release experiment was continued, the matrix swelled from the surface, and this led to the rapid disintegration of the matrix, resulting in the increase of release rate. At the later stage of the release experiment, the decreased amount of insulin in the matrix after cycle affected the decreasing release rate.

## CONCLUSIONS

New pH/temperature-sensitive polymer systems with transitions resulting from both polymer–polymer and polymer–water interactions have been demonstrated and their pH/temperature-induced phase transition has been investigated. Intra/intermolecular interactions via hydrogen bond play an important role in determining the phase transition. By manipulating the

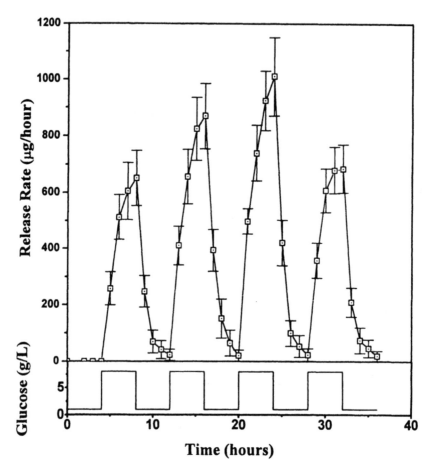

**Figure 11** Insulin release from the insulin-loaded matrix in response to alternating change of glucose concentration.

pH/temperature-responsiveness of polymers, glucose-controlled insulin release and thermosensitive permeation of hydrocortisone have been accomplished. Although these systems are far from a practical application, the concept reported here may be used in the design of stimulus-modulated drug delivery system.

## REFERENCES

1. Kopecek, J., Vacik, J., and Lim, D. *J. Polym. Sci.*, 1971, 9, 2801.
2. Chung, D. W., Higuichi, H., Maeda, M., and Inoue, S. *J. Am. Chem. Soc.*, 1986, 108, 5823.

3. Siegel, R. A. and Firestone, B. A. *Macromolecules,* 1988, 21, 3254.

4. Brannon-Peppas, L. and Peppas, N. A. *Biomaterials,* 1990, 11, 635.

5. Yuk, S. H., Cho, S. H., and Lee, H. B. *J. Controlled Release,* 1995, 37, 69.

6. Heskins, M. and Guillet, J. E. *J. Macromol. Sci.-Chem.,* 1986, A2(8), 1441.

7. Hoffman, A. S., Afrassiabi, A. and Dong, L. C. *J. Controlled Release,* 1986, 4, 213.

8. Chen, G. H. and Hoffman, A. S. *Nature,* 1995, 373, 49.

9. Bae, Y. H., Okano, T. and Kim, S. W. *Makromol. Chem., Rapid Commun.,* 1987, 8, 481.

10. Ilman, F., Tanaka, T. and Kokufuta, E. *Nature,* 1991, 349, 400.

11. Irie, M. and Kunwatchakun, D. *Macromolecules,* 1986, 19, 2476.

12. Irie, M. *Macromolecules,* 1986, 19, 2890.

13. Irie, M. and Suzuki, T. *Makromol. Chem., Rapid Commun.,* 1987, 8, 607.

14. Suzuki, A. and Tanaka, T. *Nature,* 1990, 346, 345.

15. Ricka, J. and Tanaka, T. *Macromolecules,* 1984, 17, 2916.

16. Horbett, T. A., Ratner, B. D., Kost, J. and Singh, M. A bioresponsive membrane for insulin delivery, in: Anderson, J. M. and Kim, S. W. (Eds.), *Recent Advances in Drug Delivery Systems,* Plenum Press, New York, NY, 1984, p. 209.

17. Horbett, T. A., Kost, J. and Ratner, B. D. Swelling behavior of glucose sensitive membrane, in: Shalaby, S. W., Hoffman, A. S., Ratner, B. D. and Horbett, T. A. (Eds.), *Polymers as Biomaterial,* Plenum Press, New York, NY, 1984, p. 193.

18. Kokufuta, E., Zhang, Y.-Q. and Tanaka, T. *Nature,* 1991, 351, 302.

19. Kataoka, K., Miyazaki, H., Okano, T. and Sakurai, Y. *Macromolecules,* 1994, 27, 1061.

20. Stayton, P. S., Shimoboji, T., Long, C., Chilkoti, A., Chen, G. H., Harris, J. M. and Hoffman, A. S. *Nature,* 1995, 378, 472.

21. Kim, S. W. and Bae, Y. H. Smart drug delivery system, in: Sam, A. P. and Fokkens, J. G. (Eds.), *Innovation in Drug Delivery: Impact on Pharmacotherapy,* The Anselmus Foundation, Houten, The Netherlands, 1995, p. 112.

22. Cho, S. H., Jhon, M. S., Yuk, S. H. and Lee, H. B. *J. Polym. Sci. B: Polym. Phys.,* 1997, 35, 595.

23. Yuk, S. H., Cho, S. H. and Lee, S. H., *Macromolecules,* 1997, 30, 6856.

24. McCormick, C. L., Nonaka, T. and Johnson, C. B. *Polymer,* 1988, 29, 371.

25. Feil, H., Bae, Y. H., Feijen, J. and Kim S. W. *Macromolecules,* 1993, 26, 1259.

26. Abe, K. and Koide, M. *Macromolecules,* 1977, 10, 2496.

27. McCormick, C. L., Blackmon, K. P. and Elliott, D. L. *Polymer* 1986, 27, 1976.

28. Russell, G. A., Hiltner, P. A., Gregonis, D. E., deVisser, A. C. and Andrade, J. D., *Polym. Sci.,: Polym. Phys Ed.,* 1980, 18, 1271.

# CHAPTER 4

# Design of Polymeric Systems for Targeted Administration of Peptide and Protein Drugs

EMO CHIELLINI[1], ELISABETTA E. CHIELLINI[2]
FEDERICA CHIELLINI[1], ROBERTO SOLARO[1]

## INTRODUCTION

PEPTIDE and protein drugs are assuming an ever-growing interest as a novel and effective class of therapeutic agents [1–5]. Their use is, however, limited by the rapid clearance they experience from body compartments, which can lead to the administration of exceedingly high doses in order to maintain an acceptable therapeutic level. These may cause toxic effects, which could force a stop to continuing therapy. Possible solutions to avoid these drawbacks are represented by the realization of smart systems that allow site specific administration of protein drugs embodied into a polymeric core or a coatings shell impervious to body immunological responses.

In this respect, we have been involved in research activities aimed at developing nanoparticle delivery systems based on bioerodible/biodegradable polymeric matrices for targeted administration of protein drug such as $\alpha$-interferon (IFN$\alpha$). In the present paper we report a strategy that was adopted for the design of polymeric matrices suitable to establish nondenaturing interactions with protein components. In addition, these conditions should be susceptible to provide formulations amenable to the production of nanoparticles, holding on the surface structural moieties with site-specific recognition activity. The parenchymal liver cells were the selected organ, the disease to be treated was hepatitis C and the protein drug selected was human leucocyte $\alpha$-interferon (IFN$\alpha$).

[1]Department of Chemistry and Industrial Chemistry, University of Pisa, via Risorgimento 35, 56126 Pisa, Italy.
[2]VectorPharma International SpA, via del Follatoio 12, 34148 Trieste, Italy.

The data presented here constitute part of the results attained in the development of a research project funded by the European Community (EC) and performed in cooperation with three independent European companies [6]. Hybrid polymeric materials based on intimate blends of human serum albumin (HSA), alkyl hemiesters of alternating copolymers of maleic anhydride (MAn), and vinyl ethers of monomethoxyoligoethylene glycols (PEGVE) were selected as biocompatible matrices for the formulation of the nanoparticles.

## EXPERIMENTAL

The preparation and characterization of alternating copolymers of Maleic anhydride (MAn) and poly (ethylene glycol-vinyl ether) as well as their chemical conversions to provide various alkyl hemiesters (Scheme 1) have been described elsewhere [7]. The matrices are quoted as PAM$mn$, where $m$ represent the number of oxyethylene units in R and $n$ the number of carbon atoms in R. Human serum albumin (HSA) was provided by Istituto Sierovaccinogeno Italiano SpA, Italy.

Multihydroxyl containing β-cyclodextrins (βCD) were prepared by grafting glycidyl ethers of glycerol monoacetonide and of pentitol (xylitol, arabitol, and ribitol) diacetonides on βCD under experimental conditions described elsewhere [8]. The controlled removal of the protecting acetonyl groups was carried out under experimental conditions not affecting the

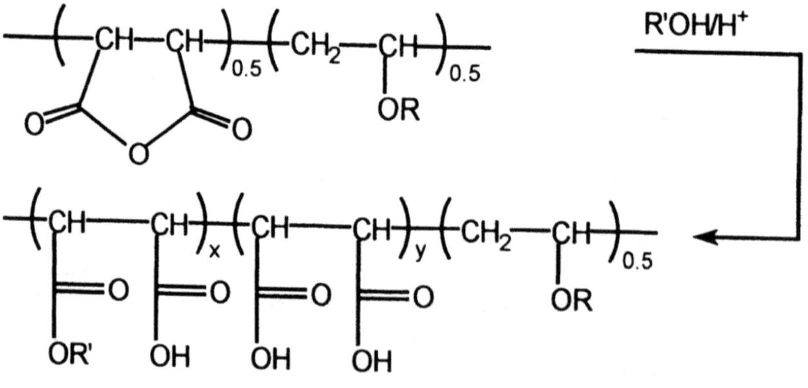

R = $-(CH_2CH_2O)_mCH_3$  (m= 1-4)
R' = $-(CH_2)_nH$  (n = 1-4)
x + y = 0.5

Scheme 1 Synthetic polymeric matrices used in the formulation of nanoparticles.

structure of the heptaglucosyl toroidal backbone of the native βCD [9]. The average degree of grafting was determined by NMR coupled to GPC analysis. Analogous conditions were adopted in the further succinoylation of the multihydroxylated βCD obtained after complete removal of the protecting acetonyl moieties [10]. Polar lipids such as digalactosyl glycerol (DGDG), the corresponding fully hydrogenated derivative (HDGDG), and dihexosyl-ceramide sphingolipid were provided by Scotia LipidTeknik (Sweden).

The hybrid matrices from the various PAM*mn* and blends of HSA-IFNα or HSA-myoglobin (MYO) in which MYO was used as a model substitute for IFNα, comprised in some cases of perhydroxylated βCD prepared according to a nondenaturing technique operating at low temperature [11]. Total protein assay, water absorption and weight loss, biocompatibility tests including *in vitro* platelet aggregation, complement activation, *in vivo* tests for acute toxicity, and acute thromboembolia were carried out according to procedures already described [8].

The nanoparticles were prepared according to two different procedures:

(1) *Slow solvent evaporation:* as a typical example, 313 mg of polymer matrix PAM14 and 228 mg of DGDG lipid were added under stirring to 10 mL of CHCl$_3$. Once a homogenous solution was obtained, 20 mg of HSA loaded with IFNα were added under stirring. After 15 min the resulting dispersion was added dropwise to 100 mL of water (acidified with HCl to pH 5.0) under stirring with an ultratorrex mixer at 8800 rpm. Once the addition was complete, the CHCl$_3$ was slowly evaporated at 45°C under constant stirring. After evaporation of all the organic solvents the dispersion was stored at 4°C.

(2) *Controlled coprecipitation:* as a typical example, a solution containing 200 mg of the hemiester PAM14 dissolved in 4 mL of 1/4 water/ethanol mixture was added dropwise from a syringe to a solution consisting of 40 mg of HSA loaded with IFNα and 500 mg of βCD grafted with glycidyl di*iso*propylidenearabitol in 10 mL of double-distilled water at room temperature under vigorous magnetic stirring. A Water/ethanol dispersion containing 3 mL of water and 20 mg of DGDG was added to the former suspension. The suspension was maintained under stirring for another 60 min and then stored at 4°C. In some cases the ultimate suspension was submitted to yophylization and the suspension regenerated when needed.

In both procedures purification of the nanoparticles was carried out either by means of filtration of the suspension over a 0.2 μm cellulose acetate filter or by Eppendorf centrifuging for 15 min followed by resuspension in water up to a final suspension of the nanoparticles in PBS buffer (pH 7.2) or in physiological solution. Nanoparticles loaded with a fluorescent marker were also prepared by using HSA containing 9 fluorescein *iso*cyanate groups per molecule (HSA-FITC). The distribution of the particle size was performed by means of a Zetasizer 4-Malvern Instrument by using a ZET5110 cell assembled in an orthogonal mode and a multimodal data elaboration.

Biological characterization of the nanoparticles was carried out by monitoring *in vitro* interactions with hepatocytes isolated from rat liver [12]. Haemagglutination inhibition test of erythrocytes with ricine agglutinin (RCA$_{120}$) was carried out [13]. The fluorescence of the hepatocytes incubated with the FITC-labeled nanoparticles was determined by means of a FACS Star Becto-Dickinson instrument.

### RESULTS AND DISCUSSION

The design of new injectable dosage forms for the administration of peptide or protein drugs for the realization of the targeted release systems in the nanoscale range requires three fundamental ingredients:

(1) A compatible polymeric matrix soluble in water or water/organic solvents provided with structural functionality suitable to interact with protein drugs and protein stabilizers without any adverse effects.
(2) A low or high molar mass stabilizing component whenever the polymer matrix of choice does not exert a stabilizing effect on the trapped protein drug.
(3) A structurally defined component able to comply with the requirements quoted in 1 and 2 and to establish interactions with specific receptor sites of the organ taken as the target for the drug treatment.

A series of carboxyl containing bioerodible polymeric materials, characterized by modulated functionality and hydrophobic/hydrophilic balance, was prepared both on a lab-scale and in the pilot plant. Procedures were setup as amenable for scaled-up productions. Those materials displayed a high versatility to combine with proteins in different proportion and to provide hybrid bioerodible matrices without any adverse effect on protein structure and activity.

The synthetic polymeric components as well as their combinations with proteins such as human serum albumin (HSA), bovine serum albumin (BSA), human serum albumin/α-interferon mixtures (HSA-IFNα) and myoglobin (MYO) did not give any negative response to *in vitro* and *in vivo* biocompatibility tests, such as platelet aggregation, complement activation, acute toxicity, and acute thromboembolic potential.

The choice of protein stabilizer stemmed from the necessity of burying the hybrid formulate in a water pool structured by a strongly hydrophilic coating shell. This appears to be an extremely efficient option allowing for the minimization of opsonizing effects on foreign bodies eventually injected in the bloodstream.

Multihydroxyl containing monomeric or oligomeric β-cyclodextrins (βCD) such as those attained by grafting with glycidyl ethers of protected polyols (glycerol and pentitols) appeared rather promising components for their amphiphilic character, connected to the presence of an hydrophobic pocket and an external hydrophilic shell with an amplified number of hydroxyl groups.

Targeting of the new dosage form to the asialoglycoprotein receptors located on the liver hepatocytes was achieved by the inclusion of a galactolipid, such as digalactosylglycerol (DGDG), the corresponding fully hydrogenated derivative (HDGDG), and dihexosyl-ceramide sphingolipid, in the hybrid formulation.

## PREPARATION OF MICROSPHERE SUSPENSIONS

Using two different procedures, based, respectively, on slow solvent evaporation[14] and on coprecipitation, a new proprietary method was developed during the fulfillment of a research project funded by the European Community (EC) under the Brite-Euram Program[15,16], were tested for the preparation of microspheres.

### By Slow Solvent Evaporation

An organic solution of synthetic polymer and lipid, containing a dispersion of the protein, was added under vigorous stirring to a dilute aqueous solution at pH 5. The resulting oil-in-water emulsion was converted to a nanoparticle suspension by slow evaporation of the volatile organic solvent. The strong tendency of nanoparticles to coalesce during solvent removal constituted the major difficulty connected with this procedure.

Polymer/lipid and polymer/protein ratios, spanning a rather wide range, were adopted to highlight their influence on nanosphere sizes and physical stability of the dispersions.

The slow solvent evaporation technique afforded milky nanosphere dispersions. No effect of the polymer/protein/lipid weight ratios on the dispersion stability was detected, at least in the investigated range.

Dynamic light scattering analysis of the dispersions highlighted a heterogeneous dimensional distribution that in most cases were bimodal (Figure 1). Depending on the suspension chemical composition, particles with an average diameter ranging between 0.1 and 3.4 μm were obtained.

SEM micrographs of lyophilized samples show the presence of particle dispersion of comparable size embedded in a homogeneous matrix

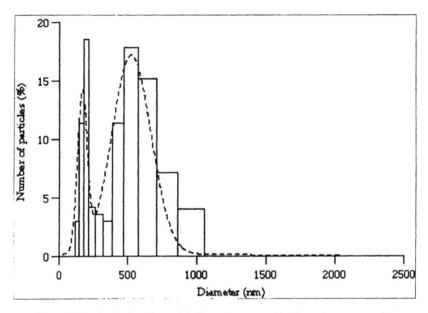

**Figure 1** Size distribution of a particle dispersion prepared by slow solvent evaporation.

(Figure 2), consisting of residual polymer, protein, and lipid not incorporated in the microspheres.

### By Co-precipitation

By taking into account the difficulties connected to the slow solvent evaporation technique, a new procedure for nanoparticle preparation by co-precipitation was set up. This technique does not use chlorinated solvents and energetic mixing, which are both known to cause appreciable protein denaturation[17,18].

The coprecipitation technique was based on the dropwise addition of a synthetic polymer solution, in a solvent mixture, into an aqueous protein solution under magnetic stirring. The progressive interaction between the water insoluble polymer and the protein gave rise to the microsphere formation. The glycolipid was then added as an aqueous dispersion to the nanoparticle suspension. No sedimentation was observed after several weeks of storage at room temperature.

A series of nanoparticle suspensions was prepared according to this technique by varying protein concentration, polymer/lipid ratio, and type of solvent used to dissolve the polymer in order to establish the best experimental conditions.

Better results, as far as dispersion homogeneity and stability are concerned, were obtained when the *n*-butyl hemiester (PAM14) of the

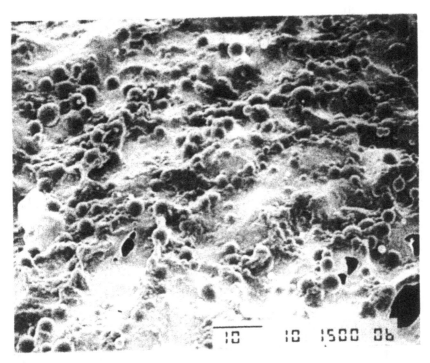

**Figure 2** SEM micrograph (1,500×) of a lyophilized particle dispersion prepared by slow solvent evaporation.

AnM,/Peg1VE copolymer and a 10:1:2 polymer/lipid/protein weight ratio was used.

Dimensional analysis of the different suspensions by dynamic light scattering indicated that the nanospheres had an average diameter of 0.15 μm and a polydispersity index of 0.1–0.3 (Figure 3). The use of β-cyclodextrins functionalized with the glycidyl ether of di*iso*propylideneribitol (GDR-βCD) as a dispersion stabilizer caused a slight increase of the average particle size to 0.2–0.3 μm.

Morphological analysis of the nanoparticle suspensions by SEM showed a homogeneous distribution of spheroidal particles with diameters less than 1 μm embedded in a continuous matrix (Figure 4) consisting of polymeric material not incorporated within the microspheres.

## MICROPARTICLE PURIFICATION AND CHARACTERIZATION

The nanoparticle suspensions were centrifuged three times at 8,000 rpm for 15 minutes. After each centrifugation the pellets were resuspended in double-distilled water. However, SEM analysis of the pellets showed extensive aggregation–fragmentation of the microspheres (Figure 5), that

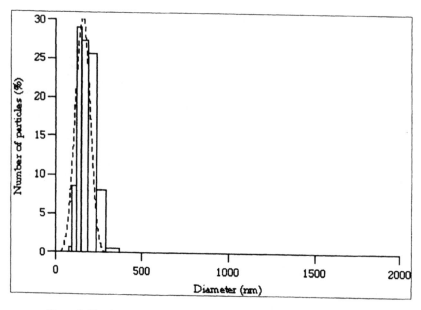

**Figure 3** Size distribution of a particle dispersion prepared by co-precipitation.

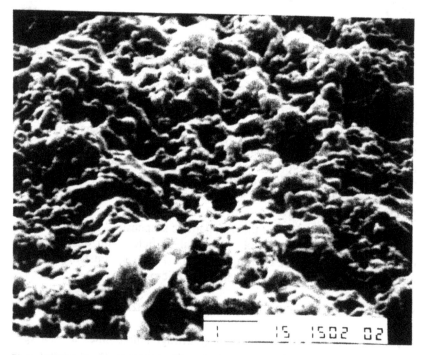

**Figure 4** SEM micrograph (15,000×) of a lyophilized particle dispersion prepared by co- precipitation.

**Figure 5** SEM micrograph (1,000×) of a centrifuged particle dispersion prepared by co-precipitation.

was likely due to the mechanical stress connected with the centrifugation process. On the other hand, a homogeneous distribution of almost spherical microparticles completely free from the embedding polymer matrix was obtained from samples prepared by using a protein solution containing 5% of functionalized βCD (Figure 6). Molecular modeling of the three-dimensional structure of IFNα and its docking with βCD provided an indication of the role played by the modified βCDs[19–21].

Reconstitution of the original dispersion was easily attained by resuspending the pellet either in water or in phosphate buffer solution.

The presence of exposed galactosyl residues on the microsphere surface was determined by an *in vitro* hemagglutination inhibition test. This test is based on the agglutination of blood red cells induced by ricine, a lectin from *Ricinus communis* characterized by a strong affinity towards galactose[13]. Galactosyl groups present on the surface of centrifuged nanoparticle dispersions containing DGDG effectively inhibited the hemagglutination process by competitively interacting with ricine receptors.

In order to test the ability of galactose labeled microspheres to actively target hepatocytes, some preliminary experiments were carried out by flow cytofluorimetry (FACS)[22]. Experiments were performed on rat hepatocyte primary cultures by using microparticle suspensions containing fluoresceinated human serum albumin (HSA-FITC). The best results were obtained with suspensions containing a combination of digalactosylglycerol (DGDG) and the corresponding hydrogenated derivative (HDGDG).

Preliminary information gained on the *in vivo* biodistribution of double-radiolabeled nanoparticles recorded in rabbits nicely confirmed the *in vitro* test carried out on a primary cell line of rat hepatocytes.

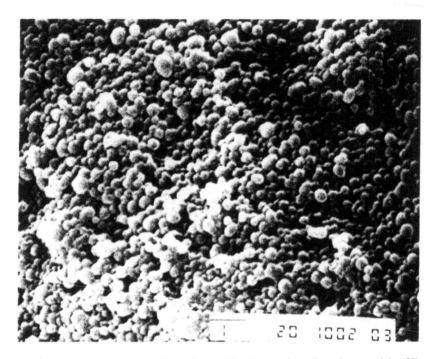

**Figure 6** SEM micrograph (10,000×) of a centrifuged nanosphere dispersion containing 5% of functionalized βCD

## CONCLUSIONS

Within the framework of a cooperative research project founded by the EC, a procedure has been developed for the formulation of injectable bio-erodible nanoparticles with fairly uniform size distribution and high environmental stability that appear well suited for targeted administration of protein and peptide drugs.

The prepared nanoparticles can be stored in lyophilized form and resuspended in physiological solutions prior to administration. The procedure developed on the laboratory scale has been found amenable to scale-up productions.

## REFERENCES

1. Lee, V. H. L. *Petide and Protein Drug Delivery,* Marcel Dekker, New York, 1991.
2. Auolus, K. L. and Raub, T. J. *Biological Barriers to Protein Delivery,* Plenum, New York, 1993.
3. Cohen, S. and Bernstein, H. *Micro-Particulate Systems for the Delivery as Proteins and Vaccines,* Marcel Dekker Inc., New York, 1996.

4. Sugiyama, K. and Oku, T. *Polym. J.,* 1995, 27(2), 179.
5. Schröder, U. and Mosbach, K. *US Pat.* 4,501,726, 1985.
6. Brite-Euram Project BE-7052, Contract BRE2.CT94. 0530 *Bioerodible-Biodegradable Polymeric Matrices for Targeted Protein Drug Release.*
7. Chiellini, E., Solaro, R., Leonardi, G., and Giannasi, D. *J. Bioact. Compat. Polym.,* 1992, 7, 161.
8. Chiellini, E., Solaro, R., Leonardi, G., Giannasi, D., Lisciani, R., and Mazzanti, G. *J. Bioact. Compat. Polym.,* 1992, 22, 273.
9. Solaro, R., D'Antone, S., Bemporad, L. and Chiellini, E. *J. Bioact. Compat. Polym.,* 1993, 8, 236.
10. Chiellini, E. E. *PhD thesis,* University of Brescia, Italy, 1996.
11. Chiellini, E., Solaro, R., Leonardi, G., Lisciani, R. and Mazzanti, G. *Eur. Pat. Appl.* 0509968A1, 1992.
12. Adachi, N., Maruyana, A., Ishihara, T. and Akaike, T. *J. Biomater. Sci. Polym. Edn.,* 1994, 6(5), 463.
13. Yoshioka, H., Ohmura, T., Hasegawa, M., Hirota, S., Makino, M. and Kamiya, M. *J. Pharm. Sci.,* 1993, 82(3), 273.
14. Arshady, R. *Polym. Eng. Sci.,* 1990, 30, 915.
15. Carlsson, A., Chiellini, E. E., Chiellini, F., Cowdall, J., Davies, J., Mazzanti, G., Roberts, M., Söderlind, E. and Solaro, R. *Ital. Pat. Appl.* RM97A0418, 1997; *PTC* IT98/00/92, 1998.
16. Carlsson, A., Chiellini, E., Chiellini, E. E., Cowdall, J., Davies, J., Mazzanti, G., Roberts, M., Söderlind, E. and Solaro, R. *Ital. Pat. Appl.* RM97A0417, 1997.
17. Lee, V. H. In *Peptide and Protein Drug Delivery,* Lee, V. H., Ed., Marcel Dekker, New York, 1991, 1.
18. Cleland, J. L. and Jones, A. J. S. *US Pat. Appl.* US/94/01666; *WO Pat. Appl.* 94/19020, 1994.
19. Miertus, S., Tomasi, J., Mazzanti, G., Chiellini, E. E., Solaro, R. and Chiellini, E. *Int. J. Biol. Macromol.,* 1997, 20, 85.
20. Miertus, S., Frecer, V., Chiellini, E., Chiellini, F., Solaro, R. and Tomasi, J. *J. Incl. Phenom. Mol. Recogn. Chem.,* 1998, 32, 23.
21. Miertus, S., Nair, A. C., Frecer, V., Chiellini, E., Chiellini, F., Solaro, R. and Tomasi, J. Modelling of β-cyclodextrin with L-α-aminoacids residues. *J. of Inclusion Phenomena and Marcrocyclic Chemistry,* 1999, 34, 69–84.
22. Chance, J. T., Larsen, S. A., Pope, V., Measel, J. W. and Cox, D. L. *Cytometry,* 1995, 22(3), 232.

# CHAPTER 5

# Drug Release from Ionic Drugs from Water-Insoluble Drug-Polyion Complex Tablets

NANDINI KONAR[1]
CHERNG-JU KIM[1]

## INTRODUCTION

IN the past, swellable polyelectrolyte gel matrices of both anionic and cationic types have been investigated for their drug release properties [1–3]. Such materials release drugs over an extended period of time by an ion-exchange mechanism but usually provide square root of time ($\sqrt{t}$) drug release kinetics with a tailing toward the end of release. In gel matrix systems, square root of time kinetics are the most common unless there is the increase of diffusion rate with time (time-dependent diffusion coefficient) in which case it is possible to obtain linear release kinetics [4]. However, the total release time is very short [5]. The degree of crosslinking as well as the nature of the charged functional groups controls the release kinetics by affecting the degree of swelling of the gel matrix.

The effects of the above two factors may be eliminated by designing a carrier that is not crosslinked, containing strongly charged functional groups on a sufficiently high molecular weight polymer that would decrease the dissolution rate enough to permit prolonged drug release. Binding an oppositely charged drug moiety to such a carrier would provide a material that possesses all the positive qualities of swellable polyelectrolyte gels while curtailing the negative aspects. Nujoma and Kim [6] demonstrated this by synthesizing a strongly anionic carrier poly(sulfopropyl methacrylate potassium-co-methyl methacrylate) (PSPMK/MMA)

[1]Temple University, School of Pharmacy, 3307 N. Broad St., Philadelphia, PA 19140, USA.

and binding it to cationic drugs. Zero-order release kinetics and high drug loadings (>40 wt%) were reported. In addition, Konar and Kim [7] found analogous results by synthesizing a strongly cationic carrier poly(trimethylammonium ethyl methacrylate chloride-*co*-methyl methcrylate) (PTMAEMC/MMA) and binding it to anionic drugs. Similar results were observed with other strongly cationic and anionic systems, poly(trimethylammonium acrylate chloride-co-methyl methacrylate (PTMAEAC/MMA), poly(methacrylamido propyl trimethyl ammonium chloride-co-methyl methacrylate (PMAPTAC/MMA), and poly(acrylamido-methyl-propanesulfonate sodium co-methyl methacrylate) (PAMPS/MMA) [8,9]. In the above systems the dissolution rate of the polymer was greatly reduced by copolymerizing the ionogenic monomer with a hydrophobic monomer such as methyl methacrylate. The resulting materials were found to dissolve very slowly but completely and proved to be excellent erodible carriers for prolonged drug release without significant tail effects. Also, the drugs ionically bound to the strongly charged groups could be dissociated only in an ionic medium, and the release kinetics were more dependent on the properties of the carrier and the bound drug than on those of the external environment.

In this study, we report the release properties of two new polyelectrolyte materials poly(acrylamido-methyl-propanesulfonate) (PAMPS) and poly (diallydimethyl ammonium chloride) (PDADMAC), which were used as anionic and cationic carriers, respectively, for oppositely charged drugs. These polymers proved to be very promising and practical as erodible carriers for controlled drug delivery as they are available commercially. Binding ionic moieties to the linear polymer backbone can be done by a simple mixing process.

## EXPERIMENTAL

All chemicals were used as received. PDADMAC and PAMPS were obtained from Aldrich Chemical Co. (Milwaukee, WI). Diclofenac sodium, sodium sulfathiazole, labetalol HCl, propranolol HCl, verapamil HCl, and diltiazem HCl were purchased from Sigma Chemical (St. Louis, MO). Dextrose USP was obtained from Amend Co. (Irvinton, NJ). Water was distilled and deionized using a Nanopure® purification system (Fischer Scientific, Fair Lawn, NJ). Simulated intestinal fluid was prepared using a 0.01M phosphate buffer (sodium phosphate monobasic and potassium phosphate dibasic) at pH 7 and 5.5 with different amounts of NaCl to vary the ionic strength. Simulated gastric fluid (pH 1.5) was prepared with concentrated HCl with different amounts of NaCl to vary the ionic strength.

## PREPARATION OF DRUG-POLYMER COMPLEXES

An excess amount of aqueous drug solution (greater than 1.5 times the mole ratio of drug to the polymer) was added to an oppositely charged aqueous polymer solution to obtain a precipitated drug–polymer complex that was then thoroughly washed free of soluble components and dried. PDADMAC was complexed in this manner with anionic drug sodium salts (diclofenac and sulfathiazole), whereas PAMPS was complexed with the cationic drugs (labetalol · HCl, propranolol · HCl, verapamil · HCl, and diltiazem · HCl). The dried drug–polymer complexes were pulverized in a mortar and pestle, incorporating 20% dextrose as a binder, and compressed into tablets (approximately 2.0 mm thickness) using a 9.5-mm diameter die and a flat punch in a Carver press (Wabash, IN) under a compression force of 5,000 lbs.

## POLYMER CHARACTERIZATION

The polymers were characterized using an aqueous gel permeation chromatographic (GPC) method to determine the molecular weight, using a 2 × PL Aquagel-OH mixed bed column (300 × 7.5 mm) (Polymer Laboratory, Amherst, MA), 80% 0.01 M phosphate buffer containing 0.3 M sodium nitrate and 20% methyl alcohol as the mobile phase, Spectra pump (Thermo Separation Systems, Fremont, CA), and a flow rate of 1 mL/min. A calibration curve using sodium sulfonated polystyrene molecular weight standards (Scientific Polymer Products, Ontario, NY) was constructed using a refractive index detector (Shimadzu, Kyoto, Japan). The infrared spectra were recorded on a Perkin-Elmer Model 1600 FTIR spectrometer by the KBr pellet method to observe the complexation between drug and polymer. The samples were scanned at a resolution of 16 cm$^{-1}$ from 500 to 4,000 cm$^{-1}$.

## DRUG RELEASE KINETICS

The release kinetics from the tablets of the drug–polymer complexes were carried out in buffered release media containing 0.01 M phosphate and NaCl ranging from 0.2 M to 0.02 M at 37°C by the USP basket method at 100 rpm. Drug release was monitored on a HP 8452A diode-array spectrophotometer at 250, 306, 306, 270, 278, 278, and 274 nm for sodium diclofenac and sulfathiazole, labetalol · HCl, propranolol · HCl, verapamil · HCl, and diltiazem · HCl, respectively.

The linearity of drug release was assessed by fitting the release data to the phenomenological equation [10]:

$$\frac{M_t}{M_\infty} = kt^n \qquad (1)$$

where $M_t$ and $M_\infty$ are the amounts of drug released at time $t$ and the total amount of drug in a tablet, respectively, and $k$ and $n$ are a constant and a release exponent, respectively. The dissociation/erosion mechanism of the drug release kinetics were evaluated using the following equation [6]:

$$\frac{M_t}{M_\infty} = 1 - \left(1 - \frac{k_e t}{C_o r_o}\right)^2 \left(1 - \frac{2k_e t}{C_o l}\right) \tag{2}$$

where $k_e$, $C_o$, $r_o$, and $l$ are the dissociation/erosion rate constant, the initial drug concentration in a tablet, the tablet radius, and the tablet thickness, respectively.

## RESULTS AND DISCUSSION

When an excess drug solution (to saturate all the binding sites) was poured into an aqueous polymer solution, a water-insoluble complex formed, and precipitated as a white, agglomerated mass, which was collected, washed, and dried. As counter ions in a buffered medium diffused into the drug–polyelectrolyte complexes, the drug–polymer bonds dissociated as depicted in Scheme 1, releasing the bound drug into the surrounding water. The drug-free polymer dissolved, leaving no residue. Evidence of complex formation between drug and polymer was furnished by FTIR studies. Shown in Figure 1 is the FTIR spectrum of diclofenac–PDADMAC complex. The peak around $1,576 \text{ cm}^{-1}$ in the spectrum of the complex is due to the carboxylate group of the diclofenac ion involved in complexation with the amine group of the polymer. Similar results have been observed with diclofenac-PTMAEMC/MMA complex [7]. The spectra of drug–PAMPS are shown in Figure 2. All drug–PAMPS complexes as well as polymer showed peaks around $1,654 \text{ cm}^{-1}$, due to the carbonyl stretch of amide bonds. Also, both polymer and all complexes showed peaks in the range $600\text{–}700 \text{ cm}^{-1}$ which are attributed to the C-S linkage. The peaks in the range $1,451\text{–}1,474$ $\text{cm}^{-1}$ in the complexes are of strong intensity and are due to C-N stretching. Peaks from $1,518\text{–}1,497 \text{ cm}^{-1}$ have not been observed in polymer but appear in the complexes.

Investigations with commercial ion-exchange resins have been carried out for several decades and are generally regarded as old technology. The potential of ion-exchange resins for sustained delivery of drugs was recognized as early as 1950, when Sanders and Srivasta reported the absorption of quinine by a carboxylic acid ion-exchange resin [11]. Subsequently, the work of several other investigators [12] promoted interest in developing sustained release dosage forms based on ion-exchange resins, as reflected in the numerous patents that appeared in the 1960s and the 1970s [13]. Of

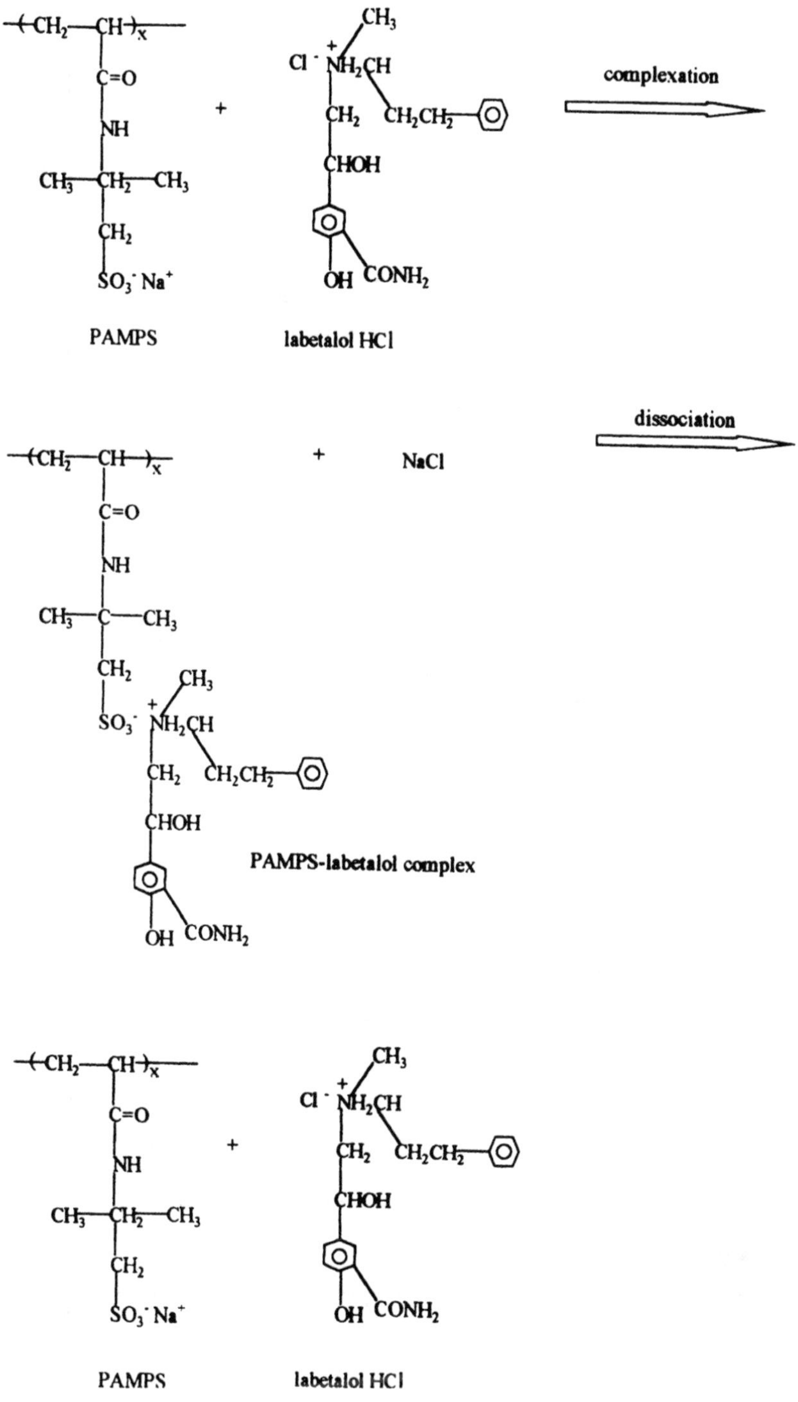

**PAMPS**        **labetalol HCl**

complexation

dissociation

$+$        NaCl

**PAMPS-labetalol complex**

**PAMPS**        **labetalol HCl**

**Scheme 1** Complexation and dissociation of labetalol HCl with PAMPS.

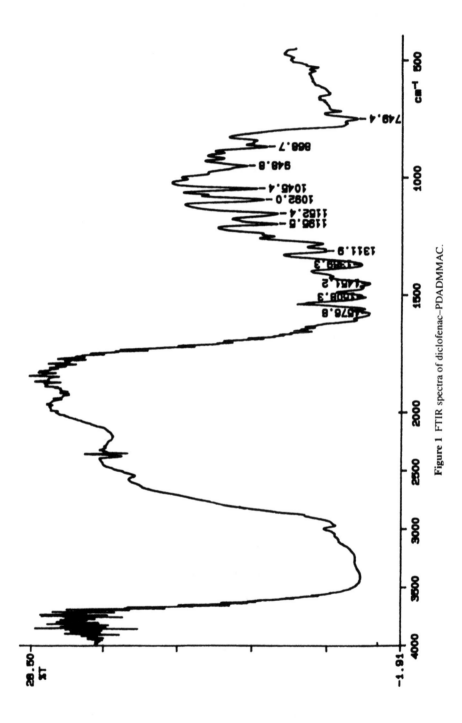

**Figure 1** FTIR spectra of diclofenac–PDADMMAC.

74

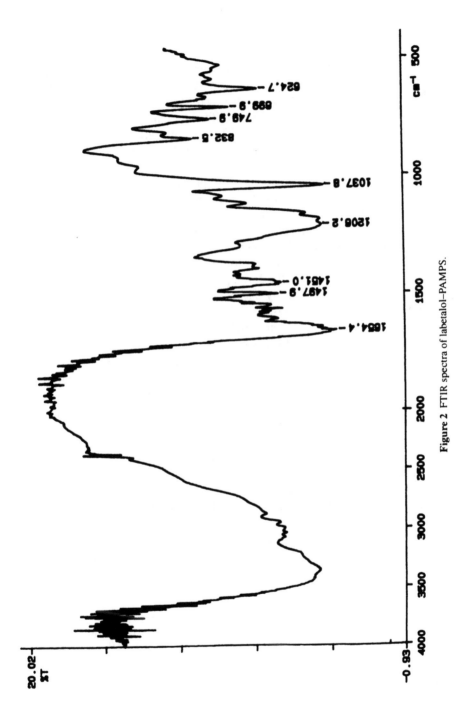

**Figure 2** FTIR spectra of labetalol–PAMPS.

late, there appears to be a renewed interest in the research and development of ion-exchange resin drug delivery systems [14,15], and several different approaches have been discussed in literature. However, they still involved the use of crosslinked systems, which are prone to the limitations of square root of time kinetics, and coating by different techniques effected further prolongation of drug release [16]. The use of water-soluble polyelectrolytes made possible the elimination of the drawbacks associated with crosslinked systems. In previously reported studies [6–9], water-soluble polyionic carriers of low solubility were prepared by copolymerizing an ionogenic monomer with an appropriate composition of methyl methacrylate (MMA). MMA imparted sufficient hydrophobicity to the polyelectrolyte so that it slowly but completely dissolved without any gel composition that contributes to swelling. The commercial availability of suitable (water-soluble polyelectrolytes) materials for application in this type of system marks an important point in the development of ion-exchange polymers for sustained drug delivery. Slowly dissolving, high molecular weight, water-soluble polymers such as PDADMAC and PAMPS preclude the necessity of synthesis and characterization, which have been done by the manufacturer.

The previously described water-soluble polyionic carriers were synthesized in the laboratory and, although promising in terms of drug release characteristics, they had to be characterized as they were new materials. Since the success of these systems has been well established based on our original polymer design, the same principle can be extended to commercially available material, which we neither have to prepare nor characterize. Since PAMPS and PDADMAC are homopolymers, higher drug loading was possible as the entire polymer chain was available for binding with an oppositely charged drug. Drug loading for both polymers exceeds 60 wt% for all drugs studied (Table 1). In the case of the copolymers the level of drug loading was dependent on the mole composition of the ionogenic monomer [6–9].

The effect of ionic strength of the release medium on the diclofenac release rates from PDADMAC is shown in Figure 3. It is evident that, in a

TABLE 1. Drug Loadings.

| PDADMAC | $M_w$ 400,000–500,000 | Drug Loading (wt%) |
|---|---|---|
| | Diclofenac | 64.0 |
| | Sulfathiazole | 63.7 |
| PAMPS | $M_w$ 2,000,000 | |
| | Labetalol | 65.2 |
| | Propranolol | 63.4 |
| | Verapamil | 60.1 |
| | Diltiazem | 68.4 |

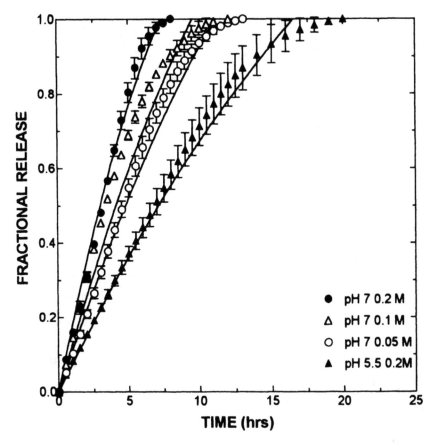

**Figure 3** Release of diclofenac from tablets of complexes with PDADMAC at pH 7.0 (0.05, 0.1, and 0.2 M NaCl) and at pH 5.5 (0.2 M NaCl).

buffered medium of 0.2 M strength (pH 7), diclofenac was released over a period of about 8 hours. As the ionic strength was reduced to 0.05 M, the release rate slowed down to about 12 hours. This effect of ionic strength on the drug release rate was also observed in the case of labetalol release from PAMPS tablets (Figure 4). The drug was released in about 8, 12, 20, and 36 hours in 0.2 M, 0.1 M, 0.05 M, and 0.02 M media, respectively. Drug release profiles of water-soluble drugs from drug-SPMK/MMA were reportedly independent of ionic strength higher than 0.1 M NaCl [6,17]. It seems that the more highly ionic polymer requires higher ionic strengths to obtain an indifferent release profile. The release profiles of labetalol in release media of different pH values (7.0 and 1.5) were similar. This is because the sulfonate ion remains strongly ionized at both pH values, and thus the drug release rates were not affected much. The pH-independent

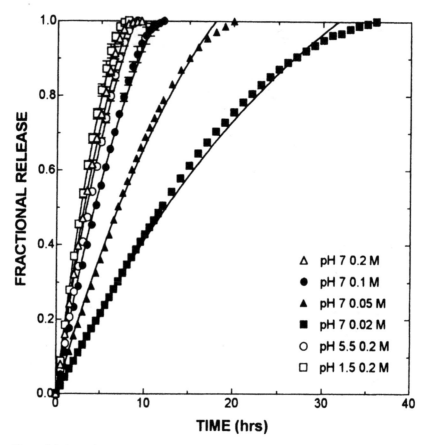

**Figure 4** Release of labetalol from tablets of complexes with PAMPS at pH 1.5 (0.2 M NaCl), pH 5.5 (0.2 M NaCl), and pH 7.0 (0.02 M, 0.05 M, 0.1 M, and 0.2M NaCl).

release characteristics of sulfonate containing systems have been reported previously [6,9]. However, diclofenac release from PDADMAC at pH 5.5 was much slower because the solubility of diclofenac is lower, its ionization is suppressed at the lower pH value; however, the ionization of the tertiary amine groups in PDADMAC were not affected by the low pH. In this case, only the effect of ionic strength of the release medium on the drug release rates from the polymer was carried out. A difference in diclofenac release from PDADMAC at different ionic strength values of the buffer medium was observed. In contrast, drug release profiles in different ionic strengths solutions were superimposable with those of quaternery amine carriers described in previous studies [7,8].

The effect of drug solubility on the release from PDADMAC is shown in Figure 5. The slightly soluble diclofenac (2.5 wt%) was released over a

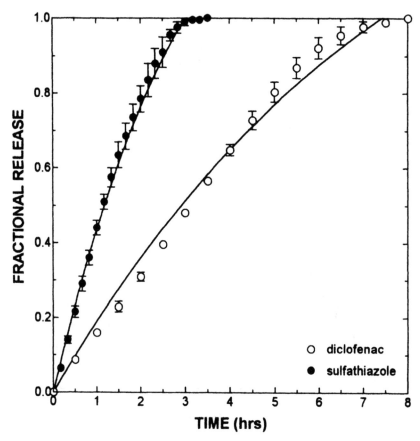

**Figure 5** Effect of drug solubility on drug release from drug–PDADMAC complex tablets at pH 7.0 (0.2 M NaCl).

period of 8 hours more slowly than sulfathiazole (24 wt%), which was released in just 3 hours. This effect of drug solubility on drug release profiles was also observed in the release of cationic drugs from PAMPS. Shown in Figure 6 are the effects of drug solubility on drug release from PAMPS. The release rates were decreased as drug solubility decreased. Diltiazem, the drug with the highest solubility in this group (66 wt%), was released the fastest (in about 5 h), whereas verapamil and propranolol which have similar solubilities (9 and 7 wt%, respectively), showed similar release durations of (about 6 h). Labetalol with the lowest solubility (1.5 wt%) was released over the longest period (8 h). The less water-soluble drug may make the drug–polymer complex more hydrophobic [9,17]. Shown in Figure 6 are the effects of the type (secondary and tertiary amine) of cationic drug on the release kinetics. Drug release profiles of propranolol

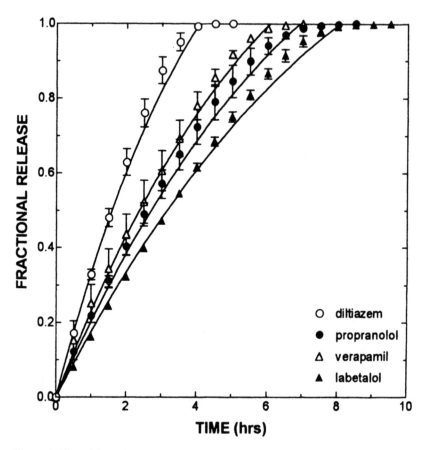

**Figure 6** Effect of drug solubility and type of amine on drug release from drug–PAMPS complex tablets at pH 7.0 (0.2 M NaCl).

and verapamil were almost superimposable. It seems that the type of amine did not influence the drug release since there is not much difference between the solubility of verapamil · HCl and propranolol · HCl. However, Nujoma et al. [17] showed that drug release from drug–PSPMK/MMA complexes prepared with verapamil · HCl was almost four times longer that that prepared with propranolol · HCl. The hydrophobicity of drug–polymer complex can be adjusted by synthesizing the copolymer less ionic (hydrophilic) and more hydrophobic monomers [9,17]. Figure 7 shows the release of labetalol from PAMPS and PAMPS/MMA. Even if the molecular weight of PAMPS/MMA has much less ($3.0 \times 10^5$) than that of PAMPS ($1.96 \times 10^6$), PAMPS/MMA provides much longer release times for all the drugs mentioned. Fro example, labetalol was released over 36 hours with PAMPS/MMA, compared to the 8 hours observed with PAMPS.

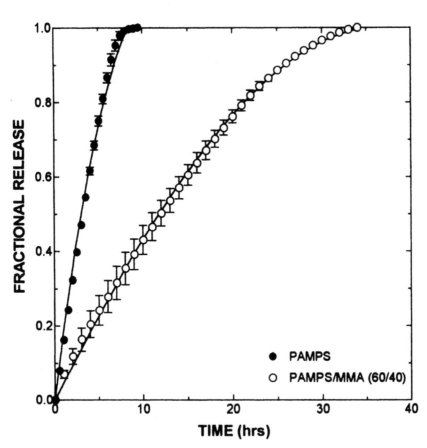

**Figure 7** Effect of polymer composition on the release of labetalol from drug-polyelectrolyte complex tablets at pH 7.0 (0.2 M NaCl).

The molecular weight was determined by an aqueous GPC method and compared with the manufacturer supplied value of $2 \times 10^6$.

The dissociation/erosion rate constant $(k_e)$ of drug–polymer complexes was determined using a nonlinear regression analysis (PRIZM®, GraphPad Sofware, Inc., San Diego, CA) of Equation (2). The results are presented in Table 2. In general, the dissociation/erosion mechanism expressed by Equation (2) represents the experimental data studied here as shown in Figures 3 to 8. Linear drug release was observed, by calculating release exponent $n > 0.85$ (up to 60% release), for a majority of drug–polymer complexes. When experimental data having $n > 0.9$ (up to 60% release) were regressed up to 80% release, the release exponent $n$ maintained $> 0.9$, as shown in Table 2. This result demonstrates excellent linearity of the drug release.

TABLE 2. Dissociation/Erosion Rate Constants and Linearity Exponents of PDADMAC and PAMPS.

| | $k_e$ (up to 90% rel) | $n$ up to 60% (w/o burst) | $n$ up to 80% (w/o burst) |
|---|---|---|---|
| **PAMPS** | | | |
| Labetalol (200 mg) | | | |
| pH 7, 0.2 M | 10.58 ± 0.16 | 0.97 ± 0.02 | 0.94 ± 0.03 |
| 0 M | 7.17 ± 0.12 | 0.97 ± 0.02 | 0.93 ± 0.02 |
| 0.05 M | 4.34 ± 0.04 | 0.97 ± 0.01 | 0.92 ± 0.02 |
| 0.02 M | 2.56 ± 0.01 | 0.95 ± 0.01 | 0.87 ± 0.01 |
| pH 5.5, 0.2 M | 9.45 ± 0.26 | 0.99 ± 0.01 | 0.98 ± 0.02 |
| pH 1.5, 0.2 M | 12.12 ± 0.20 | 0.95 ± 0.04 | 0.89 ± 0.04 |
| Propranolol (200 mg) | | | |
| pH 7, 0.2 M | 11.87 ± 0.40 | 0.87 ± 0.11 | 0.85 ± 0.08 |
| 0.1 M | 8.45 ± 0.22 | 0.89 ± 0.09 | 0.87 ± 0.06 |
| 0.05 M | 7.33 ± 0.20 | 0.87 ± 0.15 | 0.85 ± 0.08 |
| 0.02 M | 4.55 ± 0.08 | 0.94 ± 0.07 | 0.85 ± 0.05 |
| pH 5.5, 0.2 M | 12.93 ± 0.63 | 0.88 ± 0.10 | 0.88 ± 0.11 |
| pH 1.5, 0.2 M | 16.09 ± 0.19 | 0.82 ± 0.03 | 0.80 ± 0.02 |
| Verapamil (200 mg) | | | |
| pH 7, 0.2 M | 12.38 ± 0.63 | 0.79 ± 0.22 | 0.82 ± 0.14 |
| 0.1 M | 9.16 ± 0.20 | 0.98 ± 0.07 | 0.90 ± 0.06 |
| 0.05 M | 5.01 ± 0.11 | 0.93 ± 0.08 | 0.89 ± 0.05 |
| 0.02 M | 3.93 ± 0.15 | 1.01 ± 0.14 | 0.96 ± 0.10 |
| pH 5.5, 0.2 M | 11.05 ± 0.27 | 0.97 ± 0.06 | 0.96 ± 0.04 |
| pH 1.5, 0.2 M | 13.43 ± 0.23 | 0.93 ± 0.05 | 0.87 ± 0.04 |
| Diltiazem (200 mg) | | | |
| pH 7, 0.2 M | 21.96 ± 1.00 | 0.94 ± 0.03 | 0.89 ± 0.09 |
| 0.1 M | 15.66 ± 0.42 | 0.95 ± 0.08 | 0.94 ± 0.06 |
| 0.05 M | 11.75 ± 0.35 | 0.98 ± 0.01 | 0.97 ± 0.02 |
| 0.02 M | 9.37 ± 0.25 | 0.99 ± 0.05 | 0.98 ± 0.03 |
| pH 5.5, 0.2 M | 20.34 ± 0.73 | 0.95 ± 0.07 | 0.91 ± 0.08 |
| pH 1.5, 0.2 M | 24.61 ± 0.60 | 0.99 ± 0.12 | 0.88 ± 0.07 |
| Labetalol + Verapamil (200 mg) | | | |
| Labetalol | 4.89 ± 0.10 | 0.97 ± 0.04 | 0.93 ± 0.03 |
| Verapamil | 6.05 ± 0.10 | 0.92 ± 0.03 | 0.89 ± 0.03 |
| **PDADMAC** | | | |
| Diclofenac | | | |
| pH 7, 110 mg, 0.2 M | 7.24 ± 0.18 | 1.03 ± 0.06 | 1.02 ± 0.04 |
| 0.1 M | 5.89 ± 0.12 | 0.96 ± 0.05 | 0.87 ± 0.04 |
| 0.05 M | 4.74 ± 0.10 | 1.04 ± 0.06 | 1.00 ± 0.05 |
| pH 5.5, 110 mg, 0.2 M | 2.87 ± 0.06 | 0.95 ± 0.06 | 0.93 ± 0.06 |
| Sulfathiazole | | | |
| pH 7, 110 mg, 0.2 M | 36.39 ± 0.86 | 1.03 ± 0.08 | 0.92 ± 0.07 |
| pH 7, 110 mg, 0.1 M | 32.44 ± 0.82 | 1.02 ± 0.07 | 0.99 ± 0.05 |
| pH 7, 110 mg, 0.05 M | 29.60 ± 0.56 | 0.99 ± 0.03 | 0.99 ± 0.02 |
| pH 7, 180 mg, 0.2 M, 1.3 mm | 49.93 ± 0.89 | 0.99 ± 0.05 | 0.96 ± 0.04 |

continued

TABLE 2. *continued*

| | | PDADMAC | |
|---|---|---|---|
| | $k_e$ (up to 90% rel) | *n* up to 60% (w/o burst) | *n* up to 80% (w/o burst) |
| pH 7, 180 mg, 0.2 M, 2 mm | 37.26 ± 0.64 | 0.98 ± 0.06 | 0.93 ± 0.05 |
| pH 7, 150 mg, 0.2 M | 35.36 ± 0.43 | 0.99 ± 0.05 | 0.94 ± 0.03 |
| pH 7, 150 mg, 0.1 M | 32.13 ± 0.57 | 0.93 ± 0.04 | 0.89 ± 0.04 |
| pH 7, 150 mg, 0.05 M | 30.37 ± 0.33 | 0.94 ± 0.05 | 0.87 ± 0.03 |
| pH 1.5, 150 mg, 0.2 M | 38.14 ± 1.31 | 1.21 ± 0.08 | 1.07 ± 0.07 |

The effect of tablet thickness on release profiles from the drug–polymer complexes has been discussed previously [6–9]. The influence of tablet thickness on drug release profiles is again evident (Figure 8), which shows drug release from a tablet containing equal mixtures (by weight) of two different drug–polymer complexes (labetalol–PAMPS and verapamil–PAMPS with 20% dextrose). Even though these tablets have half the quantity of labetalol–PAMPS compared to those formulated with PAMPS–labetalol only, the total release times are not very different in either case. This is because the two types of tablets, although differing in composition, have the same thickness. The total dissociation/erosion path length remains the same. Thus, tablet thickness and not drug loading is the driving force for drug release. An attempt was made in vain to release anionic and cationic drugs simultaneously from oppositely charged drug–polymer pairs in a single tablet. If this were done, both polymers (anionic and cationic) once released from the bound drugs would immediately complex with each other, forming a nonerodible system that would trap any undissociated complex.

## CONCLUSIONS

Commercially available polymeric materials have been utilized to develop controlled release tablets, from which drug release may be well characterized by the dissociation/erosion mechanism. The nature of the bound drugs was identified to be among the factors affecting drug release. Being homopolymers, the hydrophobicities were considerably less than those of the copolymers reported earlier, resulting in a less prolonged effect on the release kinetics compared to those of the latter. The release properties of the homopolymeric carriers may be further enhanced if they were available

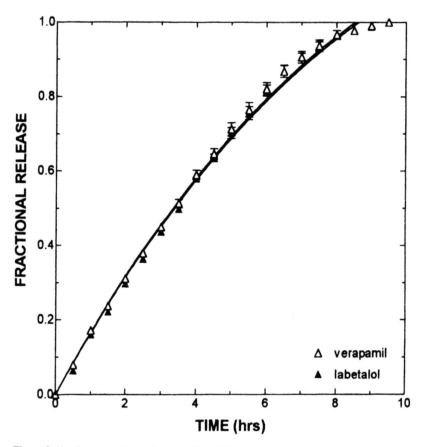

**Figure 8** Simultaneous release of verapamil and labetalol from a single tablet of drug-PAMPS complexes at pH 7.0 (0.2M NaCl).

in higher molecular weights. However, the advantage is that much higher drug loadings are possible in the case of the homopolymers.

## REFERENCES

1. Irwin, W. J., McHale, R., and Watt, P. J. *Drug Dev. Ind. Pharm.*, 1990, 16, 883.
2. Raghunathan, Y., Amsel, L, Hinsvark, O., and Bryant, W. *J. Pharm. Sci.*, 1981, 70, 379.
3. Burke, G. B., Mendes, R. W., and Jambhekar, S. S., *Drug Dev. Ind. Pharm.*, 1986, 12, 713.
4. Lee, P. I. in *Controlled Release Technology: Pharmaceutical Applications* (P. I. Lee and W. R. Woods, Eds.), ACS Symp. Ser. No. 348, ACS, Washington, DC, 1987, 1.

5. Kim, C. J. *Chemtech,* August, 1994, 36.

6. Nujoma, Y. N. and Kim, C. J. *J. Pharm. Sci.,* 1996, 85, 1091.

7. Konar, N. and Kim, C. J. *J. Pharm. Sci.,* 1997, 86, 1339.

8. Konar, N. and Kim, C. J. *J. Appl. Polym. Sci.,* 1998, 69, 263.

9. Konar, N. and Kim, C. J. *J. Control. Rel.,* 1999, 57, 141.

10. Ritger, P. I. and Peppas, N. A. *J. Control. Rel.,* 1987, 5, 23.

11. Sanders, L. and Srivasta, R. *J. Chem. Soc.,* 1950, 2915.

12. Schlichting, D. A. *J. Pharm. Sci.,* 1962, 51, 134.

13. Keating, J. W. US Pat. 527, 130 (1955), 528, 346 (1956), 3,143,465, 1964.

14. Kelleher, W. J. and Carpanzano, A. E. US Pat. 4,265,910, 1990.

15. Jani, R. and Harris, R. G. US Pat. 4,911,920, 1990.

16. Motycka, S., Newth, C. J. L., and Narin, J. G. *J. Pharm. Sci.,* 1985, 74, 643.

17. Nujoma, Y. N., Kim, C. J., and Chern, R. T. in *Materials for Controlled Release Applications.* McCullough, I. and Shalaby, S. (Eds.), ACS Symp. Ser., ACS, Washington, DC, 1998, 67.

# CHAPTER 6

# Polychelating Amphiphilic Polymers (PAP) as Key Components of Microparticulate Diagnostic Agents

VLADIMIR P. TORCHILIN[1]

## IMAGING PRINCIPLES AND MODALITIES

CLINICAL diagnostic imaging requires an appropriate signal intensity to differentiate certain structures or areas of interest from surrounding tissues. According to the physical principles applied, currently used imaging modalities include gamma-scintigraphy, magnetic resonance, computed tomography, and ultrasonography. Attenuations (i.e., the ability of a tissue to absorb a certain signal, such as X-rays, sound waves, radiation, or radiofrequencies) of different tissues differ; however, usually this difference is not sufficient for clear discrimination between normal and pathological areas. Besides, nonenhanced imaging techniques are useful only when relatively large tissue areas are involved in the pathological process. To solve the problem and to achieve a sufficient attenuation, contrast agents are used. These are the substances that are able to absorb certain types of signal (irradiation) much stronger than surrounding tissues. Different chemical nature of reporter moieties used in different modalities and different signal intensity (sensitivity and resolution) require various amounts of a diagnostic label to be delivered into the area of interest.

In many cases, contrast agent-mediated imaging is based on the ability of some tissues (i.e., macrophage-rich ones) to absorb particulate substances. This process is particle size dependent and relies on a fine balance between particles small enough to enter the blood or lymphatic capillaries yet large enough to be retained within the tissue.

[1]Department of Pharmaceutical Sciences, School of Pharmacy, Bouve College of Health Sciences, Northeastern University, Mugar Building 312, 360 Huntington Avenue, Boston, MA 02115, USA.

Nuclear medicine (gamma-scintigraphy) and magnetic resonance (MR) are widely used in contemporary clinical practice to obtain images of pathological areas. Gamma-scintigraphy and MR imaging both require a sufficient quantity of radionuclide or paramagnetic metal to be associated with a microparticulate carrier. The tissue concentration that must be achieved for successful imaging varies between diagnostic moieties and is relatively high in the case of MRI ($10^{-4}$ M). For this reason it was a natural progression to use microparticulate carriers for the efficient delivery of contrast agents to areas of interest. Liposomes and micelles are of a special interest among these diagnostically significant particulates [1]. Liposomes, microscopic artificial phospholipid vesicles, and micelles, amphiphilic compound-formed colloidal particles with hydrophobic core and hydrophilic corona, draw an attention because of their easily controlled properties and good pharmacological characteristics. Pursuing different *in vivo* delivery purposes, one can easily change the size, charge, and surface properties of these carriers simply by adding new ingredients to the starting mixture before liposome or micelle preparation and/or by variation of preparation methods.

The use of liposomes for the delivery of imaging agents has a long history [2,3]. The ability of liposomes to entrap different substances into both the aqueous phase and the liposome membrane compartment makes them suitable for carrying diagnostic moieties used with all imaging modalities. Phospholipid liposomes, if introduced into the circulation, are rapidly sequestered by reticuloendothelial system, and the sequestration is almost independent on their size, charge, and composition. To increase liposome accumulation in the "required" areas, the use of targeted liposomes or immunoliposomes has been suggested [4]. However, a majority of the immunoliposomes still ended up in the liver, which was usually insufficient time for the interaction between the target and targeted liposome. This is especially true in cases when a target of choice has a diminished blood supply (ischemic or necrotic areas) or the concentration of the target antigen is very low. In both cases much better accumulation can be achieved if liposomes can stay in the circulation longer. Different methods have been suggested to achieve long circulation, including coating the liposome surface with inert, biocompatible polymers, such as poly(ethylene glycol) (PEG), which form a protective layer over the liposome surface and slows down the liposome recognition by opsonins and subsequent clearance [5,6]. Long-circulating liposomes are now widely used in biomedical *in vitro* and *in vivo* studies and have even found their way into clinical practice [7]. The use of PEGylated liposomes and micelles as carriers for contrast agents now forms an important area of research [2,8].

## LOADING OF LIPOSOMES AND MICELLES WITH CONTRAST AGENTS

Shown in Figure 1 are the principal schemes for micelle and liposome formation and loading with various reporter moieties that might be covalently or noncovalently incorporated into different compartments of these particulate carriers. Although micelles may be loaded with a contrast agent only into the core in the process of micelle assembly, liposomes may incorporate contrast agents in both the internal water compartment and the bilayer membrane.

Both gamma-scintigraphy and MR imaging require a significant quantity of radionuclide or paramagnetic metal to associate with the liposome. There are two very general approaches most often used and efficient to prepare liposomes for gamma- and MR imaging. First, the metal is chelated with a water-soluble chelate (such as diethylene triamine pentaacetic acid or DTPA) and then entrapped in the aqueous interior of the liposome [9]. Alternatively, DTPA or a similar chelating compound may be chemically derivatized by the incorporation of a hydrophobic group, which can anchor the chelating moiety on the liposome surface during or after liposome preparation [10]. Different chelators and different hydrophobic anchors have been tried in the preparation of [111]In, [99m]Tc, Mn, and Gd liposomes [10–12]. Low-molecular-weight water-soluble paramagnetic probes leak from the liposomes upon contact with body fluids, which destabilizes the liposomal membranes. Moreover, it has been shown that, when high concentrations of Gd-DTPA are encapsulated inside liposomes for better enhancement, the relaxivity of the compound might be lower than the nonencapsulated Gd-DTPA complex, probably because of decreased residence lifetime of the water molecules inside the vesicles [13]. Membranotropic chelating agents—DTPA-stearylamine [14] and DTPA-phosphatidyl ethanolamine [15]—have a polar head, containing chelated paramagnetic atom, and a lipid moiety, which anchors the metal–chelate complex in the liposome membrane. This approach has been shown to be far superior in terms of the relaxivity then the liposome-encapsulated paramagnetic ions [13] due to the decrease in the rotational correlation times of the paramagnetic moiety rigidly connected with relatively large particle. Liposomes with membrane-bound paramagnetic ion also demonstrated reduced leakage in the body. Membranotropic chelates are also suitable for micelle incorporation (they anchor in the hydrophobic micelle core) and may serve to load micelles with heavy radiometals (Figure 1).

## DIAGNOSTIC LIPOSOMES AND MICELLES WITH PAPs

To improve the efficacy of liposomes and micelles as contrast carriers the quantity of carrier-associated reporter metal (such as Gd or [111]In) must be

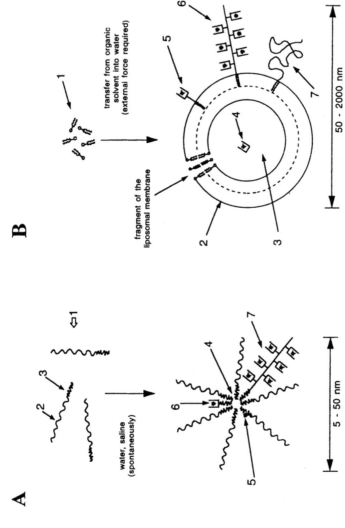

**Figure 1** Schematic structures of micelle and liposome, their formation and loading with a contrast agent. (a) A micelle is formed spontaneously in aqueous media from an amphiphilic compound (1) that consists of distinct hydrophilic (2) and hydrophobic (3) moieties. Hydrophobic moieties form the micelle core (4). Contrast agent (asterisk; gamma- or MR-active metal-loaded chelating group, or heavy element, such as iodine or bromine) can be directly coupled to the hydrophobic moiety within the micelle core (5), or incorporated into the micelle as an individual monomeric (6) or polymeric (7) amphiphilic unit. (b) A liposome can be prepared from individual phospholipid molecules (1) that consists of a bilayered membrane (2) and internal aqueous compartment (3). Contrast agent (asterisk) can be entrapped in the inner water space of the liposome as a soluble entity (4) or incorporated into the liposome membrane as a part of monomeric (5) or polymeric (6) amphiphilic unit (similar to that in case of micelle). Additionally, liposomes can be sterically protected by amphiphilic derivatization with PEG or PEG-like polymer (7) [1].

increased to enhance the signal intensity. We have tried to solve this problem by using "so-called" polychelating amphiphilic polymers, or PAPs [16]. The basic idea behind the new generation of microparticulate contrast agents was to increase the number of chelated metal atoms attached to a single lipid anchor and to incorporate them into the liposomal membrane or into the micelle core. This process could drastically increase the number of bound reporter metal atoms per vesicle and, in turn, decrease the dosage of administered lipid without compromising the image signal intensity.

To increase the load of liposomes and micelles with reporter metals, we designed a new family of amphiphilic single-terminus modified polymers containing multiple chelating groups that could be incorporated into the hydrophobic domains of liposomes and micelles. The approach is based on the use of CBZ-protected polylysine (PL) with a free terminal amino group, which is derivatized into a reactive form with subsequent deprotection and incorporation of DTPA residues. This was initially suggested by us for heavy metal load on proteins and antibodies [17].

Initially, to couple multiple reporter groups to a single protein (antibody) molecule, we used a polymer chain carrying many chelator moieties as side groups [17,18]. These chelating moieties provided high-affinity binding for a wide range of heavy metal ions, e.g., Fe, In, Tc, Re, Ga, Gd, Mn, Eu, Y, Bi, At, and Sm. The polymeric backbone for the attachment of multiple chelating moieties has to contain a sufficient number of reactive groups, such as amino, carboxy, aldehyde, or SH groups. Both natural and synthetic polymers can be used for this purpose; poly(L-lysine; PL) which contains multiple free amino groups, is most often used. To couple the chelator to a PL, the reactive intermediate of the chelator is usually used, containing mixed anhydrides, cyclic anhydrides, $N$-hydroxysuccinimide esters, tetrafluorophenyl esters, or isothiocyanate. Chelating polymer–antibody conjugation involves the preparation of a chelating polymer, containing a single terminal reactive group capable of interaction with an activated antibody [17,18]. Because of steric hindrances, the modification of amino groups in PLL with bulky chelator groups is never complete, and residual free amino groups can be easily reached with succinic anhydride. The latter reaction is important in providing the polymer with total negative charge that *in vivo* has to diminish the nonspecific capture of label-loaded conjugates with non-target cells carrying slight net negative charge.

Using N-terminus modified polylysine, we developed a synthesis for an amphiphilic polychelator, $N,\alpha$-(DTPA-polylysyl)glutaryl phosphatidyl ethanolamine (DTPA-PL-NGPE). This polychelator was incorporated into the liposomal membrane and micelle core during liposome or micelle preparation. This system sharply increased the number of chelated Gd atoms attached to a single lipid anchor. This increased the number of bound reporter metal atoms per vesicle and decreased the dosage of an administered

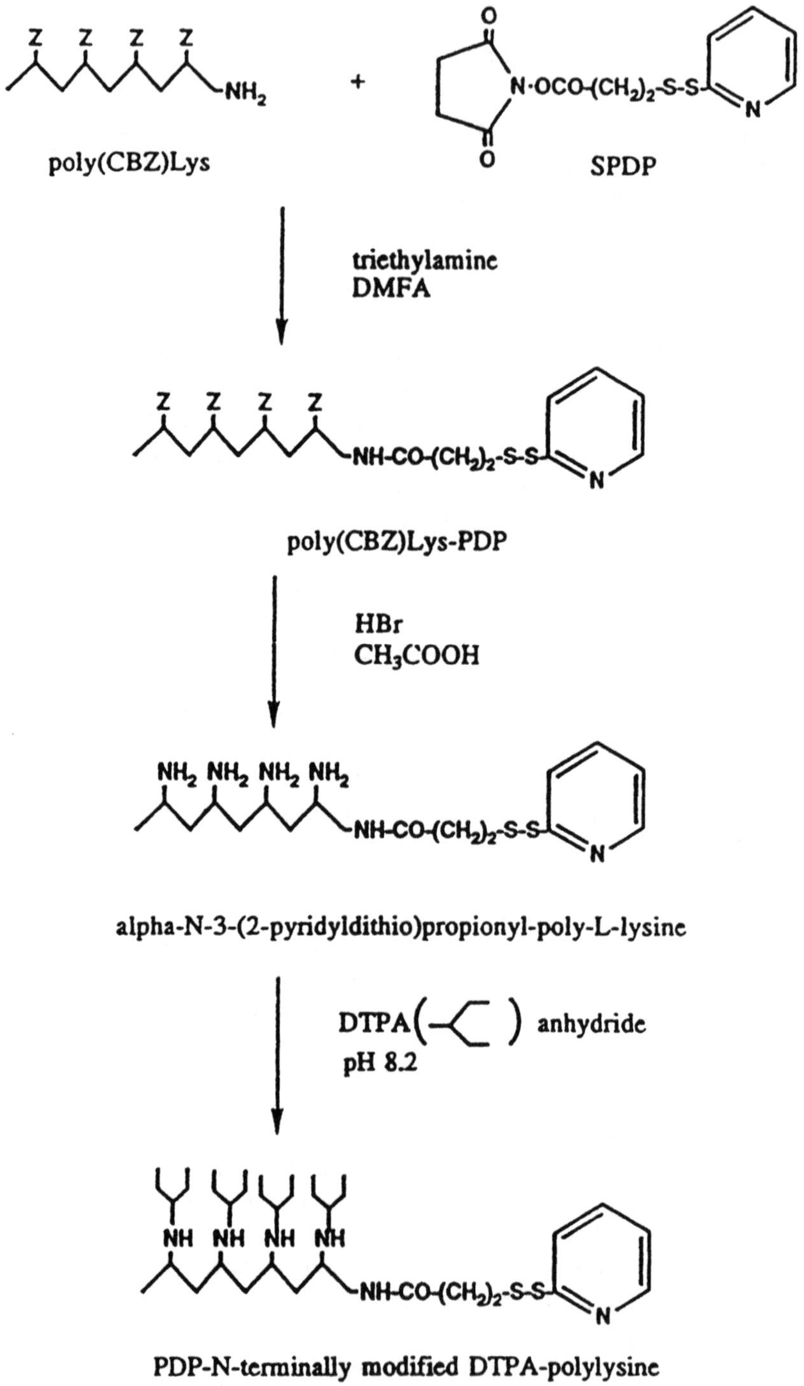

**Figure 2** Chemistry of the polymeric chelates used for loading liposomes and micelles with multiple reporter metal atoms. (a) Synthesis of a single terminus-PDP-activated chelating polymer (DTPA-polylysine) starting from CBZ-protected polylysine and SPDP.

lipid without compromising the image signal intensity. Described in Figure 2 is the chemistry of a typical single-terminus activated chelating polymer preparation and the attachment of a hydrophobic anchor.

A typical protocol for the DTPA-PL-NGPE synthesis and loading with Gd (to prepare contrast agent for MR imaging) is given below. NGPE (25 mg) was activated with $N,N'$-carbonyldiimidazole (25 mg) in the presence of $N$-hydroxysuccinimide (11.4 mg) for 16 h at room temperature. At this point, $\epsilon$, $N$-carbobenzoxy PL (100 mg, Mw 3,000 Da) and triethylamine (10 μL), were added to the initial mixture and the reaction was allowed to proceed for

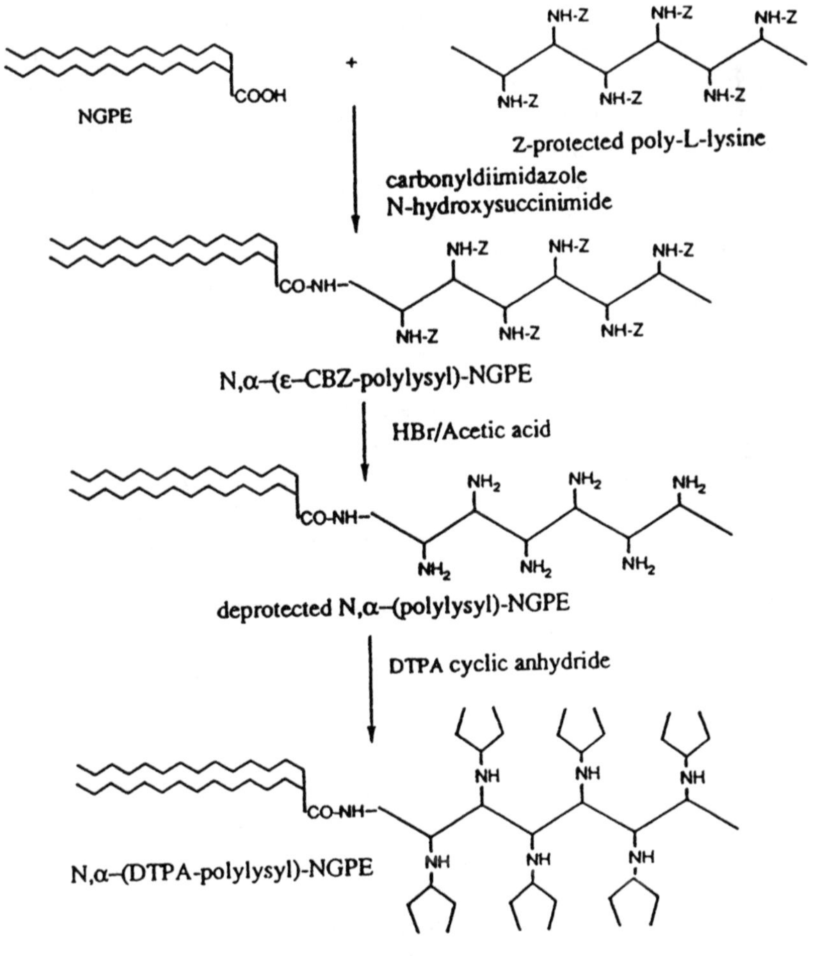

**Figure 2 (continued)** (b) Synthesis of amphiphilic DTPA-PL-NGPE consisting of hydrophilic DTPA-polylysyl moiety and hydrophobic $N$-glutaryl phosphatidyl ethanolamine moiety [1].

another 5 h. TLC demonstrated full conversion of NGPE to $N,\alpha$-($\epsilon$-CBZ-PL)-NGPE. This compound was precipitated with water, washed, and freeze-dried. $N,\alpha$-($\epsilon$-CBZ-PL)-NGPE (67.4 mg) was dissolved in 3 mL of 30% HBr in glacial acetic acid, and the reaction mixture stirred for 2 h at room temperature. The deprotected $N,\alpha$-PL-NGPE was precipitated with dry ethyl ether, washed with ether, and freeze-dried. The $N,\alpha$-PL-NGPE (37 mg) was suspended in a 1:1 chloroform:methanol mixture and reacted with DTPA anhydride (100 mg in 2 mL of dimethylsulfoxide) in the presence of 5 mL of triethylamine for 16 h at room temperature with stirring. At this time succinic anhydride (100 mg in 0.2 mL dimethylsulfoxide) was added to block the remaining polymer amino groups. The reaction mixture was dialyzed against deionized water and freeze-dried. $GdCl_3 \cdot 6H_2O$ (150 mg in 0.25 mL of 0.1 M citrate, pH 5.3) was added to 25 mg of DTPA-PL-NGPE suspended in 2 mL of dry pyridine. After 2 h incubation at room temperature, the reaction mixture was dialyzed against water and freeze-dried.

Elemental analysis has revealed that the Gd-DTPA-PL-NGPE contains ca. 40% (w/w) Gd, which corresponds to 8–10 metal atoms per single lipid-modified polymer molecule assuming a molecular weight of 3,500–4,000 Da. Since we have used poly-$\epsilon$-CBZ-L-lysine with a polymerization degree 11 for the synthesis of the polychelator, theoretically, one could introduce up to 11 metal atoms if all the polylysine e-amino groups were modified with DTPA residues. This would be superior to one metal atom per one lipid molecule previously used as the amphiphilic chelator Gd-DTPA-PE and Gd-DTPA-SA [14,15,19] probes. The higher Gd content provided better relaxivity and, consequently, to greater MR signal intensity (assuming the Gd tissue concentration does not exceed millimolar range).

The NGPE anchor grafted with a short-chain chelating polymer forms a "coat" of chelating groups around the liposomal membrane or within the micelle's corona (Figure 1). Metal atoms chelated by these groups are directly exposed to the both interior and exterior water environment and have better access to the adjacent tissue water protons. This should lead to a corresponding enhancement of the vesicle contrast properties. As a result, amphiphilic polychelator-containing liposomes and micelles should have a higher relaxation influence on water protons compared with conventional preparations with the same phospholipid content. To prove this, we have performed proton relaxivity measurements for the different liposomal preparations each containing 3% mol of the individual amphiphilic Gd-containing probe (Figure 3). The results demonstrated that polychelator-containing liposomes have higher relaxation influence on water protons compared with conventional liposomal preparation with the same phospholipid content. Clinically, this means that one can considerably reduce the total lipid dose of the contrast material required for the diagnostic procedure without decreasing the MR signal intensity.

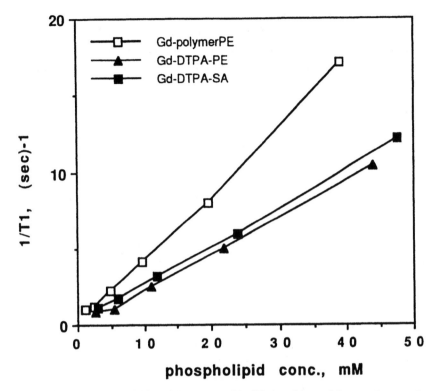

**Figure 3** Molecular relaxivities of liposomes with different Gd-containing membranotropic chelators. Liposomes (egg lecithin:cholesterol:chelator = 72:25:3) were prepared by consecutive extrusion of lipid suspension in HEPES buffered saline, pH 7.4, through the set of polycarbonate filters with pore size of 0.6, 0.4, and 0.2 mm. Liposome final size was between 205 and 225 nm. Gd content determination was performed by Galbraith Laboratories, Inc. The relaxation parameters of all preparations were measured at room temperature using a 5-MHz RADX nuclear magnetic resonance proton spin analyzer. The relaxivity of liposomes with polymeric chelators is noticeably greater because of the larger number of Gd atoms bound to a single lipid residue [16].

Micelles formed by self-assembled amphiphilic polymers (such as PEG-phosphatidyl ethanolamine) can also be loaded with amphiphilic PL-based chelates carrying diagnostically important metal ions such as [111]In and Gd [20]. The final preparations are quite stable and serve as fast and efficient agents for scintigraphy or MR imaging.

## EXPERIMENTAL VISUALIZATION WITH PAP-CONTAINING MICROPARTICULAR CONTRAST AGENTS

To prove the efficiency of the Gd-DTPA-PL-NGPE liposomes as MR contrast agents, we performed a lymph node visualization in rabbits using subcutaneous administration of 20 mg of egg PC/cholesterol (75;25) liposomes (215 nm) containing 5% mol of amphipathic polychelator. The liposome suspension (0.5 mL; 20 mg total lipid) was administered s.c. into the right forepaw of a sedated New Zealand rabbit. MR images were acquired for 2 h at 5–10 min intervals using a 1.5 Tesla GE Signa MRI instrument. Along with the preinjection image, transverse slice MR scans were taken using a fat suppression mode and $T_1$-weighted pulse sequence with TR = 500 msec and TE = 12 msec. Slice thickness was 2 mm. These rabbit transverse MRI scans (Figure 4) demonstrates that axillary and subscapular lymph nodes can be seen on the scans taken only 5 min after injection. This fast response should be valuable for MR lymph node visualization in clinical circumstances. This was additionally proved by the fast and informative diagnostic visualization of a VX2 human sarcoma in a rabbit's popliteal lymph node (Figure 5). Similar liposome-tumors were clearly seen 10 minutes postinjection (the time between contrast agent administration and image aquisition for other lymphotropic imaging agents is in a range of several hours).

To investigate the biological behavior and determine the sites of accumulation of Gd-DTPA-PL-NGPE-liposomes upon intravenous administration, the liposomes were additionally labeled with [111]In by transchelation from an In-citrate complex and injected into the rabbit with subsequent label distribution monitoring by a γ-camera. The radioscintigram taken 2 h post i.v. injection demonstrated characteristic liposome accumulation sites in the liver, spleen, and bone marrow.

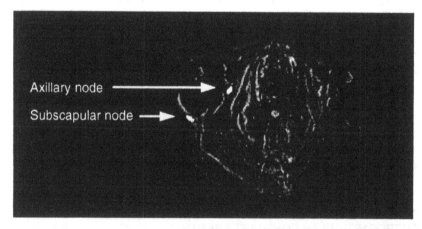

**Figure 4** Transverse scan of axillary and subscapular lymph nodes in a rabbit 5 min postinjection of Gd-containing liposomes. Liposomes (egg lecithin:cholesterol:Gd-poly-NGPE = 70:25:5, 20 mg total lipid) were injected subcutaneously into the forepaw of anesthesized rabbit in 0.5 mL of HEPES-buffered saline. Images were acquired by using a 1.5 Tesla GE Signa MRI scanner operated at fat suppression mode and $T_1$-weighted pulse sequence [16].

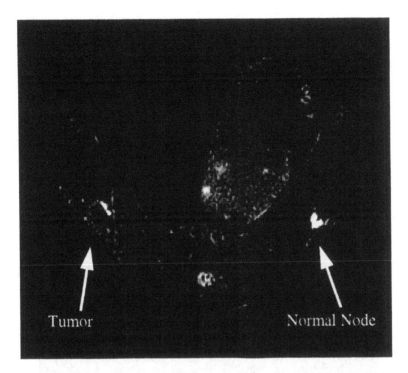

**Figure 5** MR visualization of VX$_2$ human sarcoma in rabbit popliteal lymph node with Gd-loaded PAP-containing liposomes. Ten minutes postinjection time liposomes (PC:CH:Gd-PAP = 70:25:5 molar ratio, 20 mg total lipid) were injected subcutaneously into the forepaw of a anesthesized rabbit in 0.5 mL of HEPES-buffered saline. The Gd in the polychelate-modified liposomes provides an image much faster than in case of liposomes loaded with Gd with monomeric chelators (within minutes compared to several hours). The abnormal lymph node (left) can easily be distinguished from the normal node (right) following the deficit of the liposomal uptake ("filling defect"). Images were acquired using 1.5 Tesla GE Signa MRI scanner operated at fat suppression mode and T$_1$-weighted pulse sequence [2].

Similar experiments with PEG-phosphatidyl ethanolamine mixed micelles with a core-incorporated amphiphilic [111]In- or Gd-loaded chelating agent PAP demonstrated fast and efficient gamma and MR visualization of different compartments of the lymphatic system. Upon subcutaneous administration, the micelles penetrate the lymphatics and effect visualization (Figure 6). Micelles mostly stay within the lymph fluid rather than accumulate in the nodal macrophages (because of protective effect of surface PEG fragments) and rapidly move via the lymphatic pathway.

Thus, amphiphilic polylysine-based chelating polymers can be successfully used as a key component of liposomal and micellar imaging agents in scintigraphy and MR imaging. This new class of polylysine-derived amphiphilic chelating agents described could be used broadly in other diagnostics and therapy applications.

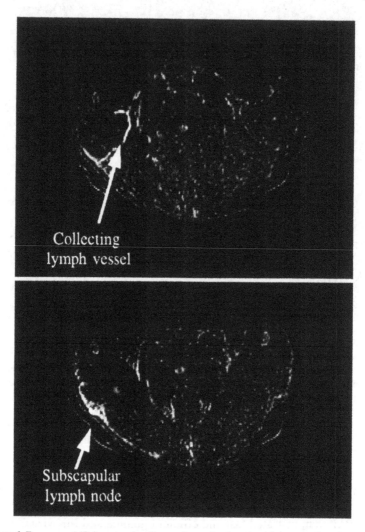

Collecting
lymph vessel

Subscapular
lymph node

**Figure 6** Transverse MR images of axillary-subscapular lymph node area in the rabbit 4 min after s.c. administration of PEG (5 kDa) -phosphatidyl ethanolamine micelles containing core-incorporated Gd-loaded amphiphilic chelate DTPA-phosphatidyl ethanolamine. The dose was 0.5 μmol Gd per injection site. Fast and clear visualization of both lymph vessel (1) and lymph node (2) was achieved. Images were acquired by using a 1.5 Tesla GE Signa MRI scanner operated at fat suppression mode and $T_1$-weighted pulse sequence [20].

98

## REFERENCES

1. Torchilin, V. P. Pharmacokinetic considerations in the development of labeled liposomes and micelles for diagnostic imaging. *Quat. J. Nucl. Med.* 1997, 41, 141–153.
2. Torchilin, V. P. Liposomes as delivery agents for medical imaging. *Mol. Med. Today* 1996, 2, 242–249.
3. Seltzer, S. E. The role of liposomes in diagnostic imaging. *Radiology* 1989, 171, 19–21.
4. Torchilin, V. P. Liposomes as targetable drug carriers. *CRC Crit. Rev. Ther. Drug. Carriers Syst.* 1985, 2, 65–115.
5. Klibanov, A., Maruyama, K., Torchilin, V., and Huang, L. Amphipathic polyethyleneglycols effectively prolong the circulation time of liposomes. *FEBS Lett.* 1990, 268, 235–237.
6. Torchilin, V. P. and Trubetskoy, V. S. Which polymers can make nanoparticulate drug carriers long-circulating? *Adv. Drug Deliv. Rev.* 1995, 16, 141–155.
7. Lasic, D., and Martin, F., editors. *Stealth Liposomes.* Boca Raton, FL: CRC Press, 1995.
8. Trubetskoy, V. S., and Torchilin, V. P. Polyethyleneglycol based micelles as carriers of therapeutic and diagnostic agents. *STP Pharm. Sci.* 1996, 6, 79–86.
9. Tilcock, C., Unger, E., Cullis, P., and MacDougall, P. Liposomal Gd-DTPA: preparation and characterization of relaxivity. *Radiology* 1989, 171, 77–80.
10. Kabalka, G. W., Davis, M. A., Holmberg, E., Maruyama, K., and Huang, L. Gadolinium-labeled liposomes containing amphiphilic Gd-DTPA derivatives of varying chain length: Targeted MRI contrast enhancement agents for the liver. *Magn. Res. Imaging* 1991, 9, 373–377.
11. Schwendener, R. A., Wuethrich, R., Duewell, S., Westera, G., and Von-Schulthess, G. K. Small unilamellar liposomes as magnetic resonance contrast agents loaded with paramagnetic manganese-, gadolinium-, and iron-DTPA-stearate complexes. *Int. J. Pharm.* 1989, 49, 249–259.
12. Jaggi, M., Khar, R., Chauhan, U., and Gangal, S. Liposomes as carriers of technetium-99m glucoheptonate for liver imaging. *Int. J. Pharm.* 1991, 69, 77–79.
13. Tilcock, C. Liposomal paramagnetic MR contrast agents. In: Gregoriadis, G., editor. *Liposome Technology,* 2nd ed. Boca Raton, FL: CRC Press, 1993, vol 2, 65–87.
14. Kabalka, G., Buonocore, E., Hubner, K., Davis, M., and Huang, L. Gadolinium-labeled liposomes containing paramagnetic amphipatic agents: targeted MRI contrast agent for the liver. *Magn. Res. Med.* 1988, 8, 89–95.
15. Grant, C., Karlik, S., and Florio, E. A liposomal MRI contrast agent: Phosphatidyl ethanolamine-DTPA. *Magn. Res. Med.* 1989, 11, 236–243.
16. Trubetskoy, V. S., and Torchilin, V. P. New approaches in the chemical design of Gd-containing liposomes for use in magnetic resonance imaging of lymph nodes. *J. Liposome Res.* 1994, 4, 961–980.
17. Slinkin, M. A., Klibanov, A. L., and Torchilin, V. P. Terminal-modofoed polylysine-based chelating polymers: highly efficient coupling to antibody with minimal loss in immunoreactivity. *Bioconj. Chem.* 1991, 2, 342–348.
18. Torchilin, V. P., and Klibanov, A. L. The antibody-linked chelating polymers for nuclear therapy and diagnostics. *CRC Crit. Rev. Ther. Drug Carriers Syst.* 1991, 7, 275–308.

19. Tilcock, C., Ahkong, Q. F., Koenig, S. H., Brown 3rd, R. D., Davis, M., and Kabalka, G. The design of liposomal paramagnetic MR agents: Effect of vesicle size upon the relaxivity of surface-incorporated lipophilic chelates. *Magn. Res. Med.* 1992, 27, 44–51.

20. Trubetskoy, V. S., Frank-Kamenetsky, M. D., Whiteman, K. R., Wolf, G. L., and Torchilin, V. P. Stable polymeric micelles: Lymphangiographic contrast media for gamma scintigraphy and magnetic resonance imaging. *Acad. Radiol.* 1996, 3, 232–238.

# CHAPTER 7

# Bioconjugation of Biodegradable Poly (lactic/glycolic acid) to Protein, Peptide, and Anti-Cancer Drug: An Alternative Pathway for Achieving Controlled Release from Micro- and Nanoparticles

TAE GWAN PARK[1]

## INTRODUCTION

BIODEGRADABLE poly(D,L-lactic-co-glycolic acid) (PLGA) was chemically conjugated to two model protein drugs, lysozyme and a tryptophan derivative. One of the two terminal functional groups, hydroxyl and carboxylic acid, on the PLGA chains was activated by suitable coupling agents and then reacted to appropriate functional groups on the protein molecules. Cleavable ester and noncleavable amide linkages were readily formed between PLGA and drugs. The conjugates were formulated as injectable microspheres while maintaining high loading efficiency and retention due to the limited water solubility of the resultant drug–PLGA conjugates. The drug release was controlled by the degradation rate of the conjugated PLGA chain, permitting a linear drug release over an extended period without showing an initial burst effect. This conjugation concept can be further extended to an anticancer drug, doxorubicin. The doxorubicin–PLGA conjugates were formulated to provide enhanced doxorubicin loaded nanoparticles for the purpose of passive targeting. Additionally, superior sustained release capability was observed to free compared doxorubicin encapsulated in PLGA nanoparticles.

[1]Department of Biological Sciences, Korea Advanced Institute of Science and Technology, Taejon 305-701, South Korea.

## CONTROLLED RELEASE FROM PLGA MICROSPHERES

Biodegradable aliphatic polyesters such as poly(L-lactic acid) and its copolymers with D-lactic and glycolic acid of poly(D,L-lactic-co-glycolic acid) [PLGA] have been extensively used as microparticulate carriers for sustained releases of various small molecular weight drugs, peptides, and proteins [1]. In particular, various PLGA polymers having different molecular weights and compositions of lactic/glycolic acid ratio have been commercially available, which degrade in a wide range of time intervals suitable for tailoring the drug release period [2]. Since PLGAs are biocompatible and biodegradable polymers, they are ideal for injectable and implantable delivery systems. Various drugs have been encapsulated into PLGA microspheres in the size range between 1 and 100 μm in diameter. Injectable formulation of PLGA microspheres containing leutenizing hormone (LH-RH) analogs has been one of the more successfully commercialized drug delivery products [3].

There have been a variety of methods for encapsulating drugs into microspheres such as solvent evaporation, solvent extraction, phase separation, spray drying, and expansion in supercritical gas [4,5]. The controlled release of drugs from the PLGA microspheres has been known to be mainly governed by diffusion in the early stage and subsequent matrix degradation enhanced diffusion in the later stage, typically showing a triphasic release profile [6].

Hydrophobic drugs such as steroids can be directly incorporated into the polymer phase of PLGA microspheres by using a single oil-in-water (O/W) emulsion solvent evaporation technique by direct dissolution of drug molecules in an organic phase. On the other hand, hydrophilic drugs such as antibiotics, peptides, and proteins are routinely encapsulated within microspheres by a double emulsion, water-in oil-in water ($W_1/O/W_2$) solvent evaporation method [7]. Water-soluble drugs are first dissolved in $W_1$ phase, emulsified into an organic phase (O), and then re-emulsified in $W_2$ phase. For the encapsulation of hydrophilic drugs by using the double-emulsion solvent evaporation method, however, an initial rapid release of incorporated drug molecules within a short duration, known as a burst effect, was often observed [8]. Additionally, drug encapsulation efficiency and loading percent within the microspheres are highly variable depending on the formulation parameters because of the tendency of the initially entrapped drug in $W_1$ phase to escape to the outer $W_2$ phase during the second emulsification procedure. Thus, it has been difficult to achieve a prolonged release profile with a minimized burst release over a desired period longer than a week. The burst effect is thought to be caused primarily by the microporous channels present within PLGA microspheres. The microporous channels are likely to be generated during freeze drying; they are inevitably developed as major water escape routes for primary inner emulsion water droplets contained in $W_1$ phase. The interconnected water filled channels might be major pathways responsible for the rapid initial release upon incubation of the microspheres in the medium. Therefore, in most cases, the resultant morphology of PLGA microspheres plays a critical role in determining the release profile of hydrophilic drugs.

## CONJUGATION OF PLGA TO DRUGS

Polymer–drug conjugation approach has been extensively studied for the past decade. The best known example is poly(ethylene glycol) conjugated therapeutic proteins for the purpose of long circulation as well as reduced immunogenicity in the body [9,10]. Additionally, water-soluble polymer–doxorubicin conjugates based on poly(N-(2-hydroxypropyl)methacrylamide) has been extensively investigated and they are now under clinical trials [11,12]. Another promising approach is to conjugate doxorubicin to an amphiphilic block copolymer composed of poly(ethylene glycol) (PEG) and poly($\alpha,\beta$-aspartic acid), which leads to a polymeric micelle structure in an aqueous solution [13,14]. The above two anticancer drug–polymer conjugates tend to deliver the conjugated doxorubicin to a tumor site in a passive targeting manner.

Biodegradable PLGA is an aliphatic polyester that has no available functional groups in its backbone for the drug conjugation. In order to introduce the backbone functionality, amino acid residues such as L-lysine were incorporated in the polymer chain during ring opening copolymerization [15,16]. Although these new biodegradable polymers are expected to have primary amino functional groups on the polymer backbone for conjugating drugs and other biospecific ligands, they might show different biodegradation behavior compared to the unmodified PLGA. Most of the commercially available PLGA is prepared by ring opening polymerization and has a hydroxyl group at one end and a long-chain alcohol (dodecyl alcohol)-blocked carboxylic acid at the other end. They are called capped PLGA. The PLGA synthesized by direct polycondensation or by modified ring opening polymerization without using dodecyl alcohol is called uncapped PLGA, and has one hydroxyl group at one end and one carboxylic acid at the other end. These two end functional groups can be used to conjugate appropriately functionalized hydrophilic drugs, polypeptides, and proteins via degradable and nondegradable linkages. The advantage to conjugating PLGA to hydrophilic drug molecules is the enhancement of the overall hydrophobicity of the conjugates, thereby increasing drug partitioning into a nonpolar medium. It is expected that the PLGA–drug conjugates could be directly dissolved in an organic phase routinely used for the formulation of nano- and microspheres. The conjugates can be incorporated into particulate drug carriers using a single W/O emulsion technique with high drug loading. This novel formulation strategy is also postulated to exhibit a unique drug release mechanism from microspheres. The conjugated drug cannot be released from the microspheres until the conjugated PLGA polymer chain gradually degrades and reaches to a critical molecular weight at which the water solubility of drug–PLGA oligomer conjugate is sufficient. Thus, the drug release rate is expected to solely depend on the chemical degradation rate of PLGA chain, which permits the controlled liberation of the conjugated drug in a form of water-soluble oligomer PLGA-conjugated drug. In other words, the drug release rate would be proportional to the mass erosion profile of PLGA, which is a linear function of time with an initial lag time. The conjugates of PLGA to lysozyme, a model protein, and doxorubicin were synthesized via various coupling reactions. The PLGA conjugates were directly incorporated into microspheres and nanoparticles. The schematic diagram of drug release behavior from microspheres is shown in Figure 1.

## CONJUGATION OF PLGA TO LYSOZYME

Lysozyme ($M_w$. 14,300) was chosen for PLGA conjugation because lysozyme can be molecularly dissolved in DMSO, a polar organic solvent, in which PLGA could be also dissolved [17]. By using dicyclohexyl carbodiimide (DCC) as the coupling agent, PLGA-lysozyme conjugate

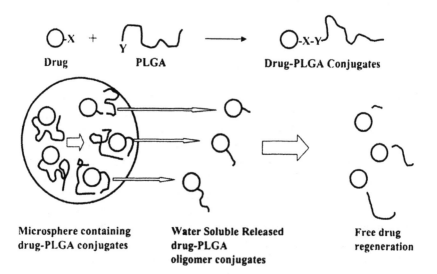

**Figure 1** A schematic representation of PLGA–drug conjugation approach.

was synthesized as shown in Figure 2 [18]. Since the carbodiimide conjugation process was known to readily occur in basic conditions, lysozyme was lyophilized at pH 9 prior to the dissolution in DMSO to preserve charge states of various amino acid residues present in the protein in the organic solvent similar to those in the aqueous solution prior to the lyophilization (pH-memory effect) [19]. The degree of PLGA–lysozyme conjugation, determined by a fluorescamine assay, indicated that out of seven lysozyme primary amine groups an average of one amine group was conjugated to PLGA ($M_W$ 8,600) molecule [20]. The PLGA–lysozyme conjugate was soluble only in DMSO and insoluble in other solvents such as water, methanol, dimethylformamide, and methyelne chloride. The microspheres encapsulated with the PLGA-conjugated lysozyme were prepared by a single oil-in-water (O/W) emulsion method using a mixture of DMSO and methylene chloride as an oil phase to dissolve the PLGA-conjugated lysozyme. The above solution was directly emulsified in poly(vinyl alcohol) (PVA)/PBS (phosphate buffered saline, 0.1 M NaCl) solution.

Micropsheres were successfully prepared using the PLGA–lysozyme conjugate by a single emulsion method. Encapsulation efficiency of lysozyme

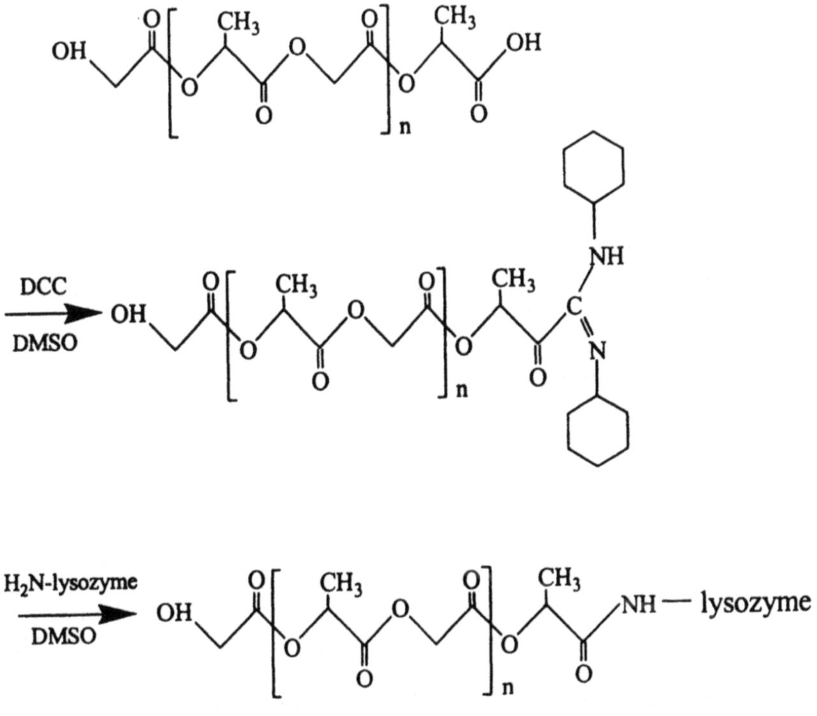

**Figure 2** Conjugation chemistry of PLGA to lysozyme.

within the microspheres was 99.9%, indicating that the PLGA–lysozyme conjugate by encapsulation produced high loading efficiencies owing to its limited water solubility in water. The SEM pictures of microspheres showed spherical shaped microspheres with smooth and nonporous surfaces. A cross-sectional view of the two microspheres revealed that internal structure was slightly porous as discussed in our previous report [21].

The release profiles of lysozyme from the microspheres demonstrated an initial low burst and subsequent accelerated release. However, a large fraction of the lysozyme was not released even after 56 days of incubation. The low cumulative release percent relative to the initial loading amounts was attributed to the ionic interaction between lysozyme and PLGA in the early incubation stage and noncovalent aggregation and nonspecific adsorption in the later incubation stage [22,23]. Although lysozyme was successfully conjugated to PLGA, the inherent lysozyme stability problems affected the release and obscurred the conjugation effect. Thus, we decided to conjugate a model amino acid derivative to PLGA.

## CONJUGATION OF PLGA TO AN AMINO ACID DERIVATIVE

A model amino acid derivative, N-(9-fluorenylmethoxycarbonyl-N-butoxycarbonyl-L-tryptophan [Fmoc-Trp(Boc)] was conjugated to a hydroxyl terminal end group of PLGA via a biodegradable ester bond linkage [24]. In the case of lysozyme–PLGA conjugation, the amide linkage formed between primary amino groups in lysozyme and carboxylic acid groups in PLGA is not desirable for the drug conjugation since it is not easily cleaved. Two different PLGAs, 50/50 and 75/25 lactic/glycolic acid compositions, were used to conjugate Fmoc-Trp(Boc) as shown in Figure 3. Various coupling agents and bases in different combinations were tested to activate the carboxylic acid group on Fmoc-Trp(Boc). Among them, bromo-trispyrrolidinophosphonium hexafluorophosphate (pyBrop)/triethylamine combination resulted in the largest conjugation percent (63%). An active ester form of Fmoc-Trp(Boc) was then reacted with a hydroxyl group of PLGA 50/50 in methylene chloride. For the formation of microspheres, a single emulsion method was used. Fmoc-Trp(Boc)-PLGA conjugates [PLGA 50/50 (formulation A), PLGA 75/25 (formulation B), and PLGA 75/25 and 50/50 mixture (formulation C), respectively] were dissolved in a cosolvent mixture of DMSO and methylene chloride. Free Fmoc-Trp(Boc) was encapsulated within PLGA 50/50 microspheres (formulation D) as control. The microspheres (formulations A, B, and C) containing the Fmoc-Trp(Boc)-PLGA conjugates demonstrated almost 100% encapsulation efficiencies due to their limited solubility in water, whereas the control microspheres (formulation D) encapsulated with unconjugated Fmoc-Trp(Boc) had only a 20.1% en-

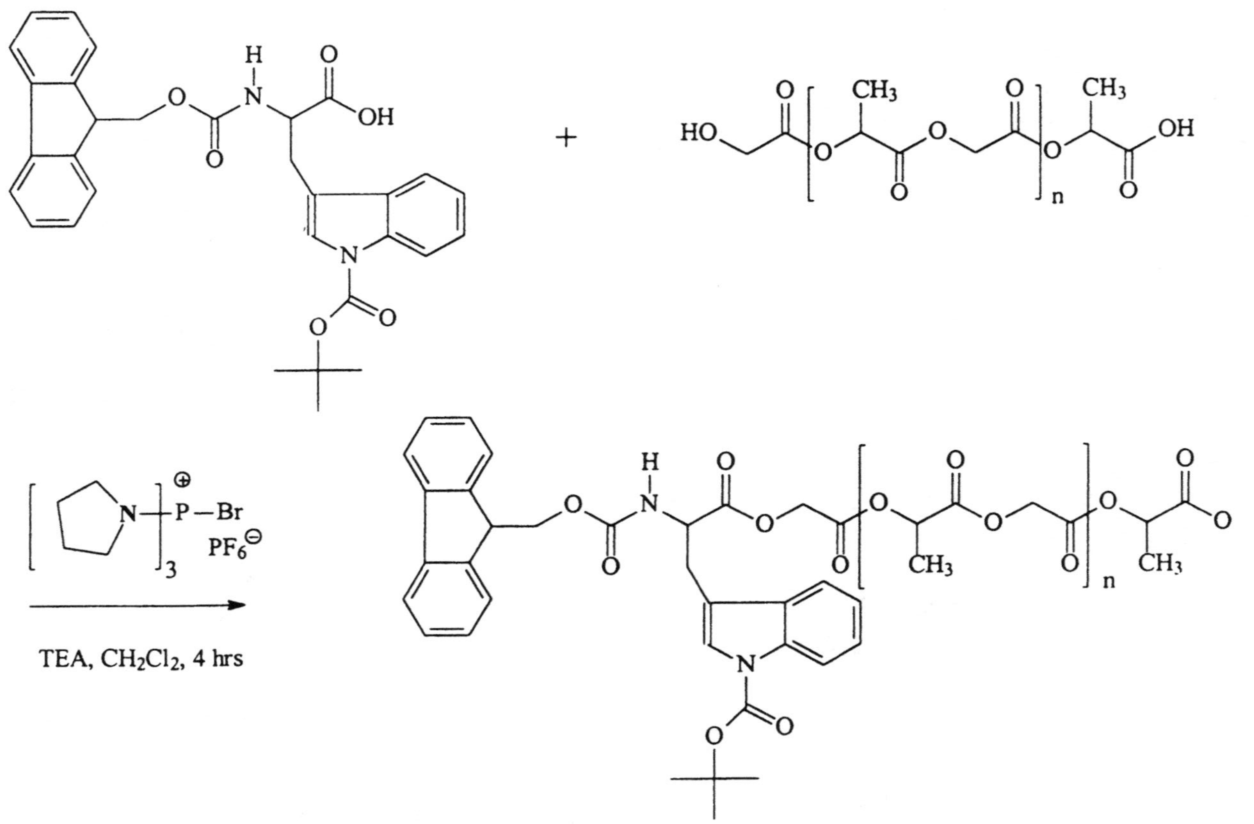

**Figure 3** Conjugation chemistry of PLGA to Fmoc-Trp(Boc) (adapted with permission from the publisher [24]).

capsulation efficiency due to the diffusion of the moderately water soluble
Fmoc-Trp(Boc) into the aqueous phase during the formulation. SEM pictures
indicated smooth surface morphology with an average diameter ranging
8.33–8.76 μm. Shown in Figure 4 are release profiles of Fmoc-Trp(Boc)-
PLGA oligomer conjugates incubated in medium from various microsphere
formulations. It can be seen that the microspheres conjugated with Fmoc-
Trp(Boc) exhibit constant release profiles for over 20 days with an initial,
short lag time period, whereas the microspheres containing unconjugated
Fmoc-Trp(Boc) show a rapid release in the initial incubation stage, resulting
in the early release termination within 5 days. This is a typical type of release
kinetic pattern from PLGA microspheres encapsulated with moderately hydrophilic drugs, which is mainly caused by the fast diffusion of encapsulated
free drug molecules through aqueous interconnecting channels in the microspheres generated upon hydration [25]. Release kinetic rates of the drug
conjugated with PLGA can be judiciously controlled by appropriately se-

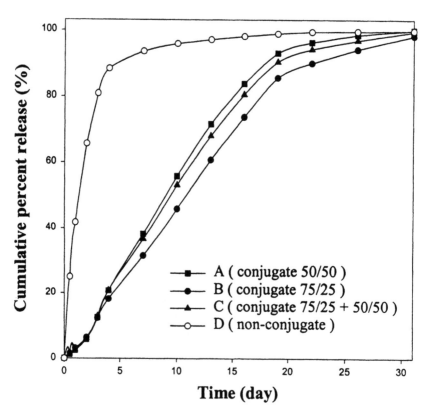

**Figure 4** Release profiles of Fmoc-Trp(Boc)–PLGA oligomer conjugates from microspheres
(adapted with permission from the publisher [24]).

lecting various PLGA polymers that degrade at different rates. Therefore, molecular weight and/or copolymer composition of PLGA to be conjugated are important variables in controlling the release rate in the present system. The above results indicate that different hydrolytic scission rates of polymer backbone composed of PLGA 50/50 and 75/25 are mainly responsible for the controlled liberation of Fmoc-Trp(Boc)-PLGA oligomer conjugates. PLGA chain segments are required to hydrolyze to reach a critical molecular weight, of about 1,000–1,100 to gain sufficient water solubility [26]. A lag time was generally observed in the mass erosion profile of PLGA microspheres, whereas the molecular weights of PLGA chains decrease continuously without exhibiting any lag time [27]. Thus, it is expected that drug release kinetic rate in the present PLGA–drug conjugate system is determined by the mass erosion rate of PLGA polymer matrices.

## CONJUGATION OF PLGA TO DOXORUBICIN

The conjugation approach can be extended to the fabrication of nanoparticles encapsulated with hydrophilic drugs. In contrast to the microspheres having their average diameters above 1 μm, it has been difficult to load sufficient amounts of water soluble drugs within the nanoparticle matrix for sustained release purposes [28]. The most widely used method is a spontaneous emulsion–solvent diffusion technique, which in most cases the drug molecules are weakly adsorbed on the nanoparticle surface and then rapidly released upon incubation. The conjugation of PLGA to drugs can be utilized for the formation of nanoparticles containing sufficient drug-loading amounts and sustained release capability.

Doxorubicin conjugated to water soluble polymers or incorporated into the core region of polymeric micelles in order to achieve passive targeting has been used as a potent anticancer drug for the treatment of various solid tumors. The conjugation of PLGA to moderately water soluble doxorubicin is expected to generate limited water-soluble conjugated doxorubicin, thereby permitting fabrication of nanoparticles with high loading amounts of doxorubicin within the matrix. PLGA-doxorubicin conjugates can be made via two synthetic routes. In doxorubicin, there are two kinds of functional groups in the structure: one primary hydroxyl group and one primary amino group. Both of them can be utilized for conjugation to the terminal end groups of PLGA.

In the first attempt, PLGA was conjugated to the primary amino group of the sugar moiety of doxorubicin by a carbamate linkage that is not cleavable. A terminal hydroxyl group of PLGA was activated by *p*-nitrophenol chloroformate and then reacted to the primary amino group in the sugar moiety. The reaction scheme is shown in Figure 5. In the second approach,

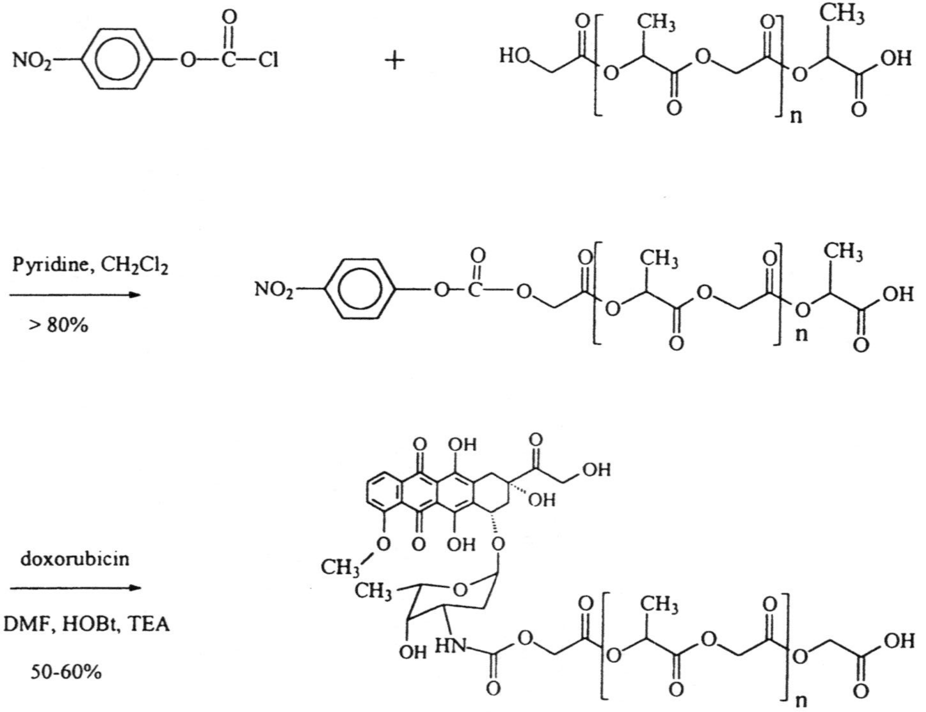

**Figure 5** Conjugation of PLGA to doxorubicin via a noncleavable carbamate linkage (adapted with permission from the publisher [24]).

110

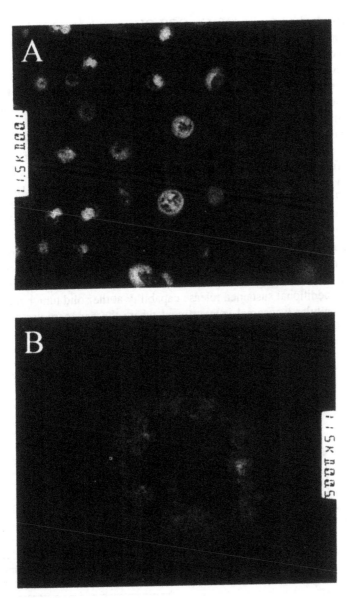

**Figure 6** Transmission electron microscopic picture of nanoparticles containing PLGA-doxoru-bicin conjugates. Scale bar is 2 μm in (A) and 200 nm in (B) (adapted with permission from the publisher [24]).

a primary hydroxyl group of -C=OCH₂OH group in the cyclohexyl ring
of doxorubicn is the primary reaction site for the conjugation, although
other secondary and tertiary hydroxyl groups are present in the structure.
The second conjugation approach generates a cleavable ester linkage between doxorubicin and PLGA, which has a distinctive advantage for re-
generation of intact doxorubicin after release. Nanoparticles 200–300 nm
in size based on the PLGA–doxorubicin conjugate were readily formed by
spontaneous emulsion and solvent diffusion as shown in Figure 6. The
loading efficiency was near 100% in contrast to very low loading percentage of free doxorubicin loading within PLGA. When incubated in buffer
solution, the doxorubicin was released in a sustained manner as shown in
Figure 7, confirming that water-soluble doxorubicin–PLGA oligomer conjugates were primarily released and then they were further degraded to re-
generate free doxorubicin. The cytotoxic results determined by using
HepG2 cells showed comparable IC$_{50}$ values between free doxorubicin
and the released doxorubicin–PLGA oligomer conjugates. The direct conjugation of doxorubicin to PLGA can be potentially formulated in nanoparticles with additional sustained release capability at the solid tumor site or in
the intracellular lysosomal compartment where the nanoparticles are normally located after endocytosis [29].

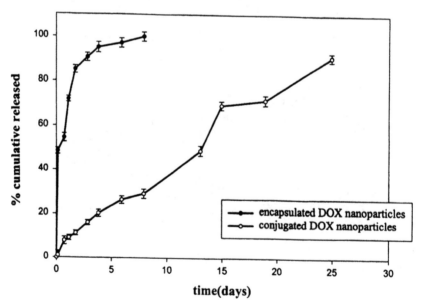

**Figure 7** Release profiles of doxorubicin from the two nanoparticles: free doxorubicin encapsulated nanoparticles and PLGA–doxorubicin loaded nanoparticles (adapted with permission from
the publisher [24]).

## CONCLUSION

The conjugation approach of PLGA to drugs can be applied to the formulation of micro- and nanoparticulates that often require high drug-loading amounts and efficiency as well as controllable drug release patterns. This new concept could be widely utilized for precisely controlled release of various hydrophilic drugs in the future.

## REFERENCES

1. Cleland, J. R. and Langer, R. *Formulation and Delivery of Proteins and Peptides,* ACS Symposium Series, vol 567, ACS, Washington DC, 1994, pp. 1–21.
2. Lewis, D. D. *Biodegradable Polymers as Drug Delivery Systems,* Chasin, M. and Langer, R., Eds., Marcel Dekker, New York, 1990, pp. 1–41.
3. Ogawa, Y., Okada, H., Yamamoto, M., and Shinamoto, T. *Chem. Pharm. Bull.,* 1998, 36, 2576–2588.
4. Crotts, G. and Park, T. G. *J. Controlled Release,* 1995, 35, 91–105.
5. Wang, H. T., Schmitt, E., Flagnagan, D. R., and Linhardt, R. J. *J. Controlled Release,* 1991, 17, 23–32.
6. Kenley, R. A., Lee, M. O., Mahoney II, T. R., and Sanders, L. M. *Macromolecules,* 1987, 20, 2398–2403.
7. Cohen, S., Yoshioka, T., Lucarelli, M., Hwang, L. H., and Langer, R., *Pharm. Res.,* 1991, 8, 713–720.
8. Alonso, M. J., Cohen, S., Park, T. G., Gupta, R. K., Siber, G. R., and Langer, R. *Pharm. Res.,* 1993, 10, 945–953.
9. Abuchowski, A., McCoy, J. R., Palczuk, N. C., van Es, T., and Davis, F., *J. Biol. Chem.,* 1977, 252, 3582.
10. Zalipsky, S. and Lee, C. *Poly(ethylene glycol) Chemistry,* Harris, J. M. Ed., Plenum Press, New York, 1992, pp. 347–370.
11. Duncan, R. *Anti-cancer Drugs,* 1992, 3, 175.
12. Putnam, D. and Kopecek, J. *Adv. Polymer. Sci.,* 1995, 122, 55.
13. Kataoka, K., Kwon, G. S., Yokoyama, M., Okano, T., and Sakurai, Y. *J. Controlled Release,* 1994, 24, 119.
14. Kataoka, K., *Controlled Drug Delivery,* Park, K., Ed., ACS, Washington, DC, 1997, pp. 49–71.
15. Barrera, D. A., Zylstra, E., Lansbury, P. T., Langer, R., *J. Am. Chem. Soc.* 1993, 115, 11010–11011.
16. Shakesheff, K. M., Cannizzaro, S. M., and Langer, R. *J. Biomater. Sci., Polymer Edn.,* 1998, 9, 507–518.
17. Chin, J. T., Wheeler, S. L., and Klibanov, A., *Biotech. Bioeng.,* 1994, 44, 140–145.
18. Bodanszky, M., *Principles of Peptide Synthesis,* Springer-Verlag, Berlin, pp. 36–58, 1984.
19. Costantino, H. R., Griebenow, K., Langer, R., and Klibanov, A. M. *Biotech. Bioeng.,* 1997, 53, 345–348.

20. Stocks, S. J., Jones, A. J. M., Ramey, C. W., and Brooks, D. E. *Analytical Biochemistry,* 1986, 154, 232–234.
21. Crotts, G. and Park, T. G. *J. Controlled Release,* 1997, 44, 123–134.
22. Crotts, G., Sah, H., and Park, T. G. *J. Controlled Release,* 1997, 47, 101–111.
23. Park, T. G., Lee, H. Y., and Nam, Y. S., *J. Controlled Release,* 1998, 55, 181–191.
24. Oh, J. E., Nam, Y. S., Lee, K. H., and Park, T. G. *J. Controlled Release,* 1999, 57, 269–280.
25. Blanco-Prieto, M. J., Fattal, E., Gulik, A., Dedieu, J. C., Roques, B. P., and Couvreur, P. *J. Controlled Release,* 1997, 43, 81–87.
26. Park, T. G. *J. Controlled Release,* 1994, 30, 161–173.
27. Park, T. G., Lu, W., and Crotts, G. *J. Controlled Release,* 1995, 33, 221–232.
28. Couvreur, P., Roblot-Treupel, L., Poupon, M. F., Brasseur, F., and Puisieux, F. *Adv. Drug. Deliv. Rev.,* 1990, 5, 209–230.
29. Yoo, H. S., Oh, J. E., Lee, K. H., and Park, T. G. *Pharmaceutical Research,* 1999, 16, 1114–1118.

# CHAPTER 8

# Elastin–Mimetic Protein Networks Derived from Chemically Crosslinked Synthetic Polypeptides

R. ANDREW McMILLAN[1]
VINCENT P. CONTICELLO[1]

## INTRODUCTION

POLYMER hydrogels have enormous importance in the biomedical field, in which they are employed as prosthetic devices, soft contact lenses, drug delivery agents, and matrices for cell encapsulation and tissue engineering [1]. The structure of the polymer network dramatically affects biomedically important materials properties, e.g., the porosity and solute permeability of the matrix [2]. However, most polymer networks have a statistical distribution of crosslink sites along the polymer backbone as a consequence of random incorporation of bifunctional or crosslinkable monomers into the polymer chain during synthesis. Consequently, most gels have heterogeneous architectures with a distribution of microenvironments and pore sizes within the network [3]. Regulation of the porosity and substrate permeability of synthetic hydrogels is an important consideration in the design of materials for drug delivery and bioartificial organs, as selectivity in solute permeability is directly related to the success of these devices [2].

We have prepared a synthetic protein polymer based on repeat sequence **Lys-25** to investigate the effect of uniformity of crosslink placement on the physical properties of a polymer hydrogel (Figure 1). The design of **Lys-25** reflects two essential structural requirements for formation of polymer hydrogels: (1) a flexible, hydrated (polyamide) backbone and

[1]Department of Chemistry, Emory University, 1515 Pierce Drive, Atlanta, GA 30322, USA.

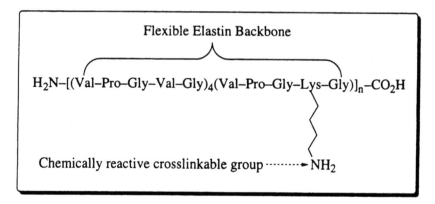

**Figure 1** Design of a crosslinkable amino acid sequence based on the elastin–mimetic repeat **Lys-25.**

(2) a chemically reactive, crosslinkable functional group, specifically, the amino substituent of the lysine side chain. The amino acid sequence of **Lys-25** is modeled on the pentapeptide repeat of bovine elastin (Val-Pro-Gly-Val-Gly) [4]. The central "Pro-Gly" element of this repeat adopts a type II reverse turn structure, which forms a flexible helix on tandem repetition known as the "β-spiral" [5]. Urry et al. have accumulated a large body of experimental evidence in support of the β-spiral structure based on structural investigations of sequence variants of the elastin polypentapeptide [(Val-Pro-Gly-Xaa-Gly)n][6]. These studies suggest that position 4 in the elastin pentapeptide repeat can tolerate wide substitution, including most of the natural, proteinogenic amino acids, without disruption of the β-spiral structure [7].

Elastin-mimetic protein polymers [(Val-Pro-Gly-Xaa-Gly)$_n$] undergo reversible temperature-dependent, hydrophobic assembly in aqueous solution [8]. This process, known as the inverse temperature transition, converts the polypeptide from a soluble, expanded state to a collapsed, viscoelastic state upon passing through a critical temperature $T_t$. This transition involves the controlled folding and assembly of the hydrophobic sidechains in the polypeptide. This process results in desolvation and aggregation of the chains into a dense, protein-rich coacervate. The temperature of this transition depends on the balance between polar and apolar moieties in the polypeptide sequence. Consequently, factors that perturb this balance can strongly attenuate the value of the $T_t$. In a chemosynthetic series of elastin analogues [(Val-Pro-Gly-Xaa-Gly)$_n$], $T_t$ increases with increasing polarity of the substituent at position 4 in the repeat [7]. An amino acid hydrophobicity scale has been constructed based on the effect of the substituent in position 4 on $T_t$ for the series of elastin analogues [(Val-Pro-Gly-Xaa-Gly)$_n$] [7,9]. The elastic response of these polymers to changes in environmental stimuli such as pH, temperature, and ionic strength of the medium can be predictably altered by variation of the residue in position 4 of the repeat sequence [6].

Elastin–mimetic protein polymers have been fabricated into elastic networks primarily via γ-radiation-induced, radical crosslinking of the material in the coacervate state [10]. Although effective, this method cannot produce polymers gels of defined molecular architecture, i.e., specific crosslink position and density, due to the lack of chemoselectivity in radical reactions. In addition, the ionizing radiation employed in this technique can cause material damage, and the reproducibility of specimen preparations may vary between different batches of material. In contrast, the ε-amino groups of the lysine residues in polymers based on **Lys-25** can be chemically crosslinked *under controllable conditions* into synthetic protein networks *(vide infra)*. Elastic networks based on **Lys-25** should contain crosslinks at well-defined position and density, determined by the sequence of the repeat, *in the limit of complete substitution* of the amino groups.

## RESULTS AND DISCUSSION

### SYNTHESIS AND CHARACTERIZATION OF POLY(LYS-25)

Protein polymers based on **Lys-25** were prepared by recombinant DNA (rDNA) technology and bacterial protein expression. The main advantage of this approach is the ability to directly produce high molecular weight polypeptides of exact amino acid sequence with high fidelity as required for this investigation. In contrast to conventional polymer synthesis, protein biosynthesis proceeds with near-absolute control of macromolecular architecture, i.e., size, composition, sequence, topology, and stereochemistry. Biosynthetic poly(α-amino acids) can be considered as model uniform polymers and may possess unique structures and, hence, materials properties, as a consequence of their sequence specificity [11]. Protein biosynthesis affords an opportunity to completely specify the primary structure of the polypeptide repeat and analyze the effect of sequence and structural uniformity on the properties of the protein network.

A synthetic DNA cassette was designed that encoded the repeat sequence **Lys-25** (Figure 2). This DNA monomer was synthesized, sequenced, and self-ligated to afford a mixture of DNA concatamers. A 3,000 base pair concatameric gene was iolated from this mixture, which encoded a polymer of **Lys-25** with a molecular mass of approximately 90 kD. Bacterial expression of the concatameric gene afforded a protein of the expected molecular mass that accumulates to high levels in *E. coli* (Figure 3).

| Val | Pro | Gly | Val | Gly | Val | Pro | Gly | Val | Gly | Val | Pro | Gly |
|-----|-----|-----|-----|-----|-----|-----|-----|-----|-----|-----|-----|-----|
| GTA | CCG | GGT | GTT | GGC | GTT | CCG | GGT | GTA | GGT | GTG | CCA | GGC |
| CAT | GGC | CCA | CAA | CCG | CAA | GGC | CCA | CAT | CCA | CAC | GGT | CCG |

| Val | Gly | Val | Pro | Gly | Val | Gly | Val | Pro | Gly | Lys | Gly |
|-----|-----|-----|-----|-----|-----|-----|-----|-----|-----|-----|-----|
| GTT | GGT | GTA | CCG | GGT | GTT | GGC | GTA | CCA | GGC | AAG | GGC |
| CAA | CCA | CAT | GGC | CCA | CAA | CCG | CAT | GGT | CCG | TTC | CCG |

**Figure 2** DNA monomer design for elastin–mimetic repeat sequence **Lys-25** (the upper strand of the duplex represents the coding sequence of the monomer in the 5'-3' direction).

This polypeptide could be purified to homogeneity by either affinity chromatography or repetitive precipitation. Expression of the concatameric gene under large-scale, batch fermentation conditions afforded an unoptimized yield of 64 mg purified polypeptide per liter of culture in LB medium. The biosynthetic protein migrated as a single tight band by polyacrylamide gel electrophoresis, which suggests a uniform polypeptide composition. The details of these procedures have been reported separately [12].

The identity of the 90-kD protein polymer was confirmed by N-terminal amino acid sequence analysis of proteolytic fragments derived from cleavage with Endoproteinase Lys-C. The major cleavage product was subjected to 25 cycles of automated Edman degradation, which yielded a the sequence GVPGVGVPGVGVPGVGVPGVGVPGK for the peptide fragment. This product corresponds to the major fragment expected for digestion of poly(**Lys-25**) with endoproteinase Lys-C, which cleaves specifically after the carboxy terminus of lysine residues in the polypeptide. The identity of this cleavage fragment is supported by the $[M+H]^+$ peak at 2092.3 m/z (expected: 2093.5 m/z) in the MALDI-TOF mass spectrum of this protein fragment. Amino acid compositional analysis of the

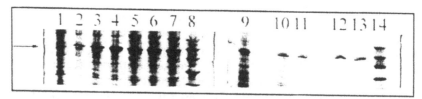

**Figure 3** Biosynthesis and purification of 90-kD elastin analogue analyzed by denaturing polyacrylamide gel electrophoresis (10–15% gradient, visualized by silver staining). Lanes 1–7; time course of target protein expression at 0, 30, 60, 90, 120, 150, and 180 minutes after induction. Lane 9; soluble lysate of induced *E. coli* expression strain BLR(DE3)pRAM1. Lanes 10–13; protein fractions obtained from immobilized metal affinity chromatography of the lysate on nickel–NTA agarose (imidazole gradient elution). Lanes 8, 14; protein molecular weight standards of 50, 75, 100, and 150 kD.

purified polypeptide is also consistent with its formulation as a polymer of **Lys-25** [12].

The [1]H- and [13]C-NMR spectroscopic data support the proposed primary structure of poly(**Lys-25**). The amide carbonyl resonances are particularly informative as these signals are well resolved in the [13]C-NMR spectrum of poly(**Lys-25**) (Figure 4). An amide carbonyl resonance is observed at 174.9 ppm for poly(**Lys-25**) that does not appear in the spectrum of poly(Val-Pro-Gly-Val-Gly) [13]. The position and relative intensity of this resonance are consistent with a lysine amide carbonyl group within a peptide bond [14]. Moreover, the resonances of the amide carbonyl groups for other residues in the pentapeptide repeat are split due to the substitution of a lysine residue at position 4 in every fifth pentapeptide in **Lys-25**. In addition, the absence of splitting in amide carbonyl group of valine in position 4 (174.5 ppm) supports this assignment, as this residue is replaced by lysine in the fifth pentapeptide of the **Lys-25** repeat. The presence of other resonances attributable to the lysine residue can be detected in the [1]H- and [13]C-NMR spectra of the **Lys-25** polymer at levels commensurate with its

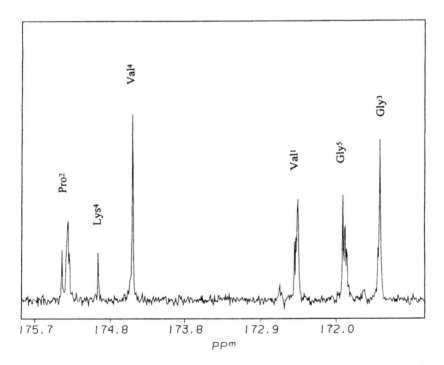

**Figure 4** Detail of the amide carbonyl resonances in the [13]C-NMR spectrum (100 MHz) of poly(**Lys-25**) in 70% $H_2O$/30% $D_2O$ solution. The spectrum was recorded on a Varian INOVA 400 NMR spectrometer. Chemical shifts were referenced to external tetramethylsilane.

percent incorporation into the repeat sequence. For example, the positions of the $\epsilon$-H (2.97 ppm) and $\epsilon$-C (40.1 ppm) resonances of the lysine side chain can be identified in the respective $^1$H- and $^{13}$C-NMR spectra of poly(**Lys-25**) and agree substantially with expected chemical shifts for this residue [14].

The **Lys-25** polymer undergoes hydrophobic assembly and precipitation in aqueous solution as the temperature is raised, analogous to the chemosynthetic elastin analogues described by Urry [8]. The position of the inverse temperature transition $(T_t)$ depends strongly on the charge state of the ionizable lysine residue as depicted in the temperature-dependent turbidimetric profiles of poly(**Lys-25**) in water and 0.1 N NaOH solution (Figure 5). The inverse temperature transition shifts from 28°C in 0.1 N NaOH with the lysine residue in the free amine state to 75°C at neutral pH with the lysine residue in the ammonium ion state. The hydrophobic assembly is completely reversible as the polypeptide precipitate redissolves on lowering the temperature below the $T_t$. The observed shift in $T_t$ with the charge state of the lysine residue coincides with the predicted effect of increased residue polarity on the self-assembly of elastin-mimetic protein

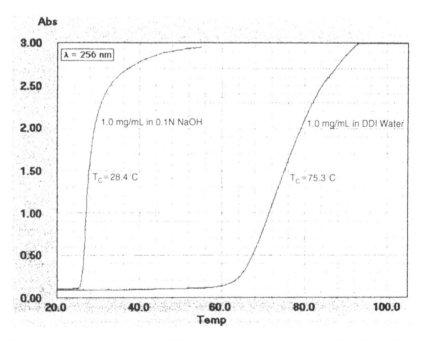

**Figure 5** Temperature-dependent turbidimetry profiles for solutions of poly(**Lys-25**) at pH 7.0 and 11.0. These measurements were performed on a Pharmacia Biotechnology Ultrospec 3000 UV/visible spectrophotometer equipped with a programmable Peltier cell and temperature control unit.

polymers. Urry et al., have described the chemical synthesis of a related series of random co-polypentapeptides [Ile-Pro-Gly-(Val/Lys)-Gly]$_n$, with fractional lysine content $(f_K)$ in position 4 ranging from 0.06 to 1.00 [15]. The aqueous p$K_a$s of the lysine amino groups in these protein polymers spanned from 8.18 to 9.60, with an observed p$K_a$ value of 9.11 at an $f_K$ of 0.22. This lysine content closely corresponds to fractional content $(f_K = 0.2)$ at position 4 in the **Lys-25** repeat. A similar p$K_a$ value for the lysine amino group would be expected for aqueous solutions of poly(**Lys-25**), as indicated by the shift in the inverse temperature transition from pH 7.0 to 11.0.

## PREPARATION AND CHARACTERIZATION OF CROSSLINKED POLY(LYS-25)

Synthesis of the elastin-mimetic protein gels is based on selective intermolecular reaction between the amino groups of the lysine residues and a bifunctional crosslinker. The choice of crosslinker has a critical impact on the properties of the polymer gel since it becomes incorporated into the network in the process. Crosslinking of lysine residues in biomolecules has been accomplished using a variety of reagants, including glutaraldehyde, bifunctional imidoesters, and, $N$-hydroxysuccinimide esters of bifunctional carboxylic acids [16]. Although glutaraldehyde has been used extensively as a crosslinking agent, neither its reaction chemistry nor the chemical identity of the resulting crosslinks is well defined. Bis(imidoesters) are also used widely for fixing biological specimens through reaction with lysyl residues; however, these reagents form positively charged amidinium ions upon reaction with amino groups. The formation of a highly charged network may have detrimental effects on the elastic properties of the material, especially since most naturally occurring elastic materials do not have an appreciable net charge. Bifunctional $N$-hydroxysuccinimide esters may be the crosslinker of choice for the preparation of synthetic polypeptide networks for several reasons: (1) well-defined reaction chemistry, (2) production of stable, uncharged amide crosslinks, and (3) commercial availability of reagents with different spacer composition and length between the reactive endgroups.

The crosslinking process involves nucleophilic displacement of the $N$-hydroxysuccimide group of the crosslinker by the amino group of a lysine residue on the polypeptide (Figure 6). This reaction is rapid and irreversible at ambient temperature under neutral to slightly basic conditions (pH 7–9). If the bifunctional molecule reacts with lysine residues from two different polypeptide chains, an intermolecular crosslink will form between the chains. As the degree of crosslinking increases, gelation of the solution occurs with formation of a protein network of defined main chain sequence and crosslink composition. Intermolecular reaction with the crosslinker should be favored at high concentrations of the polypeptide,

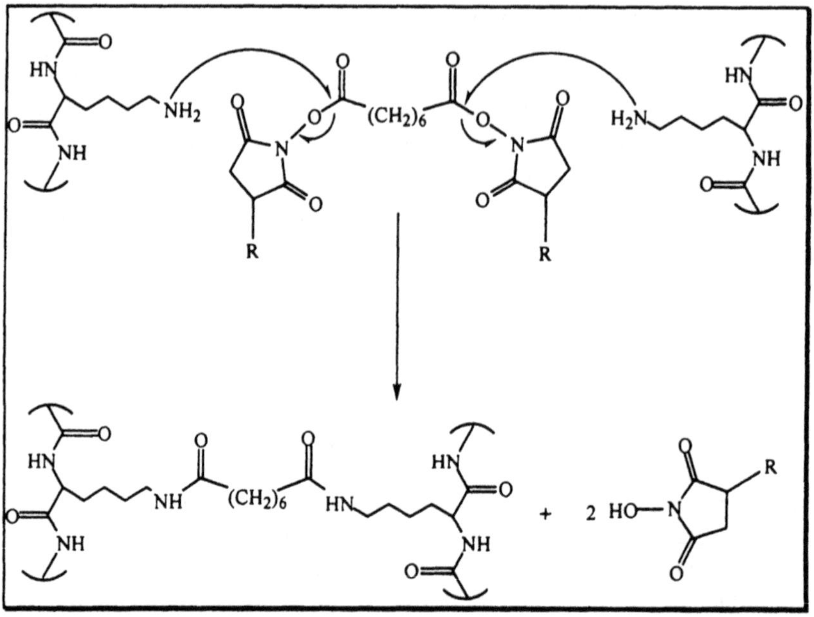

**Figure 6** Mechanism of intermolecular crosslink formation in lysine-containing polypeptides via condensation with bis($N$-hydroxysuccinimidyl)suberates ($R$ = H in DMSO reactions; $R$ = $SO_3^- Na^+$ in aqueous reactions).

although intramolecular crosslinking may also occur to various extents depending upon the particular reaction conditions.

The **Lys-25** polymer was crosslinked into a gel by treatment of a 5% (w/v) solution of the polypeptide in either aqueous phosphate buffer (pH 9) or anhydrous dimethylsulfoxide with a bis($N$-hydroxysuccinimidyl) derivative of suberic acid. The reagents were mixed at 4°C, and the solutions were warmed to 25°C. After approximately 10 minutes, the solutions formed mechanically stable gels that could be physically manipulated. The gels formed in DMSO were optically transparent in contrast to the opaque gels formed in aqueous solution.

Hydrophobic assembly and collapse occurred upon warming the the aqueous gel from 4°C to 37°C similarly to that observed for solutions of noncrosslinked poly(**Lys-25**). This process was completely reversible as the collapsed material expanded to its original volume on cooling back to 4°C. However, this process could not be followed turbidimetrically due to the opacity of the material. Since the crosslinking reaction involves the conversion of a charged, hydrophilic amine group into an uncharged, hydrophobic amide group, a large decrease in the $T_t$ would be expected for the water-swollen protein gel vis-à-vis the protein polymer in aqueous solution. The decrease in chain polarity and its effect on $T_t$ may account for local precipitation and the opacity of the protein gel at 25°C. In addition, the hydrophobic hexamethylene spacer in the crosslinker may further decrease the polarity of the network with respect to the poly(**Lys-25**) protein.

The success of the network formation depends on the efficiency of the reaction between the lysine groups and the *N*-hydroxysuccinimide (NHS) esters. Residual amine groups alter the responsive properties of the gel and introduce noncrosslinked defect sites into the network. The latter sites reduce the desired uniformity of the polypeptide hydrogel and increase its polarity. Optimization of the crosslinking conditions is vital for the formation of gels with uniform crosslink density and well-defined responsive behavior. The content of residual amino groups in the polymer gel was quantitated by treatment of the gel with sulfosuccinimidyl-4-O-[4,4'-dimethoxytrityl]butyrate (sulfo-SDTB) under aqueous conditions. This reagent is commonly used for estimating the amine content on solid substrates through generation of dimethoxytrityl cation that can be spectroscopically detected ($\lambda_{max}$ = 498 nm) [17]. The residual amino groups are converted into substituted butyramide derivatives, which undergo hydrolysis under acidic conditions to release the dimethoxytrityl cation (Figure 7). On the basis of this assay, the protein gel prepared under aqueous conditions contained 12% residual amino groups, which suggests an upper limit of 88% crosslinking efficiency under the reaction conditions described above.

The morphology of the synthetic gels was investigated by high-resolution scanning electron microscopy (HRSEM). Specimens were prepared for imaging by desolvation with increasing concentrations of ethanol in the swelling solvent. The aqueous gel sample was cycled between 4°C and 40°C during each exchange step to ensure adequate perfusion of the gradient mixtures into the specimen. The ethanol in the gels was exchanged with liquid carbon dioxide in a Polaron critical point drying apparatus, and the $CO_2$ was removed as a gas above its critical temperature and pressure. The dried gels were immobilized on carbon tape and a thin (10–20 nm) film of gold/palladium (60:40) alloy was applied with an Emscope sputter coater (7.5 mA, 50 mTorr, 3 min). Specimens were imaged in the below-lens configuration with an ISI DS-130 scanning electron microscope equipped with a $LaB_6$ electron source.

The morphology of the gels depended dramatically on the the identity of the reaction solvent (aqueous phosphate buffer versus DMSO). The crosslinking reaction should generate compositionally identical networks in the ideal limit of 100% substitution of the lysine amine groups. Although both gels form highly reticulated networks in the supermicron size domain, the morphology of the submicron features differed drastically between the two materials. The protein network prepared in aqueous solution

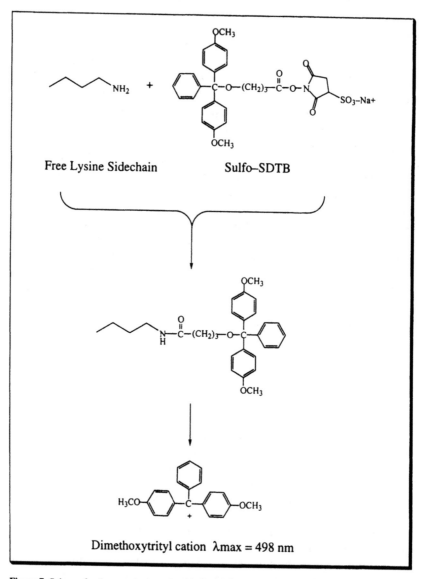

Free Lysine Sidechain        Sulfo–SDTB

Dimethoxytrityl cation $\lambda$max = 498 nm

**Figure 7** Scheme for the quantitation of residual amino groups for polypeptide gels in aqueous suspension via reaction with the chromogenic reagent sulfosuccinimidyl-4-$O$-[4,4′-dimethoxytrityl] butyrate (sulfo-SDTB).

displays a well-defined beaded or nodular morphology within the network with features on the order of several hundred nanometers (Figure 8). These features are clearly representative of the morphology of the gel as a whole since imaging at several sites within the specimen afforded nearly identical features. The protein gel prepared in DMSO has a dense filamentous morphology that more closely resembles the expected morphology for a flexible polypeptide crosslinked into a gel (Figure 9).

The nodular morphology of the aqueous gels clearly cannot arise as an artifact of sample preparation since an identical procedure was employed for both specimens. The disparity in microstructure between the two specimens presumably results from differences in interactions between the protein network and the swelling solvent. The nodular features in the hydrated gels resemble spheres that have randomly coalesced into a network. One possible hypothesis is that the charged polypeptide forms spherical micelles in aqueous solutions, which are preserved in the structure of the gel by the crosslinking reaction. Evidence in support of this hypothesis may be gleaned from an investigation of the structure of the poly(**Lys-25**) in solution under conditions relevant to the crosslinking process, although we have not yet performed these experiments.

An alternative hypothesis is that the amphiphilic character of the incompletely crosslinked network may account for the nodular morphology of poly(**Lys-25**) gels in aqueous solvents. The residual amine groups may induce a micellar structure in the protein gel under aqueous conditions. The charged ammonium groups should be localized at the interface between the aqueous solvent and hydrophobic protein gel. Local segments of the gel may adopt a spherical morphology to minimize the contact area between the two phases at the interface. Since the gels prepared in DMSO also contain unreacted amine groups, replacement of the original solvent by water may induce a nodular morphology in the material if this hypothesis is correct. Changes in the morphology of the gel upon solvent exchange should be detectable using the SEM imaging techniques described above.

## CONCLUSIONS

Protein gels have been prepared by chemical crosslinking a sequence-defined protein polymer based on elastin–mimetic repeat **Lys-25**. Both noncrosslinked proteins and protein networks display thermally responsive behavior typical of elastin analogues. The gels produced by this approach contain residual, unreacted amine groups that may influence the responsive properties and structural uniformity of the networks. Three major ongoing challenges have evolved from our initial investigation into the preparation of uniformly crosslinked protein gels from protein polymers.

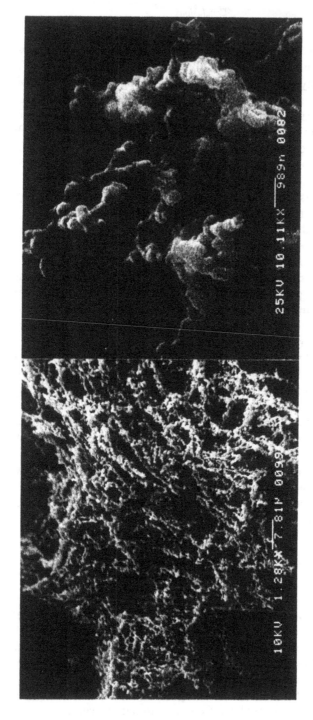

**Figure 8** Scanning electron micrographs of a protein gel prepared by reaction of poly(**Lys-25**) with a stoichiometrically equivalent amount of the crosslinker bis(sulfosuccinimidyl)suberate in aqueous phosphate buffer (pH 9.0). The micrographs in the figure depict the SEM images of gels dried under critical point conditions at two different magnifications. (SEM images courtesy of Dr. Robert Apkarian and Kevin Caran of the Integrated Microscopy and Microanalytical Facility of Emory University.)

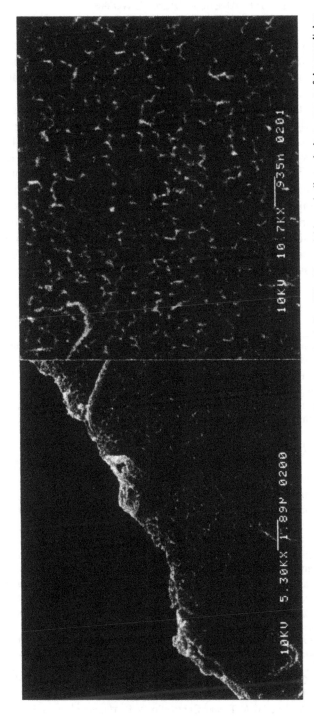

**Figure 9** Scanning electron micrographs of a protein gel prepared by reaction of poly(Lys-25) with a stoichiometrically equivalent amount of the crosslinker disuccinimidylsuberate in anhydrous DMSO. The micrographs in the figure depict the SEM images of gels dried under critical point conditions at two different magnifications. (SEM images courtesy of Dr. Robert Apkarian and Kevin Caran of the Integrated Microscopy and Microanalytical Facility of Emory University.)

127

First, the extent of crosslinking must be optimized from the observed 88% functionalization of amine groups. The preparation of uniformly crosslinked gels necessitates that the efficiency of crosslinking approaches 100% such that the majority of defect sites are removed from the network. Second, the chemical composition of the crosslinker must be matched more closely to that of the protein polymer such that changes in network polarity during the crosslinking process are minimized. A large decrease in network polarity upon crosslinking causes a decrease in $T_t$ that can result in local precipitation and microscopic inhomogeneity in the network. Finally, the origin of solvent effects on network morphology must be determined for these protein gels. The gross differences observed in network morphology between protein gels prepared in aqueous versus organic solvents may have a large effect on biomedically important properties such as solute permeability and exclusion limit. We are currently addressing these issues in our investigations.

## REFERENCES

1. Dumitrui, S. and Dumitrui-Medvichi, C. *Polymeric Biomaterials,* Dumitrui, S., Ed., Marcel Dekker, New York, 1994, pp. 3–97.
2. Gehrke, S. H., Fisher, J. P., Palasis, M., and Lund, M. E. *Ann. N.Y. Acad. Sci.* 1997, 831, 179–207.
3. Silberberg, A. in *Polyelectrolyte Gels;* Harland, R. and Prud'homme, R. K., Eds., ACS Symposium Series 480, American Chemical Society, Washington, DC, 1992, pp. 146–158.
4. Urry, D. W. *J. Prot. Chem.* 1988, 7, 1–34.
5. Wasserman, Z. R. and Salemme, F. R. *Biopolymers* 1990, 29, 1613–1628. Chang, D. K. and Urry, D. W. *Chem. Phys. Lett.* 1988, 147, 395–400.
6. Urry, D. W. *Angew. Chem. Int. Ed. Engl.* 1993, 32, 819–841.
7. Urry, D. W., Gowda, D. C., Parker, T. M. and Luan, C.-H. *Biopolymers* 1992, 32, 1243–1250.
8. Urry, D. W. *Prog. Biophys. Mol. Biol.* 1992, 57, 23–57.
9. Urry, D. W., Luan, C.-H., Harris, C. M. and Parker, T. M. *Protein-Based Materials,* McGrath, K. and Kaplan, D., Eds., Birkhauseir, Boston, 1997, pp. 133–177.
10. Urry, D. W., Parker, T. M., Reid, M. C. and Gowda, D. C. *J. Bioactive Compatible Polym.* 1991, 6, 262–282.
11. Heslot, H. *Biochimie* 1998, 80, 19–31. Ferrari, F. A. and Cappello, J. *Protein-Based Materials,* McGrath, K. and Kaplan, D., Eds., Birkhauser, Boston, 1997, pp. 37–60. Cappello, J. and Ferrari, F. *Plastics from Microbes: Microbial Synthesis of Polymers and Polymer Precursors;* Mobley, D. P., Ed., Hanser/Gardner Publications, Munich, 1994, pp. 35–92. Tirrell, J. G., Fournier, M. J., Mason, T. L., and Tirrell, D. A. *Chem. Eng. News* 1994, 72, 40–51. O'Brien, J. P. *Trends Polymer Science* 1993, 1, 228–232.
12. McMillan, R. A., Lee, T. A. T. and Conticello, V. P. *Macromolecules* 1999, 32, 3643–3648.

13. McPherson, D. T., Morrow, C., Minehan, D. S., Wu, J., Hunter, E. and Urry, D. W. *Biotechnol. Prog.* 1992, 8, 347–352.

14. Wishart, D. S. and Sykes, B. D. *J. Biomol. NMR* 1994, 4, 171–180.

15. Urry, D. W., Peng, S. Q., Gowda, D. C. and Parker, T. M. *Chem. Phys. Lett.* 1994, 225, 97–103.

16. Wong, S. S. *Chemistry of Protein Conjugation and Crosslinking.* CRC Press, Boca Raton, FL, 1991.

17. Cook, A. D., Pajvani, P. B., Hrkach, J. S., Cannizzaro, S. M. and Langer, R. *Biomaterials* 1997, 18, 1417.

# CHAPTER 9

# Genetically Engineered Protein Domains as Hydrogel Crosslinks

CHUN WANG[1]
RUSSELL J. STEWART[1]
JINDRICH KOPEČEK[1,2]

## INTRODUCTION

**H**YDROGELS are three-dimensional networks of hydrophilic polymers. They are able to swell and retain a significant portion of water without dissolving under physiological conditions. Hydrogels are biocompatible partly due to their large degree of water retention and are very useful as biomedical materials. They can be electrically neutral or charged. The equilibrium swelling of neutral hydrogels arises from the water–polymer mixing contribution to the overall free energy that is coupled to an elastic free energy contribution due to polymer expansion [1]. For charged hydrogels, ionic interactions between the charged polymer backbones and free ions in solvent also play a role in the swelling process in addition to the above two types of contribution [2].

Hydrogels can be categorized as chemical and physical gels based upon the nature of the crosslinking force [3]. Chemical gels have stable point co-valent crosslinks. Their properties are mainly dependent on the structure and properties of the primary chains and on the crosslinking density. Physical gels are three-dimensional networks where polymer chains form junction zones through noncovalent interactions. Some hydrogels undergo continuous or discontinuous changes in swelling that are mediated by external stimuli such as changes in pH, temperature, ionic strength, solvent type, electric and magnetic fields, light, and the presence of chelating

[1]Department of Bioengineering, University of Utah, Salt Lake City, UT 84112, USA.
[2]Departments of Pharmaceutics and Pharmaceutical Chemistry, University of Utah, Salt Lake City, UT 84112, USA.

131

species [4]. These are the stimuli-sensitive, or "smart" hydrogels, that have a tremendous potential as novel biomaterials for applications such as drug delivery and tissue engineering. In this overview, we discuss a new concept of hydrogel design—hybrid hydrogels consisting of synthetic polymers and engineered protein domains.

## HYDROGELS AS BIOMATERIALS

The application of hydrogels as biomaterials dates back to the landmark work of Wichterle and Lím [5], which described a rational design of hydrophilic polymer networks for biological uses. During the past several decades, hydrogels have been used as surgical sutures, artificial organs, soft tissue prosthesis, hemodialysis membranes, drug delivery systems, and soft contact lenses [6,7].

The most important ways to synthesize hydrogels are crosslinking copolymerization, crosslinking of polymeric precursors, and polymer–polymer reactions. These traditional methods of synthesis have produced numerous materials with excellent properties. However, these synthetic pathways do not permit an exact control of chain length, sequence, and three-dimensional structure. Radical polymerization usually results in a product with a distribution of different molecular weight species. In the case of hydrogels, the main problem is side reactions, such as formation of internal loops, unreacted pendant groups, and entanglements [8]. These defects or heterogeneity in the detailed structure of crosslinked polymers have profound influences on the physicochemical properties and ultimately the biological performances of these biomaterials. For example, biorecognition of ligands in hydrogels by enzymes is influenced by the structure of the ligand, equilibrium degree of swelling, and detailed structure of the network [9]. The detailed structure of hydrogels based on copolymers of $N,N$-dimethylacrylamide containing azoaromatic groups in the crosslinks depends on the method of synthesis. The structural differences result in different rates of hydrogel degradation by azoreductase in the gastrointestinal tract [8].

It appears that new ways for controlled synthesis of biomaterials are needed. The rapidly developing genetic engineering technology provides powerful tools for producing tailor-made biomaterials with predetermined three-dimensional structures. Exact control of primary structure, composition, and chain length of protein biomaterials can be achieved by manipulating the DNA sequence encoding the protein structure [10,11].

## GENETICALLY ENGINEERED BIOMATERIALS

Genetic engineering, or recombinant DNA technology [12], is being employed to produce protein-based biomaterials. First, a gene encoding the target protein is synthesized chemically. For a long or more complicated gene, it can be constructed either by stepwise ligation or polymerase chain reaction (PCR). Then the gene is ligated into a plasmid vector, a special circular DNA molecule, which is able to propagate with host cells. The production of the target protein by the host cells, most often bacteria, can be triggered or induced at any time. The expressed protein may accumulate as a soluble form in the cytoplasm, as an insoluble form in inclusion bodies, or be secreted into the periplasm of the bacteria or the culture media. One of the most convenient methods of purifying recombinant proteins is immobilized metal affinity chromatography (IMAC), which uses Ni(II) ions immobilized on agarose beads to preferentially bind target proteins containing terminal histidines.

There are distinct advantages of biosynthesis over chemical synthesis [10]. It is possible to obtain protein products with a very narrow and even uniform molecular weight distribution, whereas chemical synthesis inevitably results in a mixture of products with different chain lengths. In biosynthesis, by defining the DNA sequence of the gene and by changing the cell culture composition, exact control over the composition and stereochemistry of the target protein can be achieved. Through similar methods uncommon amino acids or amino acid analogs can also be incorporated into the protein chains [13]. By cutting and pasting gene segments through routine molecular biology procedures, different protein domains can be easily assembled, rearranged, and modified to generate new chimeras with novel or improved functions [14].

Biological synthesis of protein-based biomaterials has already created analogs of silk [15] and elastin [16], materials with more precise control of the three-dimensional structures [17], and molecules for biorecognition [18]. As the molecular basis of natural protein materials, such as that of the spider dragline silk [19], begins to unravel, rational design of artificial protein materials will be greatly facilitated [20]. This field is now witnessing tremendous advances and will probably hold the future of biomaterials development and research.

## COILED COILS

One common folding motif of proteins is the coiled coil, which is a slightly left-handed super-helix consisting of two or more right-handed $\alpha$-helices [21]. The coiled-coil motif has been found in over 200 native proteins, and this number is growing rapidly. The primary structure of the coiled-coil strands has the characteristic 4-3 (heptad) repeats. One heptad constitutes exactly two turns, each covering three and a half residues. As illustrated in Figure 1, amino acid residues in a heptad are designated as "a, b, c, d, e, f, g." Hydrophobic residues at "a" and "d" positions pack their side chains

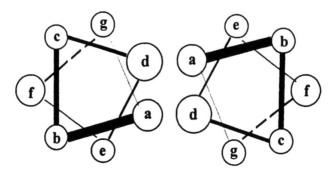

**Figure 1** A helical wheel diagram of a dimeric coiled-coil. Letters "a" through "g" denote the seven amino acid residues of a heptad repeating unit.

tightly in aqueous environment and form the stabilizing interfaces between the helices. Other residues are usually polar ones. In particular, ionic interactions between "e" and "g" residues are important in specific association among helices.

Because of its structural simplicity and instrumental role in protein function, the coiled coil is one of the most investigated protein folding motifs. Native coiled-coil sequences and their mutants have been synthesized and studied. Numerous model coiled-coil peptides have been designed *de novo* [22]. Although many theoretical questions remain unanswered, much has been learned about the sequence–structure relationship. It is even possible to design and engineer new coiled-coil sequences and structures that have never existed before [23].

The ability of coiled coils to form stable, specific oligomers has been most frequently used to engineer novel proteins. As a transplantable oligomerization module, native coiled-coil sequences of the dimeric leucine zipper have been used to construct highly avid antibodies [24,25], epitope-displaying scaffold [26], and other novel chimeric proteins with biological and therapeutic importance [27–30]. Examples of using coiled coils as engineering materials are schematically depicted in Figure 2.

## DESIGNS OF HYBRID HYDROGELS

### RATIONALE

Hybrid hydrogels are usually referred to as hydrogel systems whose components are at least two distinct classes of molecules, for example, synthetic polymers and biological macromolecules, interconnected either covalently or noncovalently. They have been of particular interest because

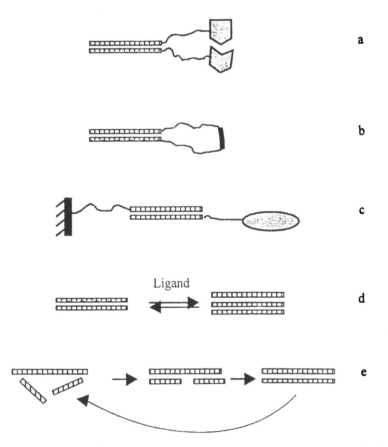

**Figure 2** Examples of coiled-coils used as engineering materials—transplantable oligomerization module (a) [24,25,27], epitope-displaying scaffold (b) [26], biorecognition element (c) [28], allosteric switch (d) [29], and self-replicating peptide (e) [30].

it is possible to combine and superimpose the properties of the component molecules onto the hydrogels. Degradation of hybrid hydrogels of *N*-(2-hydroxypropyl)methacrylamide with enzymatically degradable oligopeptide crosslinks have been studied [31]. Novel hybrid hydrogels of a thermal sensitive polymer *N*-isopropylacrylamide and lipids have been prepared and displayed interesting properties [32]. Hydrogels based upon glucose-binding protein concanavalin A as crosslinks of synthetic polymers are able to sense the concentration of external sugar molecules and release insulin [33]. Complementary oligonucleotides have also been used to crosslink water-soluble polymers, and the thermal properties of these hydrogels appear to be controlled by the different oligonucleotide sequences used [34].

We are interested in designing hybrid hydrogels assembled from water-soluble synthetic polymers and genetically engineered protein molecules. The advantage of using proteins instead of other biological macromolecules is twofold. Proteins are the machinery of life. They are extremely versatile molecules capable of serving all kinds of structural and functional purposes. By using structurally simple and well-defined protein folding motifs, such as the coiled coil, one could design hybrid hydrogels with the same versatility by adjusting the primary sequence of the protein molecules. Moreover, such an engineering endeavor is greatly facilitated by the availability of the entire arsenal of molecular genetics. At least in theory, one could be able to manipulate the DNA sequence at will and to generate unlimited varieties of protein products that have well-defined structures and properties.

The specific goal of the work described here is to utilize the self-association behavior of the coiled-coil protein domains to bring together synthetic macromolecules creating hybrid three-dimensional networks as illustrated in Figure 3. A coiled-coil protein has a string of terminal histidine residues (his-tag), which forms a complex with a Ni(II) and a nitrogen–oxygen–donor ligand on the side chain of a linear synthetic polymer. Thus the coiled-coils crosslink the water-soluble primary chains resulting in a hydrogel network. We genetically engineered two coiled-coil proteins whose secondary structures change differently in response to temperature in solution. We found that hybrid hydrogels formed by these proteins and water-soluble synthetic

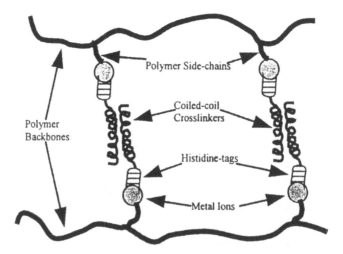

**Figure 3** Schematic illustration of a hybrid hydrogel system—genetically engineered coiled-coil protein domains used to crosslink synthetic water-soluble polymers. Divalent transition metal ions are shown to form complexes with nitrogen–oxygen–donor ligands on the synthetic polymer side chains and the terminal histidine residues in the coiled coils.

macromolecules displayed properties that were predetermined by the coiled-coil crosslinks [35].

## METAL-CHELATING COPOLYMER OF HPMA

HPMA [36] and a vinyl metal-chelating monomer $N$-($N'$,$N'$-dicarboxy-methylaminopropyl)methacrylamide synthesized [35]. Chemical structures of HPMA and DAMA are given in Figure 4. Poly(HPMA-*co*-DAMA) was prepared by free radical copolymerization in methanol with AIBN as initiator. Molecular weight distribution was determined by size exclusion chromatography and content of side-chain carboxylic group by acid–base titration.

## BIOSYNTHESIS OF COILED-COIL PROTEIN DOMAINS

The first coiled-coil protein domain was a segment of the motor protein kinesin [37]. The gene encoding the stalk region of kinesin, KS590, was extracted and subcloned. KS590 was expressed in *E. coli* and purified using metal affinity chromatography [35].

The second coiled-coil protein domain was based on a *de novo* designed sequence EK42 [38]. It has the sequence of $(VSSLESK)_6$ and was originally designed to form coiled-coil homodimers. A gene encoding EK42 was designed using the degeneracy of the genetic codons (Figure 5). After being synthesized by solid phase method, the gene was cloned. A protein containing EK42 and a tag sequence, TEK42, was expressed in *E. coli* and purified using metal affinity chromatography [35].

Structures of the two proteins were characterized using circular dichroism (Table 1) [35]. Both proteins were α-helical coiled coils. Although

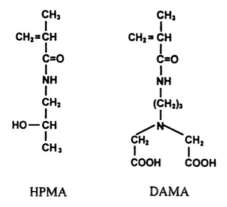

HPMA              DAMA

**Figure 4** Chemical structure of $N$-(2-hydroxypropyl)methacrylamide (HPMA) and $N$-($N'$,$N'$-dicarboxymethylaminopropyl)methacrylamide (DAMA).

5'-<u>GAT CC</u>G GTG AGT TCC CTC GAG AGC AAA GTC TCT AGC CTG GAA TCT AAA GTA
   3'-<u>GC</u> CAC TCA AGG GAG CTC TCG TTT CAG AGA TCG GAC CTT AGA TTT CAT
*BamHI*

AGT TC<u>T CTA GA</u>A AGT AAA GTT TCC AGC CTG GAA TCC AAA GTG AGC TCC CTG GAA
TCA AG<u>A GAT CT</u>T TCA TTT CAA AGG TCG GAC CTT AGG TTT CAC TCG AGG GAC CTT
   *XbaI*

AGC AAA GTC TCA AGC TTG GAA TCC AAA **TGA** <u>G</u>-3'
TCG TTT CAG AGT TCG AAC CTT AGG TTT **ACT** <u>CTT AA</u>-5'
   *EcoRI*

**Figure 5** Sequence of the designed artificial gene encoding EK42 domain. Restriction sites underlined facilitate ligation into expression vector and detection of the insert gene. The stop codon is highlighted.

KS590 displayed cooperative thermal unfolding, TEK42 remained stable in the range of 10–90°C.

## PREPARATION OF THE HYBRID HYDROGELS

As the first step, a Ni(II) metal complex of HPMA copolymer was formed and purified [35]. It was then mixed with protein solutions to form dried films, which were rehydrated to investigate the swelling profile and stimuli sensitivity.

## DYNAMIC SWELLING OF THE HYBRID HYDROGELS

Swelling profile determinations of the hydrogel films are provided elsewhere [35]. The results were given as the volume swelling ratio plotted

TABLE 1. Characterization of the Coiled-Coil Protein Crosslinkers.

|  | KS590 | TEK42 |
|---|---|---|
| Length (number of amino acids) | 255 | 72 |
| Length of coiled-coil domain (nm) | 23 ~ 25 | 6 |
| $\Theta_{222}$ (deg cm$^2$ dmol$^{-1}$) | −22086 | −13591 |
| $\Theta_{208}$ (deg cm$^2$ dmol$^{-1}$) | −20506 | −13820 |
| $\Theta_{222}/\Theta_{200}$ | 1.08 | 0.98 |
| α-helicity (%)[a] | 66 | 56 |
| α-helicity (%)[b] | 56 | 36 |
| Melting temperature (°C) | 35 (major) | 108[c] |
|  | 65 (minor) |  |

[a]Estimated based on amino acid sequences.
[b]Calculated from CD spectra using the method described in Reference [41].
[c]Estimated from thermal melting in the presence of different concentrations of guanidine hydrochloride.

against time when the gels were immersed in phosphate-buffered saline (PBS, 20 mM Na phosphate, 150 mM NaCl, pH 7.4). In chemical gels crosslinked through short covalent bonds, swelling mainly depends on the hydration of primary chains. However, in the hybrid hydrogels in this case, the contribution of hydrophilic rodlike coiled-coil proteins should also be considered.

## STIMULI-SENSITIVITY OF THE HYBRID HYDROGELS

The hydrogels crosslinked with the two proteins responded to elevating temperature the same way as the free proteins, i.e., gels containing KS590 displayed a volume transition (Figure 6), whereas gels containing TEK42 did not. The sharp yet continuous transition appeared to be the direct result of temperature-induced change in the crosslink secondary structure. We speculate that large change in hydrodrynamic volume of the coiled-coil crosslinkers might be the major driving force of the gel volume transition, as illustrated in Figure 7. The gels were also responsive to the presence of strong metal-chelating ligands, such as imidazole (Figure 8), which competed with the his-tags for Ni(II) and detached the protein crosslinkers from the gel backbones. Partial replacement off the his-tags with imidazole resulted in a decrease in gel crosslinking density, and the degree of gel

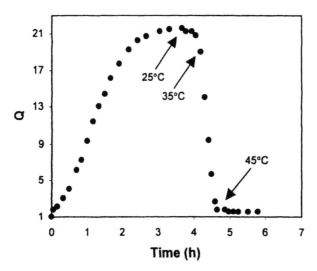

**Figure 6** Swelling of hybrid hydrogels in response to temperature. The hydrogels were prepared from HPMA-DAMA copolymers, Ni(II), and KS590 coiled coil. The gels were equilibrated in PBS (pH 7.4) at 25°C prior to an increase in temperature. Rate of heating 1°C/min plus 2 min equilibration time. Arrows indicate elevated temperatures of 25°C, 35°C, and 45°C, respectively. Modified from Reference [35].

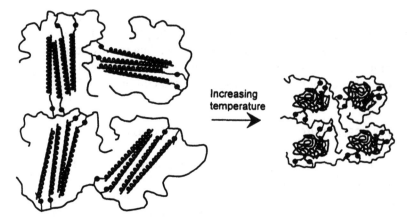

**Figure 7** Schematic diagram of the mechanism of temperature sensitivity of hybrid hydrogels.

swelling increased. When the imidazole concentration was high enough (e.g., 500 mM) to replace all the crosslinks, the gels readily dissolved.

## CONCLUSIONS AND FUTURE PERSPECTIVES

In recent years, protein-based linear [17] and block [39] copolymers have been investigated for use as novel biomaterials. Their biggest advantage over synthetic polymers is the possibility to produce macromolecules

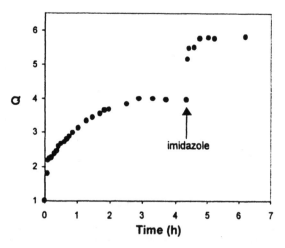

**Figure 8** Swelling of hybrid hydrogels in response to competing metal-chelating ligand. The hydrogels were prepared from HPMA–DAMA copolymers, Ni(II), and TEK42 coiled coil. The gels were swollen in PBS (pH 7.4) at 25°C. The arrow indicates the addition of 100 mM buffered imidazole.

with predetermined structure and function through genetic engineering. In this chapter, a new concept of using genetically engineered protein domains as crosslinkers of synthetic macromolecules was described. The central theme was to create highly swollen physical hybrid hydrogels on which the properties of the protein domains can be superimposed. It represents our first effort toward the goal of engineering proteins to build structurally and functionally well-defined biomaterials.

One unique feature of our design is the use of coiled-coil protein domains. In addition to temperature-induced volume transition and responsiveness to chelating agents, sensitivity toward other stimuli, such as pH, ionic strength, solvents, electric current, mechanical force, or specific recognition and binding with ligand, could also be built into the hydrogel system. This could be accomplished by incorporating protein domains with specially engineered sequences. Furthermore, certain physical properties of the hydrogels, such as viscosity, gelation temperature, elasticity, rigidity, porosity, and bioerodibility, can be similarly tailored. As one of the possible extensions of the work described here, all the "e" and "g" positions of the heptad repeats of a coiled-coil strand could be occupied by glutamic acid or by lysine residues. Electrostatic interaction between two of such strands would favor specific formation of a heterodimer. Polar residues, such as asparagine, could be inserted at the hydrophobic interface to facilitate specific alignment and orientation of the two strands and to decrease overall stability. Compared with the hybrid hydrogels formed using TEK42 as crosslinkers, gels crosslinked by such heterodimeric coiled-coil domains would be expected to have better defined pore size, better crosslinking efficiency with less intramolecular crosslinks, and a lower dissociation temperature.

Another variation in the system is to change the connection between the coiled-coil crosslinker and the synthetic polymeric backbone. Instead of using metal complexation, covalent conjugation of the protein crosslinkers and the hydrogel primary chains is highly desirable because of its stability. For example, a water-soluble polymer with pendant maleimide groups could be used to conjugate protein crosslinkers containing terminal cysteine residues. Stable thioester bonds could be formed readily at neutral pH. Thus, properties of the hydrogels would solely depend on the properties of the crosslinking coiled-coil domains.

Potential applications for the hybrid hydrogels are numerous and would include all current applications of synthetic and natural hydrogels. One specific area of application may be drug delivery. Low molecular weight drugs could easily be loaded to the gels either by physical entrapment or by covalent attachment using well-developed methodology [40]. Yet the biggest advantage of such a system is for delivering protein or peptide drugs. For example, therapeutic proteins can be fused with the coiled-coil crosslinking domain by genetic engineering technology and be incorporated into the hydrogel. Alternatively, large protein molecules could also be physically trapped into the pores of the hydrogels. The fact that no organic solvent is needed for hydrogel preparation is a major advantage of such a formulation, thus the biological activities of the protein drugs could be preserved. For targeted delivery and release of drugs, biorecognizable targeting moieties such as antibodies or signaling peptides could be incorporated into the hydrogel system to mediate recognition by specific tissue and cell types and to trigger drug release. Using coiled-coils of different length, hydrogels with well-defined pore size could be prepared to modulate the rate of drug release.

Before such a hybrid hydrogel system could be used for *in vivo* applications, the issue of immunogenecity must first be addressed. Most transition metal ions suitable for attached his-tagged proteins can have adverse effects on the body when the local concentration is high. However, there are some cases in which the presence of metal ions is crucial of achieving certain therapeutic effects. One such example is prolonged effective time of insulin in the presence of zinc(II). Hydrogels can also be tracked *in vivo* by magnetic resonance imaging, if Ga(II) is used. Another consideration is the possible presence of epitope sequence in *de novo* designed protein crosslinkers. Immunologic studies on the degradation products of the protein crosslinks should be carried out. The behavior of these exogenous protein molecules in serum must also be evaluated to ensure not only that they do not elicit adverse effects but also that they maintain desired structures and functions in such protein-rich environment.

## REFERENCES

1. Kúdela, V. in *Encyclopedia of Polym. Sci. and Technol.*, Mark, H. F. and Kroschwitz, J. I. (Eds.) Wiley, New York, 1985, p. 783.
2. Katchalsky, A., Lifson, S., and Eisenberg, H. *J. Polym. Sci.*, 1951, 7, 571.
3. Park, K., Shalaby, W. S. W., and Park, H. *Biodegradable Hydrogels for Drug Delivery*, Technomic Publishing Co., Inc. 1993.
4. Tanaka, T. in *Polyelectrolyte Gels: Properties, Preparation, and Applications*, Harland, R. S. and Prud'homme, R. K. (Eds.), American Chemical Society Symposium Series, Vol. 480, Washington DC, 1992.
5. Wichterle, O. and Lím, D. *Nature*, 1960, 185, 117.
6. Andrade, J. D. (Ed.) *Hydrogels for Medical and Related Applications*, American Chemical Society Symposium Series, Vol. 31, Washington DC, 1976.
7. Peppas, N. A. *Curr. Opin. Colloid & Interface Sci.*, 1997, 2, 531.
8. Yeh, P.-Y., Kopečková, P., and Kopeček, J. *Macromol. Chem. Phys.*, 1995, 196, 2183.
9. Kopeček, J. and Rejmanová, P. in *Controlled Drug Delivery*, Bruck, S. D. (Ed.), CRC Press, Boca Raton, Florida, 1983, Vol. I, p. 81.
10. Cappello, J. *Trends in Biotechnol.*, 1990, 8, 309.
11. McGrath, K. P., Fournier, M. J., Mason, T. L., and Tirrell, D. A. *J. Am. Chem. Soc.*, 1992, 114, 727.

12. Sambrook, J., Fritsch, E. F., and Maniatis, T. *Molecular Cloning: A Laboratory Manual*, 2nd Ed. Cold Spring Harbor Laboratory Press, 1989.

13. Deming, T. J., Fournier, M. J., Mason, T. L., and Tirrell, D. A. *Macromolecules*, 1996, 29, 1442.

14. Stinson, S. C. *C & EN*, July 16, 1990, p. 26

15. Cappello, J., Crissman, J., Dorman, M., Mikolajczak, M., Textor, G., Marquet, M., and Ferrari, F. *Biotechnol. Prog.*, 1990, 6, 198.

16. McPherson, D. T., Morrow, C., Minehan, D. S., Wu, J., Hunter, E., and Urry, D. *Biotechnol. Prog.*, 1992, 8, 347.

17. Krejchi, M. T., Atkins, E. D. T., Waddon, A. J., Fournier, M. J., Mason, T. L., and Tirrell, D. A. *Science*, 1994, 265, 1427.

18. McGrath, K. and Kaplan, D. *Mater. Res. Soc. Symp. Proc.*, 1993, 292, 83.

19. Simmons, A. H., Michal, C. A., and Jelinski, L. W. *Science*, 1996, 271, 84.

20. Tirrell, D. A. *Science*, 1996, 271, 39.

21. Lupas, A. *Trends Biochem. Sci.*, 1996, 21, 375.

22. Hodges, R. S. *Biochem. Cell Biol.*, 1996, 74, 133.

23. Harbury, P. B., Plecs, J. J., Tidor, B., Alber, T., and Kim, P. S. *Science*, 1998, 282, 1462.

24. Pack, P., Müller, K., Zahn, R., and Plückthun, A. *J. Mol. Biol.*, 1995, 246, 28.

25. Terskikh, A. V., Le Doussal, J.-M., Crameri, R., Fisch, I., Mach, J.-P., and Kajava, A. V. *Proc. Natl. Acad. Sci. USA*, 1997, 94, 1663.

26. Miceli, R. M., Myszka, D. G., Peishoff, C. E., and Chaiken, I. M., in *Molecular Diversity and Combinatorial Chemistry: Libraries and Drug Discovery*, Chainken, I. M. and Janda, K. D. (Ed.) American Chemical Society, Washington DC, 1996, p. 172.

27. Waterman, M. J. F., Waterman, J. L. F., and Halazonetis, T. D. *Cancer Res.*, 1996, 56, 158.

28. Tripet, B., Yu, L., Bautista, D. L., Wong, W. Y., Irvin, R. T., and Hodges, R. S. *Protein Eng.*, 1996, 9, 1029.

29. Gonzalez, L., Plecs, J. R., and Alber, T. *Nature Struct. Biol.*, 1996, 3, 510.

30. Lee, D. H., Granja, J. R., Martinez, J. A., Sererin, K., and GHadiri, M. R. *Nature*, 1996, 382, 525.

31. Šubr, V., Duncan, R., and Kopeček, J. *J. Biomater. Sci. Polym. Ed.*, 1990, 1, 261.

32. Tsujii, K., Hayakawa, M., Onda, T., and Tanaka, T. *Macromolecules*, 1997, 30, 7397.

33. Obaidat, A. A. and Park, K. *Biomaterials*, 1997, 18, 801.

34. Nagahara, S. and Matsuda, T. *Polym. Gels and Networks*, 1996, 4, 111.

35. Wang, C., Stewart, R. J., and Kopeček, J. *Nature*, 1999, 397, 417.

36. Kopeček, J. and Bažilová, H. *Eur. Polym. J.*, 1973, 9, 7.

37. Yang, J. T., Laymon, R. A., and Goldstein, L. S. B. *Cell*, 1989, 56, 879.

38. Graddis, T. J., Myszka, D. G., and Chaiken, I. M. *Biochemistry*, 1993, 32, 12664.

39. Petka, W. A., Harden, J. L., McGrath, K. P., Wirtz, D., and Tirrell, D. A. *Science*, 1998, 281, 389.

40. Putnam, D. and Kopeček, J. *Adv. Polym. Sci.*, 1995, 122, 55.

41. Chen, Y.-H., Yang, J. T., and Chau, K. H. *Biochemistry*, 1974, 13, 3350.

# CHAPTER 10

# Superporous Hydrogel Composites: A New Generation of Hydrogels with Fast Swelling Kinetics, High Swelling Ratio and High Mechanical Strength

KINAM PARK[1]
JUN CHEN[2]
HAESUN PARK[1]

## INTRODUCTION

A hydrogel is a crosslinked polymer network that is insoluble in water but swells to an equilibrium size in the presence of excess water. Shown in Figure 1 is a schematic description of chemical gels. In chemical gels, the polymer chains are crosslinked by covalent bonding. If the polymer chains are crosslinked by noncovalent bonding, such networks are called physical gels. The research on hydrogels started in 1960s with a land-mark paper on poly(hydroxyethyl methacrylate) by Wichterle and Lim [1]. Since then, various types of hydrogels have been synthesized and characterized due to the unique properties of hydrogels and potential applications in various areas including controlled drug delivery. Much of the work on hydrogels has been concentrated on lightly crosslinked, homogeneous homopolymers and copolymers. One of the first applications of hydrogels in controlled drug delivery was slow release of the loaded drugs from dried hydrogels exposed to an aqueous environment. For dried hydrogels to swell, water has to be absorbed into the glassy matrix of the dried hydrogels. The swelling kinetics of the dried hydrogels thus depend on the absorption of water occurring by a diffusional process and the relaxation of the polymer chains in the rubbery region. This is a slow process. Although

[1]Purdue University, School of Pharmacy, West Lafayette, IN 47907, USA.
[2]Merial Limited, WP 78-110, West Point, PA 19486, USA.

**Figure 1** Schematic description of a chemical gel. In chemical gels, hydrophilic polymer chains are crosslinked through covalent bonds. In the presence of abundant water, polymer chains try to dissolve in water and this results in swelling of the polymer network. The swelling is limited by the crosslinking of polymer chains. The dried hydrogels appear nonporous even by scanning electron microscopy.

the slow swelling of dried hydrogels has been useful in many applications, there are situations where faster swelling of dried hydrogels is desirable.

## SWELLING KINETICS OF HYDROGELS

### SLOW SWELLING OF HYDROGELS

According to the swelling kinetics of a gel [2], the characteristic time of swelling ($\tau$) is proportional to the square of the characteristic length of the gel *(L)* and is inversely proportional to the diffusion coefficient of the gel network in the solvent *(D)* as follows:

$$\tau = L^2/D$$

The characteristic length is the radius for spherical hydrogels and the thickness for hydrogel sheets. The diffusion coefficient of hydrogel networks is in the order of $10^{-7}$ cm²/sec [2,3]. A 1-mm-thick gel slab with a diffusion coefficient of $10^{-7}$ cm²/sec takes over an hour to reach 50% of the equilibrium swelling and more than 6 hours to reach 90% of equilibrium [4]. Although this slow swelling property has been useful in some applications, such as developing sustained release drug delivery systems, it may be too slow for hydrogels to be used in other applications, such as superabsorbents in baby diapers. It has been a common practice to reduce the characteristic swelling time by reducing the size of the hydrogel. The size restriction on gels however limits the useful applications of hydrogels; therefore, hydrogels with larger dimension would be highly desirable if they could swell faster [5].

## HYDROGELS WITH FASTER SWELLING KINETICS

We have been interested in developing hydrogel-based gastric retention devices. The concept is that hydrogels swell to a large size so as to be retained in the stomach. In our previous study, hydrogels were successfully used as a gastric retention device that stayed in dog stomachs for up to 60 hours [6,7]. In those studies, however, the hydrogels had to be preswollen for a few hours before being administered to the dog in order to avoid premature emptying into the intestine. Without preswelling, all the hydrogels were emptied into the intestine within 30 minutes after administration. Consequently, making hydrogels with fast swelling properties, i.e., swelling in a matter of minutes rather than hours, has been our research goal for the last several years.

Recently, we have synthesized superporous hydrogels (SPHs) that swell within minutes regardless of the size of the matrix [8–13]. Although SPHs provided faster swelling kinetics and higher swelling extent than the conventional hydrogels, the mechanical strength of the fully swollen SPHs was rather poor. Usually, mechanically strong SPHs can be made by increasing the crosslinking density, but this resulted in the loss of the high swelling property. Since making SPHs with fast swelling and high swelling ratio as well as with high mechanical strength was our goal, we have synthesized SPH composites. To understand the unique properties of SPHs and SPH composites, we compared properties of various types of hydrogels and SPHs.

## HYDROGELS, MICROPOROUS HYDROGELS, AND MACROPOROUS HYDROGELS

Hydrogels can be made by a number of methods, but one of the most widely used methods is free radical polymerization of vinyl monomers. Examples of vinyl monomers used in our laboratory are shown in Table 1. Monomers are crosslinked with divalent monomers such as $N,N'$-methylenebisacrylamide. Biodegradable crosslinking agents can also be used as long as they have bifunctionality.

Polymerization of monomers in the absence of any other solvent is called bulk polymerization. Bulk polymerization of monomers, such as hydroxyethyl methacrylate (or HEMA), leads to the production of a glassy, transparent polymer matrix that is very hard. When immersed in water, such a glassy matrix swells to become relatively soft and flexible. Although it allows the transfer of water and some low molecular weight solutes, this kind of swollen polymer matrix (i.e., hydrogel) is considered nonporous. The

TABLE 1. Vinyl Monomers Used for Making Various Hydrogels.

| Chemical Name | Monomer Structure |
|---|---|
| Acrylamide | |
| N-Isopropylacrylamide | |
| 2-Hydroxyethyl methacrylate | |
| 2-Hydroxypropyl methacrylate | |
| N-Vinyl pyrrolidinone | |
| Acrylic acid | |
| Sodium acrylate | |
| 2-Acrylamido-2-methyl-1-propanesulfonic acid | |
| 3-Sulfopropyl acrylate, potassium salt | |
| 2-(Acryloyloxy)ethyltrimethyl-ammonium methyl sulfate | |

Natural polymers modified with vinyl groups (e.g., albumin or gelatin modified with glycidyl acrylate).

pores between polymer chains are in fact the only spaces available for mass transfer, and the pore size is within the range of molecular dimensions (a few manometers or less) [14]. Under a scanning electron microscope, the surface of the dried hydrogels appears completely nonferrous. In this case, the transfer of water or other solutes is achieved by a pure diffusional mechanism [14]. This restricts the rate of absorption and to some extent the size of the species that are absorbed [15]. The homogeneous hydrogels have been used widely in various applications, especially in the controlled drug delivery area where limited diffusional characteristics are required [16].

Hydrogels are usually prepared by solution polymerization where monomers are mixed with a suitable solvent. The nature of the synthesized hydrogel, whether a compact gel or a loose polymer network, depends on the type of monomer, the amount of diluent in the monomer mixture (i.e., monomer–diluent ratio), and the amount of crosslinking agent [17]. As the amount of diluent (usually water) in the monomer mixture increases, the pore size also increases up to the micrometer ($\mu$m) range [14]. Hydrogels with the effective pore sizes in the 10–100 nm range and in the 100 nm–10 $\mu$m range are called microporous and macroporous hydrogels, respectively [14,16]. In practice, the terms "microporous" and "macroporous" are used interchangeably simply due to the fact that there are no unified definitions on micro- and macropores in hydrogels. Therefore, hydrogels with pores up to 10 $\mu$m can be described as either microporous or macroporous.

It is important to distinguish the definitions for microporous and macroporous structures in hydrogels and those in other porous materials, such as polyurethane foams. In the plastic foam area, micro- and macropores are indicated for pores less than 50 $\mu$m and pores in the 100–300 $\mu$m range, respectively [18]. One of the reasons for this difference is that the hydrogels with pores larger than 10 $\mu$m were rarely made, whereas porous plastics with pores in the 100–300 $\mu$m range are very common. Porous hydrogels with pore sizes larger than 100 $\mu$m have only recently been reported [19,20], and that is probably why the definitions on porous hydrogels are different from those in porous plastics.

Microporous and macroporous hydrogels are sometimes called polymer sponges [14]. When HEMA is polymerized at the initial monomer concentration of 45 (w/w)% or higher in water, a hydrogel of poly(HEMA) (or PHEMA) is produced with a porosity higher than homogeneous PHEMA hydrogels. These heterogeneous PHEMA hydrogels are sometimes called "sponges" in the biomedical literature [14,21]. The term sponge is not recommended, since it is better known as the "rubber sponge," which is not a hydrogel in any sense. Furthermore, the properties of rubber sponges are different from porous hydrogels.

## SUPERPOROUS HYDROGELS (SPHs)

SPHs are a new type of hydrogel that have numerous supersize pores inside them. Depicted in Panels A and B in Figure 2 are cartoons drawn based on scanning electron microscopic (SEM) pictures of dried SPHs and conventional hydrogels. As shown in Figure 2, SPHs have numerous pores

A                                         B

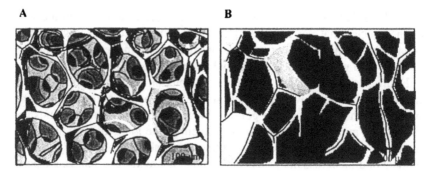

**Figure 2** Representations of a superporous hydrogel (A) and a conventional hydrogel (B) in the dried state. No porous structures are observable in conventional hydrogels by a scanning electron microscope.

while the conventional hydrogels show no pores throughout the matrix. The pore sizes in the SPHs are larger than 100 μm, usually in the range of several hundreds micrometers and can be up to the millimeter range. Most of the pores inside of the SPH are connected to form an open channel system. The size and number of the pores can be controlled by adjusting the type and amount of surfactant and gas forming agent during crosslinking polymerization. Even after drying, the pores in the SPHs remain connected to each other to form capillary channels.

It should be noted that the SPHs have distinctly different properties compared to microporous and macroporous hydrogels. First, the pore sizes in the SPHs are several hundred micrometers and can increase up to the millimeter range; this is much larger than the pores in microporous or macroporous hydrogels. Second, in contrast to conventional microporous or macroporous hydrogels, which contain a relatively small fraction of empty spaces, the SPHs can easily accommodate gas cells more than several hundred percent of the volume of the starting monomer mixtures. Third, the pores in the SPHs remain connected even after drying, and this allows the dried hydrogels to swell extremely fast.

## METHODS OF PREPARING POROUS HYDROGELS

Porous hydrogels can be prepared by a number of techniques. Those techniques are briefly described below.

### PORO(SI)GEN TECHNIQUE

Porous hydrogels can be made by preparing the hydrogels (usually polymerizing monomers) in the presence of dispersed water-soluble porosigens, which can be removed later by washing with water to leave an interconnected meshwork (i.e., porous hydrogels) [16,22]. Examples of effective porosigens are micronized sucrose [16], lactose [16], dextrin [16], sodium chloride [21], and PEG [23].

## PHASE SEPARATION TECHNIQUE

In solution polymerization, monomers are usually mixed in a diluent that is good for both monomers and polymers. If, however, the diluent is a nonsolvent for the polymer formed (e.g., PHEMA in water), the solubility of the polymers dramatically decreases as the polymerization proceeds. This results in phase separation of the polymer-rich monomer phase into droplets, which then join together to form a network filled with large spaces (i.e., heterogeneous, porous hydrogels) by the end of the polymerization process. This process is called heterogeneous solution polymerization [14,17,24]. The pore sizes of macroporous hydrogels prepared by phase separation are typically only a few micrometers. In addition, the overall porosity is very low, and this implies that the pores are not well interconnected. The major limitation of the phase separation method is that only very limited types of porous hydrogels can be prepared. In addition, there is not much control over the porosity of the gels when prepared by phase separation.

## CROSSLINKING OF INDIVIDUAL HYDROGEL PARTICLES

Individual hydrogel particles can be crosslinked to form crosslinked aggregates. Pores are formed between the hydrogel particles. Such aggregate macrostructures were prepared by initially mixing the hydrogel particles (in the range of a few hundred micrometers) with a solution of a crosslinking agent, water, and hydrophilic organic solvent such as isopropanol [25]. Pores in such structures are present between hydrogel particles, and the size of pores is much smaller than the size of particles. This approach is limited to the absorbent particles that have chemically active functional groups on the surface.

## GAS-BLOWING (OR FOAMING) TECHNIQUE

Hydrogels can be prepared in the presence of gas bubbles. In this technique the monomers are polymerized or water-soluble polymer chains are crosslinked around gas bubbles generated by a blowing agent. The gas-blowing technology has been widely used in the preparation of plastic foams from materials such as polyurethanes, rubber, and poly(vinyl chloride). The key ingredient in the foaming process is a blowing agent (or foaming agent), which is defined as any substance or combination of substances capable of producing cellular structure within a polymer matrix. Foaming agents are classified as; (a) physical foaming agents that expand when pressure is released (e.g., nitrogen and carbon dioxide) and (b) chemical foaming agents that decompose or react to form a gas (e.g., sodium bicarbonate in the presence of acid).

Recently, the gas-blowing technique was used in our laboratory to prepare SPHs [19,20]. Because this technique used is for the preparation of SPHs, they were also called "hydrogel foams." In the synthesis of SPHs by the gas-blowing technique, foaming and polymerization have to occur simultaneously [13]. For this reason, it is important to control the timing for foaming and polymerization. In the study mentioned above, inorganic carbonates, such as $Na_2CO_3$ and $NaHCO_3$, were used as the foaming agent. These inorganic carbonates have long been used safely as a gas-forming ingredient in effervescent tablets for antacids. They are safe, cheap, and easy to use.

## SUPERPOROUS HYDROGEL COMPOSITES (SPHs)

The mechanical strength of SPHs increases as the crosslinking density increases. This, however, results in reduced swelling by the hydrogels and increased brittleness. To minimize such problems, we synthesized SPH composites by mixing hydrophilic particulates in the monomer solution before polymerization. Of the many composite materials we have used, crosslinked sodium carboxymethylcellulose particulate (under the commercial name of Ac-Di-Sol®) was the most effective in increasing the mechanical strength without significantly sacrificing the swelling property. Other superdisntegrants, such as Primojel®, Explotab®, and Crospovidone®, were also effective, but not as much as Ac-Di-Sol®.

### FAST SWELLING OF SPHs AND SPH COMPOSITES

The total swelling time for a dried SPH in aqueous solution is determined by two factors: $t_1$ and $t_2$. $t_1$ is the time for water to reach all the surface of the pores in the SPHs. It is determined by the effectiveness of the capillary action in a SPH. $t_2$ is the actual swelling time of the polymer matrix, which is determined by the thickness of the cell walls and struts. Because the cell walls and struts of SPHs are very thin, they have very short characteristic swelling times. For SPHs, $t_2$ is comparable to that of a ultrathin hydrogel film. The capillary action is mainly determined by the availability of capillary channels and the wettability of the channels. Various approaches have been attempted to maintain good capillary action (i.e., to decrease $t_1$) by maintaining open intercellular channels and good surface wettability.

## APPLICATIONS

The fast swelling kinetics and high swelling property would allow SPHs and SPH composites to be used in various applications.

## SUPERABSORBENTS

Due to the excellent capillary capacity of SPHs and SPH composites, they are ideal for improving surgical pads for bleeding control and personal hygiene products such as disposable diapers and sanitary napkins. The fast water-absorbing property also allows application of the SPHs or their composites as a desiccating agents in place of silica gels. Currently, the commercially available superabsorbents are in powder form because particulate gels with only a small size can swell fast. This brings limitations in certain applications [5]. The SPH and SPH composites technique allows preparation of superabsorbent materials in any size, thickness, or shape. SPHs are easy to make and show a number of superior properties to existing hydrogel products, especially the swelling rate. These merits should greatly broaden the applications of superabsorbent polymers. The SPHs technique can be used to make a wide array of synthetic, semi-synthetic, or natural SPHs that may replace the existing superabsorbents in many applications.

## CONTROLLED DRUG DELIVERY

In the controlled drug delivery area, SPHs and SPH composites can be used as a platform for long-term oral drug delivery. Due to their fast swelling and superswelling properties, they can stay in the stomach for several hours up to more than 24 hours [8,12]. Such a long-term gastric retention time is ideal for long-term oral controlled drug delivery.

## DIET CONTROL

In the diet control area, SPHs and SPH composites can be used to control the appetite of healthy people who desire to reduce the volume of food they take. Because of the fast swelling to a very large size, SPHs and SPH composites can remain in the stomach for extended periods of time ranging from hours to days [6–8,12]. The presence of bulky SPHs and SPH composites is expected to reduce appetite and the space in the stomach and thus reduce the amount of food that can be taken. Thus, they could provide an alternative therapy for obesity.

## BIOMEDICAL APPLICATIONS

In the biomedical area, SPHs and SPH composites can be used to make various biomedical devices, such as artificial pancreas, artificial cornea, and artificial skin, articular cartilage, soft tissue substitutes, cell growth substrates in tissue engineering, burn dressings, surgical augmentation of the female breast, or hemoperfusion in blood detoxification and in the treatment of uremia.

## BIOTECHNOLOGY AREA

In the biotechnology area, they can be used in the separation of macromolecules and cells from the medium. The presence of extremely large pores makes SPHs and SPH composites are also ideal materials for chromatographic supports.

## STRUCTURAL APPLICATIONS

The low density of SPHs and SPH composites allows applications as a high-strength, light-weight structural material as well as a packaging material. They will be also good as insulators and fillers in structures with energy-sensitive applications.

## FAST RESPONSIVE STIMULI-SENSITIVE SPHs

Hydrogels that can change their volume abruptly with small changes in environmental conditions are known as "intelligent" or "smart" hydrogels. Smart hydrogels respond to changes in the environmental conditions, such as temperature, pH, solvent, electric field, specific molecules or ions, light, or pressure. Although these smart hydrogels are highly useful in various applications, the typical response time usually ranges from hours to days, and this slow response time sometimes limits the usefulness of the smart hydrogels. By making superporous smart hydrogels, the response time could be reduced to seconds or minutes.

## FUTURE OF SPHs AND SPH COMPOSITES

Although current SPHs and SPH composites have many useful properties that conventional hydrogels do not have, they can be improved further. Current SPH composites are strong enough to be dropped on the floor without breaking. They, however, can be broken by applying pressure. For example, they can be broken by squeezing them in the hand. It is scientifically challenging to make SPHs or SPH composites that have the elastic property of rubber, i.e., elastic enough to be extended several folds without breaking. Although the development of new hydrogels with more useful properties may be endless, our immediate goal is to make SPHs and SPH composites with rubberlike elasticity and strength. We know that it may not be easy, but we feel that it is possible.

## REFERENCES

1. Wichterle, O. and Lim, D. *Nature,* 1960, 185, 117–118.
2. Tanaka, T. and Fillmore, D. J. *J. Chem. Phys.,* 1979, 70, 1214–1218.
3. Kabra, B. G. and Gehrke, S. H. in *Superabsorbent Polymers,* American Chemical Society, Washington, DC, 1994, 76–86.
4. Gehrke, S. H. in *Responsive Gels: Volume Transitions II,* Springer-Verlag, New York, 1993, 81–144.
5. Knack, I. and Beckert, W. U.S. Patent 1991, 5,002,814.
6. Shalaby, W. S. W., Blevins, W. E. and Park, K. *J. Controlled Rel.,* 1992, 19, 131–144.
7. Shalaby, W. S. W., Blevins, W. E. and Park, K. *Biomaterials,* 1992, 13, 289–296.
8. Chen, J. Ph.D. Thesis, Purdue University, 1997.
9. Chen, J., Park H. and Park, K. *Proc. ACS Div. Polym. Mat. Sci. Eng.,* 1998, 79, 236–237.
10. Chen, J., Park, H. and Park, K. *Proc. Intern. Symp. Control. Rel. Bioact. Mater.,* 1998, 25, 60–61.
11. Park, K., Chen, J. and Park, H. *Polymer Preprint,* 1998, 39, 192–193.
12. Chen, J., Blevins, W. E., Park, H. and Park, K. *J. Controlled Rel.,* 2000, 64, 39–51.
13. Chen, J., Park H. and Park, K. *J. Biomed. Mater. Res.,* 1999, 44, 53–62.
14. Chirila, T. V., Constable, I. J., Crawford, G. J., Vijayasekaran, S., Thompson, D. E., Chen, Y. C., Fletcher, W. A. and Griffin, B. J. *Biomaterials,* 1993, 14, 26–38.
15. Skelly, P. J. and Tighe, B. J. *Polymer,* 1979, 20, 1051–1052.
16. Oxley, H. R., Corkhill, P. H., Fitton, J. H. and Tighe, B. J. *Biomaterials,* 1993, 14, 1065–1072.
17. Barvic, M., Kliment, K. and Zavadil, M. *J. Biomed. Mater. Res.,* 1967, 1, 313–323.
18. de Groot, J. H., Nijenhuis, A. J., Bruin, P., Pennings, A. J., Veth, R. P. H., Klompmaker, J. and Jansen, H. W. B. *Colloid and Polymer Science,* 1990, 268, 1073–1081.
19. Park, H. and Park, K. *The 20th Annual Meeting of the Society for Biomaterials* 1994, Abstract #158.
20. Park, H. and Park K. *Proc. Intern. Symp. Control. Rel. Bioact. Mater.,* 1994, 21, 21–22.
21. Kon, M. and de Visser, A. C. *Plast. Reconstruct. Surg.,* 1981, 67, 288–294.
22. Krauch, C. H. and Sanner, A. *Natur. Wissenscheften,* 1968, 55, 539–540.
23. Badiger, M. V., McNeill, M. E. and Graham, N. B. *Biomaterials,* 1993, 14, 1059–1063.
24. Dusek, K. and Sedlacek, B. *Coll. Czech. Chem. Commun.,* 1969, 34, 136–157.
25. Rezai, E., Lahrman, F. H. and Iwasaki, T. U.S. Patent 1994, 5,324,561.

# CHAPTER 11

# Structure and Solute Size Exclusion of Poly(methacrylic acid)/Poly(N-isopropyl acrylamide) Interpenetrating Polymeric Networks

JING ZHANG[1]
NICHOLAS A. PEPPAS[1]

## INTRODUCTION

HYDROGELS are hydrophilic three-dimensional networks formed by crosslinked polymer chains, entanglements, or crystalline regions. Some of the hydrogels are capable of responding to external stimuli, such as pH, temperature, ionic strength, and electric field. [1] Among these, pH-sensitive and temperature-sensitive hydrogels have been extensively investigated for possible applications in various biological and chemomechanical areas [2–7]. Poly(methacrylic acid) and poly(N-isopropyl acrylamide) are the pH-sensitive and temperature-sensitive materials that we are going to concentrate on in this chapter.

Poly(methacrylic acid) is an ionizable hydrophilic polymer. Crosslinked PMAA is insoluble but able to swell in water. Its swelling behavior is pH dependent due to the ionization/deionization of the carboxylic acid groups [8,9]. At low pH, usually pH $< pK_a$, the COOH groups are not ionized. Thus, the PMAA network is in a collapsed state. At high pH values, the -COOH groups are ionized, and the charged COO- groups repel each other; this leads to PMAA swelling.

Poly(N-isopropyl acrylamide) (PNIPAAm) is the most extensively studied temperature-sensitive polymer [10–20]. Crosslinked PNIPAAm exhibits drastic swelling transition at its lower critical solution temperature

[1]Polymer Science and Engineering Laboratories, School of Chemical Engineering, Purdue University, West Lafayette, IN 47907-1283, USA.

(LCST) of 32°C. At temperatures lower than 32°C, the gel is swollen, whereas at temperatures higher than 32°C, the gel dehydrates to the collapsed state due to the breakdown of the delicate hydrophilic/hydrophobic balance in the network structure.

In this work, we generated new PMAA/PNIPAAm IPN hydrogels with both pH and temperature sensitivities by the interpenetration of the pH-sensitive and temperature-sensitive polymer networks.

**EXPERIMENTAL**

SYNTHESIS OF THE IPNs

The temperature-sensitive poly(*N*-isopropyl acrylamide) and pH-sensitive poly(methacrylic acid) were used as the two component networks in the IPN system. Since both *N*-isopropyl acrylamide (NIPAAm) (Fisher Scientific, Pittsburgh, PA) and methacrylic acid (MAA) (Aldrich, Milwaukee, WI) react by the same polymerization mechanism, a sequential method was used to avoid the formation of a PNIPAAm/PMAA copolymer. A UV-initiated solution-polymerization technique offered a quick and convenient way to achieve the interpenetration of the networks. Polymer network I was prepared and purified before polymer network II was synthesized in the presence of network I. Figure 1 shows the typical IPN structure.

Monomer I (MAA) was dissolved in methanol and 1 mol% of crosslinking agent, tetraethyleneglycol dimethylacrylate (TEGDMA) (Polysciences, Inc., Warrington, PA), and 1 wt% of initiator, 2,2-dimethoxy-2-phenyl acetophenone (DMPA; Aldrich, Milwaukee, WI) were added. The solution was cast on glass plates equipped with spacers and reacted under an UV source with an intensity of 1 mW/cm$^2$ for 30 min. Polymer I (PMAA) was removed from the plates, washed in deionized water to remove all unreacted monomers, cut into discs, and dried in a vacuum oven.

The dried polymer network I was then swollen in monomer II/methanol solution (NIPAAm/CH$_3$OH) to equilibrium. The solution contained 1 mol% of TEGDMA and 1 wt% of DMPA. The swollen gel was UV polymerized for 10 min to form the IPN. The IPN was then washed in pH = 7.4 buffer solution to remove the unreacted monomers.

SWELLING STUDIES

The IPN hydrogels were cut into 10-mm-diameter disks and dried under vacuum at 37°C. The dried volumes were calculated using a buoyancy technique. Dry disks of each type were placed in 100 mL of buffer solutions

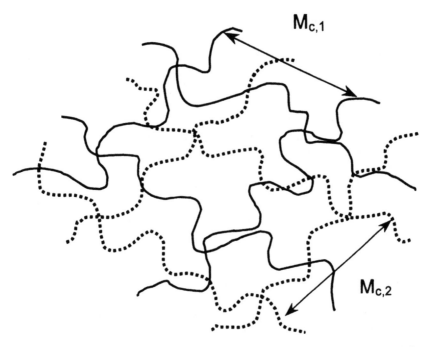

**Figure 1** Interpenetrating polymer networks, with the solid mesh representing network I and the dashed mesh representing network II. $M_c$ represents the molecular weight between the crosslinks.

ranging in pH from 3.0 to 9.0 and temperature 22 to 45°C. The ionic strength of each solution was adjusted to 0.1 M by the addition of NaCl. To determine the equilibrium swelling characteristics, the IPNs kept in the buffer solution and weighed every 24 h until the volume remained constant. The swelling solution was changed daily for each sample. The equilibrium volume swelling ratio, $Q$, was determined as the ratio of the final swollen polymer volume to the initial try polymer volume.

## THERMAL ANALYSIS

Differential scanning calorimetry (Model DSC 2910, TA Instruments, New Castle, DE) experiments were conducted on preswollen hydrogel samples, with a total weight of approximately 15 mg (polymer and water). The gels were placed into hermetically sealed pans and equilibrated at 15°C with liquid nitrogen cooling prior to heating at 2°C per minute to a final temperature of 60°C. The temperature-sensitive collapse of the hydrogel was identified as an endotherm in the thermograms. The endothermic response to the collapse of the hydrogel was caused by an increase in the system entropy as the polymer and water separate.

## DIFFUSION STUDIES

Diffusion studies were performed by using a Valia-Chen diffusion cell (Crown Glass Co., Inc., Somerville, NJ). The apparatus consisted of two half-cells, receptor cell and donor cell, with a volume of 3 mL and a side opening with a diameter of 9 mm. A magnetic stir bar was placed in each half-cell for continued agitation. Between the two half-cells a preswollen membrane was securely placed and was protected from the atmosphere to prevent evaporation of the solvent form the membrane.

To the donor cell, 3 mL of the model drug solution was added, and to the receptor half-cell 3 mL of the solvent was added. Timing of the diffusion of the solute began as soon as both half-cells were filled. At regular intervals (i.e., 15, 30, and 60 minutes) the contents of the receptor cell were moved and replaced with fresh solvent, which in this case was pH buffer solution. To ensure constant temperature of the inner cell solution, constant-temperature water was pumped through the outer half-cells.

An ultraviolet-visible light spectrophotometer (Lambda 10, Perkin Elmer, Norwalk, CT) was used to measure the characteristic absorbance of the samples taken from the receptor half-cell. Using a calibration curve derived from known concentrations of the model drugs, the concentration of each sample taken from the receptor half-cell could be determined.

Solutes with different hydrodynamic radii were used in this study to investigate the permeation behavior of the IPN membranes.

## RESULTS AND DISCUSSION

### RESULTS OF SWELLING STUDIES

Shown in Figure 2 is a typical plot of the pH influence on the swelling ratio of the IPNs. The data are in the pH range from 3.0 to 9.0, whereas physiological pH is 7.4. Similar to that of pure PMAA hydrogel, the swelling curve of the IPN exhibits a sharp transition at approximately pH 5.5. This transition is due to the electrostatic repulsion caused by the ionization of the carboxylic groups of the PMAA component at high pH, and the charged COO- groups repel each other, so that the gel will take up water and swell. The swelling ratio of the IPN is not as great as that of the pure PMAA gel even though the PMAA gel is made from the same concentration MAA solution that was used in the preparation of the IPN. This is because the incorporation of the second network increases the apparent crosslinking density of the IPN, which in turn decreases the swelling ratio and increases the mechanical strength of the IPN versus the pure PMAA. In addition, the existence of the second network decreases the carboxylic

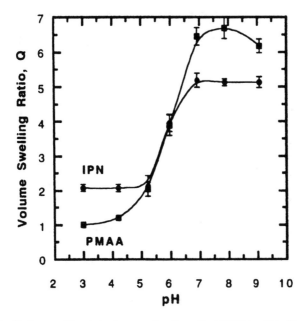

**Figure 2** Equilibrium swelling behavior as a function pH at 22°C in pH buffer solution for PMAA/PNIPAAm IPN samples (●) containing 70 mol% of PNIPAAm and pure PMAA (■) samples.

group density of the system compared to the pure PMAA hydrogel thus resulting in a lower swelling ratio.

Shown in Figure 3 is the temperature influence on the swelling behavior of the IPNs. Two IPN samples with the same PNIPAAm concentration were used in this study. One was swollen in a buffer solution with pH 4.06 while the other was swollen in pH 6.16. The temperature of the solutions was increased from 20°C to 45°C. Deswelling transitions were detected in both IPN samples. As temperature increases to the transition temperature of 32°C, the mobility of the side chains increases to break the delicate hydrophilic/hydrophobic balance in the PNIPAAm network at lower temperature. Dehydration takes place in the PNIPAAm network, which results in the subsequent aggregation of the PNIPAAm chain and lead to the deswelling of the IPNs. For the pH 4.06 sample, since the PMAA component of the IPN was in a collapsed state, the increase in temperature decreased the swelling ratio from 4.8 to 1.0. However, for the IPN sample at pH 6.16, since the PMAA component is at the swollen state, the swelling ratio decreased from 8.2 to 3.4 as temperature increased instead of going all the way to 1.0.

The ionic strength also plays an important role in the IPN swelling behavior. We have investigated the ionic strength influence on the swelling

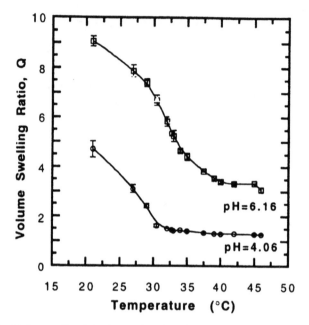

**Figure 3** Equilibrium swelling behavior as a function of temperature at pH 4.06 (○) and pH 6.16 (□) for PMAA/PNIPAAm IPNs containing 70 mol% of PNIPAAm.

behavior by using swelling agents such as NaCl solutions with different ionic strength. The temperature and pH were kept constant at pH of 7.0 and temperature of 22°C. As noted in Figure 4, the swelling ratio of PNIPAAm is almost constant as the ionic strength increases to 0.1 M (physiological state) because PNIPAAm is not ionizable. In contrast, the swelling ratio of PMAA decreased drastically as the ionic strength increased; this is due to the shielding effect caused by the counter ions at high ionic strength. The IPN sample deswells at the beginning and then flatten out when the ionic strength approaches 0.1 M. The behavior of the IPN sample is a combination of the behavior of PMAA and PNIPAAm hydrogels.

In addition to the equilibrium swelling characteristics of these IPNs, it is important to know the dynamic nature and the swelling rates that are involved in swelling. Dynamic swelling studies have been done to understand how fast the swelling response of these IPNs is to the changes in pH or temperature by keeping one parameter constant and oscillating the other parameter around the transition point. The time constants involved in the swelling and deswelling processes are very important in the characterization of the dynamics of the IPN hydrogels.

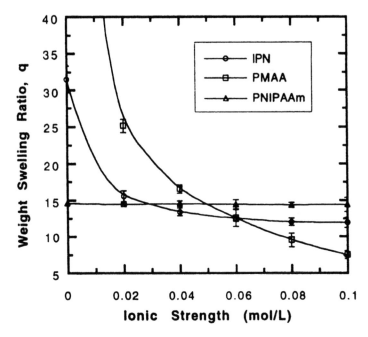

**Figure 4** Equilibrium swelling behavior as a function of ionic strength at pH 7.0, 22°C for PMAA/PNIPAAm IPN (○), PMAA (□), and PNIPAAm (△).

## RESULTS OF THERMAL ANALYSIS

DSC were conducted on PMAA/PNIPAAm IPN hydrogels swollen at different pH values. The results show that the difference in pH has great influence on the LCST transitions of the IPN hydrogels, as shown in Figure 5.

At pH 4.3, there is no significant transition detected around 32°C. Transition temperatures are detected and increase as pH increases. This is because, at low pH, the aggregation of PMAA decreases the mobility of the PNIPAAm network as well as the water uptake of the IPN, resulting in drastically lowering the temperature sensitivity of the IPN hydrogel. However, at higher pH value, the swollen PMAA allows the PNIPAAm to have a higher mobility, which makes the IPN more temperature sensitive.

Compared to the DSC diagram of pure PNIPAAm, we noticed that the LCST transitions detected in the IPNs samples were in the range of 31–32°C. There was no significant deviation from the LCST of the pure PNIPAAm, which was the major difference between a PNIPAAm-based IPN and a PNIPAAm copolymer in which the LCST will be greatly influenced by the comonomers. Thus, the formation of IPN gave a relatively independent polymer system in which each network may retain its own property.

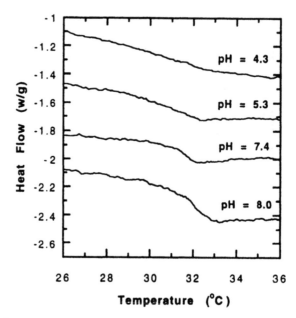

**Figure 5** Differential scanning calorimetry thermograms of PMAA/PNIPAAm IPNs at pH 4.4, 5.3, 7.4, and 8.0.

## RESULTS OF DIFFUSION STUDIES

Permeability coefficients were calculated from the data obtained in the permeation experiments using the following equation:

$$\ln\left(\frac{2C_t}{C_o} - 1\right) = \frac{2A}{V}Pt \tag{1}$$

Here, $C_t$ is the solute concentration in the receptor cell at time $t$; $C_o$ is the initial solute concentration of the donor cell; $V$ is volume of each half-cell; $A$ is the effective area of permeation; and $P$ is the membrane permeability coefficient. To determine the permeability coefficient, $P$, a plot of $-(V/2A)$ ln $[1 - 2(C_t/C_o)]$ versus time, $t$, was constructed. The linear portion yields a slope of the permeability coefficient, $P$ in cm/s.

The diffusion coefficient was calculated from the permeability coefficient, $P$, and the partition coefficient, $K_d$, as shown in the following equation:

$$D_m = \frac{Pl}{K_d} \tag{2}$$

Here, $l$ is the membrane thickness in the swollen state at constant pH and temperature, and $K_d$ is the ratio of the solute concentration in the membrane to the solute concentration the solution at equilibrium, which can be calculated from mass balance.

A significant size exclusion phenomenon was observed for the IPN membranes. Theophylline ($R_h = 1.3$ Å), proxyphylline ($R_h = 2.3$ Å), oxprenolol HCl ($R_h = 2.6$ Å), and FITC-Dextran ($R_h = 49$ Å) were used as model drugs in the diffusion study where $R_h$ denotes the hydrodynamic radius of the solute. The solute size, membrane mesh size, pH, temperature, and the affinity of the solute with the membrane can affect the permeation of the solute.

Shown in Figure 6 is a plot for the permeation of the four different model drugs through a PMAA/PNIPAAm IPN membrane containing 70 mol% of PNIPAAm in pH 7.4 buffer solution. As mentioned, the slope of each linear curve represents the permeability of each solute. As expected, higher permeability was observed for smaller model drug. From theophylline to FITC-Dextran, the permeability decreases, which is an indication of the size exclusion behavior in the IPN membrane: at the same gel swollen state, the IPNs show higher permeability for the smaller solutes.

To further investigate the pH and temperature effects on permeability, we analyzed the permeation of oxprenolol HCl under the four possible pH

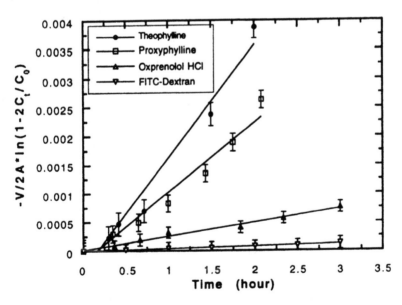

**Figure 6** Solute permeation of theophylline (O), proxyphylline (□), oxprenolol HCl (△), and FITC-Dextran (▽) through PMAA/PNIPAAm IPN. The slope represents the permeability of each solute.

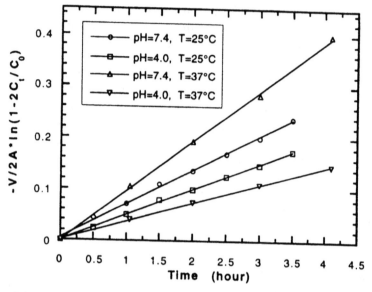

**Figure 7** Permeation behavior of oxprenolol HCl for four different pH and temperature conditions.

and temperature conditions, as shown in Figure 7. As indicated before, the four slopes of the linear curves represent four different permeabilities of the same IPN sample.

The relationships of the four permeabilities are presented in Figure 8: Here, $P$ is permeability, and $q$ is weight swelling ratio. PMAA is the pH-sensitive network I, also called primary network, and PNIPAAm is the temperature-sensitive network II, also called secondary network. From Figure 8, we note that, at low $T$ high pH, both networks are swollen, and the IPN has the highest swelling ratio. When pH decreases, the primary network collapses, and the swelling ratio drops drastically. The permeability decreases with the swelling ratio. The same observation is noted at high temperature when we reduced the pH from 7.4 to 4.0. We can see that the permeability decreases with the collapse of the primary network.

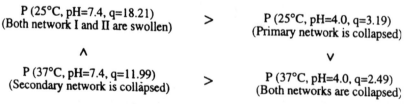

**Figure 8** The relationship of the permeabilities for oxprenolol HCl permeation.

Things are different when we change the temperature. At the state of pH = 7.4 $T$ = 37°C, the secondary network is collapsed. The swelling ratio also decreases but not as much as the previous case. The permeability at this state is higher than the state when both networks are swollen, even though it has a lower swelling ratio. This is probably because the primary network is still swollen, and the aggregation of the secondary network at higher $T$ does not decrease the swelling ratio too much. Instead, it leaves space and opens gates for the drug to permeate through. Comparing the four situations, the physiological state of pH = 7.4, $T$ = 37°C shows the highest permeability.

## CONCLUSION

Interpenetrating polymer networks of PMAA/PNIPAAm were prepared. The material exhibited both pH and temperature sensitivities. The temperature sensitivity was verified by DSC studies. The permeation studies showed a significant size exclusion phenomenon of the IPNs. The influence of pH and temperature on the IPN permeability was investigated. A hypothetical mechanism was presented to explain the high permeability at high temperature and high pH.

## REFERENCES

1. Gehrke, S. H. Synthesis, equilibrium swelling, kinetics, permeability and applications of environmentally responsive gels. *Adv. Polym. Sci.,* 1993, 110, 81–144.
2. Kost, J. and Langer, R. Equilibrium swollen hydrogels in controlled release applications, in N. A. Peppas, ed. *Hydrogels in Medicine and Pharmacy,* pp. 95–107. CRC Press, Boca Raton, FL, 1986.
3. Peppas, N. A. Other biomedical applications of hydrogels, in N. A. Peppas, Ed. *Hydrogels in Medicine and Pharmacy,* pp. 177–194. CRC Press, Boca Raton, FL, 1986.
4. Ratner, B. Biomedical applications of hydrogels: Critical appraisal and review, in D. F. Williams, Ed. *Biocompatibility of Clinical Implant Materials,* pp. 145–153. CRC Press, Boca Raton, FL, 1981.
5. Freitas, R. F. S. and Cussler, E. L. Temperature sensitive gels as extraction solvents. *Chem. Eng. Sci.,* 1987, 42, 97–103.
6. Dong, L. C. and Hoffman, A. S. Synthesis and application of thermally reversible heterogels for drug delivery. *J. Controlled Release,* 1990, 13, 21–31.
7. Kikuchi, A. and Okano, T. Temperature-responsive polymers as on-off switches for intelligent biointerfaces, in T. Okano, Ed. *Biorelated Polymers and Gels,* pp. 1–28. Academic Press, San Diego, CA, 1998.
8. Weiss, A. M., Adler, K. P., Grodzinsky, A. J., and Yarmush, M. L. Variable permeability membranes. Network structure of poly(methacrylic acid) and its relation to diffusive transport. *J. Membr. Sci.,* 1991, 58, 153–173.

9. Seno, M., Len, M. L., and Iwamoto, K. Swelling of poly(methacrylic acid) gels and adsorption of L-lysine and its polymer on the gels. *Colloid Polym. Sci.* 1991, 269, 873–879.

10. Tanaka, T. Volume phase transition in poly(*N*-isopropylacrylamide) gel. *Phys. Rev. Let.* 1978, 40, 820.

11. Schild, H. G. Poly(*N*-isopropylacrylamide): Experiment, theory and application. *Progr. Polym. Sci.* 1992, 17, 163–249.

12. Otake, K., Inomata, H., Konno, M., and Saito, S. Thermal analysis of the volume phase transition with *N*-isopropylacrylamide gels. *Macromolecules* 1990, 23, 283–289.

13. Katono, H., Maruyama, A., Sanui, K., Ogata, N., Okano, T., and Sakurai, Y. Thermo-responsive swelling and drug release switching of interpenetrating polymer networks composed of poly(acrylamide-*co*-butyl methacrylate) and poly(acrylic acid). *J. Controlled Release* 1991, 16, 215–228.

14. Gutowska, A., Bae, Y. H., Jacobs, H., Feijen, J., and Kim, S. W. Thermosensitive interpenetrating polymer networks: Synthesis, characterization, and macromolecular release. *Macromolecules,* 1994, 27, 4167–4175.

15. Lim, Y. H., Kim, D., and Lee, D. S. Drug releasing characteristics of thermo- and pH-sensitive interpenetrating polymer networks based on poly(*N*-isopropylacrylamide). *J. Appl. Polym. Sci.* 1997, 64, 2647–2655.

16. Brazel, C. S. and Peppas, N. A. Synthesis and characterization of thermo- and chemomechanically responsive poly(*N*-isopropylacrylamide-*co*-methacrylic acid) hydrogels. *Macromolecules* 1995, 28, 8016–8020.

17. Kaneko, Y., Nakamura, S., Sakai, K., Aoyagi, T., Kikuchi, A., Sakurai, Y., ,and Okano, T. Rapid deswelling response of poly(*N*-isoproplyacrylamide) hydrogels by the formation of water release channels using poly(ethylene oxide) graft chains. *Macromolecules* 1998, 31, 6099–6105.

18. Okano, T., Bae, Y. H., Jacobs, H., and Kim, S. W. Thermally on-off switching polymers for drug permeation and release. *J. Controlled Release* 1990, 11, 255–265.

19. Dong, L. C., Yan, Q., and Hoffman, A. S. Controlled release of amylase from a thermal and pH-sensitive, macroporous hydrogel. *J. Controlled Release* 1992, 19, 171–178.

20. Feil, H., Bae, Y. H., Feijen, J., and Kim, S. W. Mutual influence of pH and temperature on the swelling of ionizable and thermosensitive hydrogels. *Macromolecules* 1992, 25, 5528–5530.

# CHAPTER 12

# Thermodynamics of Water Sorption in Hyaluronic Acid and Its Derivatives

P. A. NETTI[1]
L. AMBROSIO[1]
L. NICOLAIS[1]

## INTRODUCTION

HYALURONIC acid (HA) is a naturally occurring polysaccharide that is a ubiquitous component of the extracellular matrix of higher organisms. Among the many physiological functions, HA controls the movement of fluid and macromolecules through the extracellular matrix. Since its discovery much attention has been focused on the possible biomedical application of this highly biocompatible and biodegradable polymer. Many of the biological functions of HA are a consequence of its water affinity [1]. This property, however, results in poor mechanical strength and rapid degradation that limits widespread application of HA as a biomaterial. One way to improve its mechanical properties and reduce the degradation time while maintaining its original biocompatibility is to increase the ratio of hydrophobic/hydrophilic groups on the polymer backbone. This has been achieved by esterification of the carboxyl groups of HA with therapeutically inactive or active alcohols [2]. These materials have promising potential in biomedical applications since they present enhanced mechanical strength and reduced degradation time compared to HA [3] and can be engineered into a number of medically useful configurations, for pharmacology, orthopedics, ophthalmology, and the medical–sanitary field in general. Depending on the nature of the ester group and to the degree of esterification, the polymers can be technologically processed to obtain products of

[1]Department of Materials and Production Engineering University of Naples "Federico II," and Institute of Composite Materials Technology-CNR, Piazzale Tecchio 80, 80125 Naples, Italy.

**169**

biomedical interest such as threads, films, tubes, sponges, fabrics, and microspheres [4–7].

For the development of these applications, the understanding of water sorption mechanism as well as its dependence on the structural parameters of the material is particularly important in the formulation of a suitable biomedical device. The physical properties of the hydrated material are strongly influenced both by the interactions between water and the hydrophilic segments of the polymer as well as by the organization of the water molecules within the structure. Understanding the effect of a substituting group on the water affinity of the material is a key issue to be able to engineer these materials. Therefore, the aim of this work was to explore how the dimension and chemical structure of the substituting group alters the mechanism of water sorption by HA and the consequence of this sorption on the mechanical properties of the material.

## MATERIALS

HA and three different HA esters, supplied by Fidia Advanced Biomaterials SpA (Padua, Italy), were examined. HA esters were prepared by treating a quaternary ammonium salt of HA with an esterifying agent in a suitable aprotic solvent at a controlled temperature as described in detail elsewhere [2]. In this study we analyzed the following esters: ethyl ester (HYAFF7), benzyl ester (HYAFF11), and dodecyl ester (HYAFF73; Figure 1). The materials used were both films and powders. Films were obtained by a phase inversion process from DMSO solutions as previously described [2,4].

### GRAVIMETRIC MEASUREMENTS

Water sorption was measured at 40°C at several relative pressures (0.05 $< p_w/p_w° < 0.95$), where $p_w$ is the water vapor pressure of the external phase and $p_w°$ is the water vapor pressure at the temperature of the experiment. Water sorption measurements were carried out a gravimetric apparatus based on a highly sensitive quartz spring as reported in Reference [8]. Briefly, the apparatus consists of a double-jacket sorption cell connected to a solvent reservoir and to a vacuum line. The water pressure in the sorption cell was measured by a sensitive capacitance transducer (M.K.S. BARATRON 221A, with an accuracy of ±0.5% of the reading). The water weight gain during the sorption experiment was evaluated by measuring the elongation of a quartz spring, on which polymer sample was hung, and placed in the sorption cell. The quartz spring used for these experiments

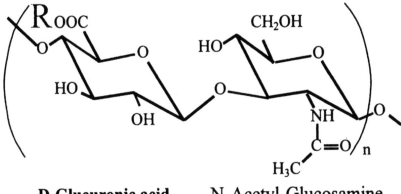

**D-Glucuronic acid**          N-Acetyl-Glucosamine

R = -H                    ⟶    Hyaluronic Acid

R = -CH₂-CH₃          ⟶    Ethyl Ester (HYAFF7)

R = ⟨O⟩                   ⟶    Benzyl Ester (HYAFF11)

R = -(CH₂)₁₁-CH₃    ⟶    Dodecyl Ester (HYAFF73)

**Figure 1** Structural formula of HA and the ethyl (Hyaff7), docecyl (Hyaff73) and benzyl (Hyaff11) esters. The esterification reaction was carried out starting from a quaternary ammonium salt of HA and the ethyl, dodecyl and benzyl alcohols, respectively, in a suitable aprotic solvent.

had a sensitivity equal to 5 mg/cm. Spring elongation was monitored using a traveling microscope.

All the samples were previously dried under vacuum overnight before starting the measurements. Double distilled and degassed water was used in all the experiments.

DENSITY MEASUREMENTS

The densities of the HA and the three esters were evaluated at 40°C by means of a Wesphail balance using *n*-hexane and chloroform as reference liquids. These two liquids were chosen because of their negligible solubility in the HA and its derivatives.

## MECHANICAL TESTS

The tensile properties of the films, both in the dry and wet state, were performed by a standard Instron testing machine Mod. 4204 at room temperature; for wet samples, the tests were carried out in physiological solution. A strain rate of 0.1 $min^{-1}$ was used for all tests, following the ASTM D-638 protocol.

## RESULTS

### SORPTION ISOTHERMS

The amounts of water absorbed as function of relative water vapor pressure (relative humidity) for HA and its esters are reported in Tables 1–4. HA absorbed the highest amount of water at all humidity levels compared to its esters. The ethyl ester (Hyaff7) absorbs more water than the other two, and the dodecyl ester (Hyaff73) absorbs more water than the benzyl ester (Hyaff11). A small percentage of water absorption hysteresis, between sorption and desorption, was found for the four different materials analyzed. No significant differences in the percentage of hysteresis was found among the HA and the three esters.

Since all the materials considered are amorphous, the differences in water sorption was exclusively due to the effect of the substituent group. It is worth nothing that, although there are large differences in the molecular weights of the substituting groups, the water sorption capacity of the ethyl

TABLE 1. Solubility Properties of Hyaluronic Acid.

| Relative Pressure Range | Water Concentration Sorption $\left[\dfrac{\text{g water}}{\text{g polymer}}\right]$ | Concentration Range Desorption $\left[\dfrac{\text{g water}}{\text{g polymer}}\right]$ | Solubility $\left[\dfrac{\text{g water}}{\text{g polymer}}\right]$ |
|---|---|---|---|
| 0.00 | 0.000 | 0.047 | |
| 0.05 | 0.038 | 0.068 | |
| 0.10 | 0.063 | 0.098 | 0.76 |
| 0.20 | 0.097 | 0.140 | 0.63 |
| 0.30 | 0.140 | 0.180 | 0.48 |
| 0.40 | 0.170 | 0.210 | 0.46 |
| 0.50 | 0.220 | 0.250 | 0.42 |
| 0.70 | 0.340 | 0.340 | 0.44 |
| 0.90 | 0.890 | 0.980 | 0.48 |
| 0.95 | 1.380 | | 0.99 |
| | | | 1.45 |

TABLE 2. Solubility Properties of Ethyl Hyaluronate (HYAFF7).

| Relative Pressure Range | Water Concentration Sorption $\left[\dfrac{\text{g water}}{\text{g polymer}}\right]$ | Water Concentration Desorption $\left[\dfrac{\text{g water}}{\text{g polymer}}\right]$ | Solubility $\left[\dfrac{\text{g water}}{\text{g polymer}}\right]$ |
|---|---|---|---|
| 0.00 | 0.000 | 0.005 | |
| 0.08 | 0.036 | — | 0.46 |
| 0.17 | 0.058 | — | 0.33 |
| 0.21 | 0.069 | — | 0.32 |
| 0.32 | 0.085 | 0.102 | 0.26 |
| 0.48 | 0.112 | 0.123 | 0.23 |
| 0.66 | 0.152 | 0.166 | 0.23 |
| 0.90 | 0.350 | 0.370 | 0.39 |

TABLE 3. Solubility Properties of Benzyl Hyaluronate (HYAFF11).

| Relative Pressure Range | Water Concentration Sorption $\left[\dfrac{\text{g water}}{\text{g polymer}}\right]$ | Water Concentration Desorption $\left[\dfrac{\text{g water}}{\text{g polymer}}\right]$ | Solubility $\left[\dfrac{\text{g water}}{\text{g polymer}}\right]$ |
|---|---|---|---|
| 0.00 | 0.000 | 0.0014 | |
| 0.08 | 0.020 | — | 0.24 |
| 0.19 | 0.035 | — | 0.18 |
| 0.21 | 0.040 | — | 0.19 |
| 0.31 | 0.050 | 0.059 | 0.16 |
| 0.49 | 0.077 | 0.088 | 0.16 |
| 0.68 | 0.120 | 0.142 | 0.17 |
| 0.92 | 0.210 | 0.213 | 0.23 |

TABLE 4. Solubility Properties of Dodecyl Hyaluronate (HYAFF73).

| Relative Pressure Range | Water Concentration Sorption $\left[\dfrac{\text{g water}}{\text{g polymer}}\right]$ | Water Concentration Desorption $\left[\dfrac{\text{g water}}{\text{g polymer}}\right]$ | Solubility $\left[\dfrac{\text{g water}}{\text{g polymer}}\right]$ |
|---|---|---|---|
| 0.00 | 0.000 | 0.001 | |
| 0.11 | 0.034 | — | 0.32 |
| 0.17 | 0.047 | — | 0.27 |
| 0.23 | 0.056 | — | 0.24 |
| 0.31 | 0.074 | 0.083 | 0.24 |
| 0.47 | 0.101 | 0.113 | 0.21 |
| 0.67 | 0.130 | 0.174 | 0.19 |
| 0.89 | 0.340 | 0.350 | 0.38 |

and dodecyl esters were found to be quite similar, whereas a large difference in water sorption capacity was found between the aliphatic (Hyaff7 and Hyaff73) and aromatic ester groups (Hyaff11).

For both HA and its esters, the solubility coefficient, defined as the ratio between water concentration in the polymer and the water relative pressure ($S = cw$/relative pressure), did not strictly increase with concentration (Tables 1–4). For example, at low water concentrations the solubility coefficient decreased with concentration, whereas an opposite trend was shown at higher water concentrations. The solubility coefficient attained a minimum relative pressure of about 0.6 for all the materials tested.

## MECHANICAL PROPERTIES

The elastic moduli of films of the ethyl and benzyl esters of HA in the dry and wet state are shown in Figure 2. In the dry state the elastic modulus of the two materials is not very different, 2.6 ± 0.1 GPa for Hyaff7 and 2.9 ± 0.16 GPa for Hyaff11. On the other hand, the elastic modulus of the two material differ significantly in the wet state, 0.026 ± 0.007 GPa for Hyaff7 and 0.53 ± 0.15 GPa for Hyaff11. The difference between the mechanical properties is specular due to the water sorption capacity of the two materials.

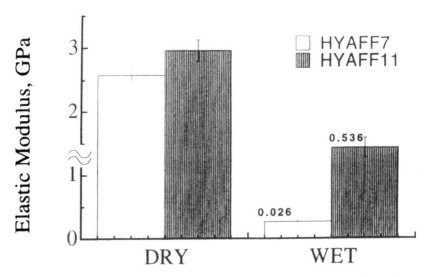

**Figure 2** Comparison among the elastic moduli of two different HA's ester in dry and wet conditions.

## ANALYSIS OF THE SORPTION ISOTHERMS

According to the Flory-Huggins theory of polymer solutions, if the mixing process were driven only by an entropic gradient (nonpolar solvent) the solubility coefficient $S_b$

$$S = \frac{v_w}{\ln(1-v_w)+v_w} \tag{1}$$

would increase as a function of the water concentration. In the case of endothermic interactions between solvent and polymer (i.e., $\chi > 0$), the solubility coefficient would be reduced. Therefore, the behavior found experimentally for the parameter $S$ (Tables 1–4) indicates that there is a strong exothermic interaction between water and the polymer at low concentration. The development of large heats of mixing at low concentration for these materials has also been proved experimentally [9,10]. As the concentration increased, the solubility decreased, indicating a reduction in the enthalpy of mixing. At higher concentrations, the solubility starts to increase as a consequence of an increase in the configurational entropy [11]. This suggests that the process of mixing between the water and the polymer occurs by two different mechanisms: at low water concentration the process is mainly driven by an enthalpic gradient, whereas entropic gradients are dominant at higher water concentration. The enthalpic driving force may be due to specific interactions between water and the polar groups on the polymer backbone, so the overall sorption process can be envisaged as occurring by two distinct mechanisms: (a) absorption of water molecules on hydrophilic groups and (b) water dissolution in the polymer matrix. Since water is absorbed onto hydrophilic groups, mainly by hydrogen bonding or weak electrostatic interactions, the absorption process can be conveniently described as a Langmuir-like isotherm:

$$v_w^i = \frac{\dfrac{\left(\dfrac{\rho_w}{\rho_p}\right)C_H ba}{(1+ba)}}{1+\dfrac{\left(\dfrac{\rho_w}{\rho_p}\right)C_H ba}{(1+ba)}} \tag{2}$$

where $a$ is the water activity, $v_w^i$ is the volume fraction of the population of water bound to the polar groups, $\rho_w$ and $\rho_p$ are the mass densities of the water and polymer, respectively, $C_H$ is the total number of polar sites on the polymer backbone, and $b$ is the affinity between water and polymer's polar groups. On the other hand, dissolution of water within the polymer matrix can be described by Flory-Huggins isotherm:

$$a = v_w^s e^{\left(v_p + \chi v_p^2\right)}$$ (3)

$v_w^s$ is the volume fraction of the population of water solubilized into the polymer matrix, $v_p$ is the polymer volume fraction. The total water volume fraction is given by the sum of the two population of water present in the polymer:

$$v_w = v_w^i + v_w^s$$ (4)

Equations (2), (3), and (4) solved together have been used to describe the sorption isotherms of HA and its esters.

Shown in Figure 3 is the model fitting of the sorption data for the HA and its esters. The mass densities of the materials (Table 5) have been used to convert the concentration to volume fraction. As shown, the model is able to satisfactorily interpolate the experimental data. The estimated parameters of the model are reported in Table 6.

The Flory-Huggins interaction parameter, $\chi$, indicates the water affinity of the polymer while $C_H$ is a measure of the fraction of bound water in the

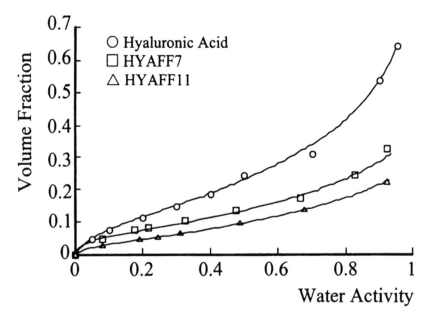

**Figure 3** Sorption isotherms for HA and its three esters. Solid lines are the model fitting of the experimental data.

TABLE 5. Density of Hyaluronic Acid and Its Esters.

| Material | Density $\frac{g}{cm^3}$ |
|---|---|
| Hyaluronic Acid | 1.308 |
| Hyaff 7 | 1.386 |
| Hyaff 11 | 1.387 |
| Hyaff 73 | 1.315 |

polymer matrix. The parameter $\chi$ is smaller than 0.5 only for the HA, indicating a complete solubility of HA in water, and increases in the order HA-Hyaff7-Hyaff73-Hyaff11. There was not a substantial difference between the parameter $\chi$ for the two aliphatic esters (Hyaff7 and Hyaff73), whereas there was a marked difference in the parameter $\chi$ between the aliphatic and the aromatic ester (Hyaff11). Different from the parameter $\chi$, the parameter $C_H$ increases in the order HA > Hyaff7 > Hyaff11 > Hyaff73.

Assuming that the number of specific interacting sites present on the polymer chain coincides with the oxygen atoms, the density of interacting sites per unit of mass of polymer ($\sigma$ = # of interacting sites/mass of polymer) is

$$\sigma = \frac{n_{O_2}}{M_W}$$

where $n_{O_2}$ and $M_W$ are the number of oxygen atoms and the molecular weight of the repeating unit, respectively. The density of interacting sites for HA and its esters are reported in Table 6. It increases in the order HA > Hyaff7 > Hyaff11 > Hyaff73. Note that, although $\sigma$ for Hyaff11 is higher than Hyaff7, the amount of bound water is less for Hyaff11 than for Hyaff7. This indicates that the number of hydrophilic groups does not correlate with the water absorption capacity of the material. The ratio $\frac{C_H}{\sigma}$ (Table 6) represents the number of bound water present for each hydrophilic group of the polymer. This ratio decreases in the order

TABLE 6. Parameters for Equations (2) and (3).

| Material | $\chi$ | $C_H$ | $b$ | $\sigma$ | $\frac{C_H}{\sigma}$ |
|---|---|---|---|---|---|
| Hyaluronic acid | 0.439 | 0.068 | 16 | 0.031 | 2.1 |
| Hyaff 7 | 0.945 | 0.036 | 44.8 | 0.027 | 1.33 |
| Hyaff 73 | 0.972 | 0.025 | 57.3 | 0.018 | 1.38 |
| Hyaff 11 | 1.127 | 0.016 | 42.3 | 0.024 | 0.66 |

HA > Hyaff7 > Hyaff73 > Hyaff11. It can be noted that the ratio for Hyaff11 is much less than that for Hyaff73, although the Hyaff11 contains more hydrophilic groups than Hyaff73.

## DISCUSSION

One of the major problem for the use of HA as a biomaterial is the loss in mechanical strength (softening) and the fast degradation time associated with hydration. By modulating the ratio between hydrophobic and hydrophilic groups it is potentially possible to engineer both the mechanical properties and degradation rate of HA. Indeed, by chemical esterification of the HA it is possible to modify water affinity (Tables 1–4) and, therefore, its water sensitivity. However, the ratio between hydrophobic and hydrophilic groups is not the only parameter controlling the water affinity of HA esters.

The water affinity is strongly affected by the chemical nature but not by the molecular weight of the substituting groups. Hyaff7 and Hyaff73 have (aliphatic) substituting groups with similar chemical structures but different molecular weight, 29 and 169, respectively. Despite a large difference in molecular weight of the substituting groups, there is no substantial difference in hydration (Table 2 and 3). Conversely, Hyaff11, which has an aromatic substituting group with an intermediate molecular weight (78) but a different chemical structure, shows hydration properties much different from the two aliphatic esters (Table 4). Although the density of the polar groups per molecule ($\sigma$) is higher for Hyaff11 than for Hyaff73, the latter absorbs more water. Taken together, all these data indicate that it's the chemical nature of the substituting group (i.e., the parameter $\chi$) more than the total number of polar groups on the molecule (i.e., parameter $\sigma$) that controls the sorption capability and, therefore, the water sensitivity of these materials. Moreover, the parameter $\sigma$, expressed as a measure of the total number of interacting sites, does not account for the availability to water molecules. The ratio $C_H/\sigma$ is a measure of this availability and decreases in the order HA > Hyaff7 > Hyaff73 > Hyaff11 (Table 6), reflecting the water sorption capability of the materials. This trend is also confirmed by the material densities (Table 5). Lower density means higher free volume and, therefore, higher availability of the polar groups to the water molecules.

Swelling capability is directly related to water affinity. It is reasonable to assume that most, if not all, of the mechanical softening, as well as the degradation of the material, is induced by the free water rather than the bound water. The absorption of water molecules onto the polar groups is expected to occur without a substantial change in the polymer structure and thus no appreciable reduction in the mechanical strength or increase in the degradation rate of the material. Since the amount of water present in our materials for relative pressure below 0.3 is mainly bound (Table 7), it is expected that the mechanical softening as well as the degradation process of the material occurs at relative pressure above 0.3. This is confirmed by the data shown in Figure 1. In our experiment, the dry state is referred as the material in equilibrium with the environmental conditions, which is at a relative humidity higher than 0.3%.

The difference in water affinity (i.e., $\chi$) between the aliphatic and the benzyl ester is reflected by the mechanical properties, (Figure 1). Therefore, the mechanical properties, as well as the resistance to degradation of HA's esters, are expected to decrease inversely to the parameter $\chi$. This suggests that it is possible to use parameter $\chi$ to engineer the mechanical strength and the degradation rate of these materials. The target parameter $\chi$ would be evaluated on the basis of the desired mechanical and degradation properties of the material. This would guide the choice of the appropriate substituting group to be used. An estimation of the parameter $\chi$ for a given substituting group can be obtained by a knowledge of $\chi$ for HA and adding to it the contribution for the given substituting group as suggested by Small [12].

## CONCLUSIONS

It is possible to improve the mechanical properties and to reduce the degradation rate of HA by esterification of the carboxyl groups with therapeutically inactive or active alcohols. Water sorption capability and thus mechanical softening and degradation rates depend upon the water affinity of these materials; a property measured by the Flory-Huggins interaction parameter $\chi$. Our data suggest that it is possible to engineer these materials by selecting appropriate substituting groups.

TABLE 7. Percent of Bound Water in Hyaluronic Acid and Its Esters.

| Relative Pressure | Percentage of Bound Water | | | |
|---|---|---|---|---|
| | HA | Hyaff7 | Hyaff73 | Hyaff11 |
| 0.01 | 75 | 85 | 83 | 75 |
| 0.10 | 55 | 58 | 52 | 44 |
| 0.15 | 48 | 49 | 42 | 35 |
| 0.20 | 42 | 42 | 35 | 29 |
| 0.30 | 33 | 32 | 26 | 21 |
| 0.50 | 21 | 20 | 16 | 13 |
| 0.80 | 11 | 10 | 8 | 7 |
| 0.90 | 9 | 9 | 7 | 5.7 |

## REFERENCES

1. Laurent, T. C. *Chem. Mol. Biol. Intercell.* 1970, 2, 703–731.
2. Della Valle, F. and Romeo, A. *New Polysaccharide Esters and Their Salts.* Italy: FIDIA SpA, 1987.
3. Zhong, S. P., Campoccia, D., Doherty, P. J., Williams, R. L., Benedetti, L., and Williams, D. F. Biodegradation of hyaluronic acid derivatives by hyaluronidase. *Biomaterials* 1994, 15, 359–365.
4. Benedetti, L. *Medical Device Technology* 1994, 11, 32–37.
5. Benedetti, L., Corvito, R., Berti, T., Pea, F., Mazzo, M., Moras, M., and Abatangelo, G. *Biomaterials* 1993, 14, 1154–1160.
6. Benedetti, L., Joshi, H. N., Goei, L., Hunt, J. A., Callegaro, L., Stella, V. J. and Topp, E. M. *New Polymeric Materials* 1991, 3, 41–48.
7. Iannace, S., Ambrosio, L., Nicolais, L., Rastrelli, A., and Pastorello, A. Thermomechanical properties of hyaluronic acid derived products. *J. Mat. Sci.: Mat. Med.* 1992, 3, 59–64.
8. Netti, P. A., Ambrosio, L., Del Nobile, M. A., Mensitieri, G., and Nicolais, L. Water transport in hyaluronic acid esters. *J. Bioactive Compatible Polymers* 1996, 11, 312–316.
9. Netti, P. A. Transport propertiues of hydrogel. Ph.D. thesis, University of Naples "Federico II," Naples, Italy, 1995.
10. Netti, P. A., Ambrosio, L., Astarita, G., DelNobile, M. A., Benedetti, L., and Nicolais, L. *Water Sorption in Hyaluronic Acid Esters.* October 19–21, Berlin, Germany, 1995, 18, 432.
11. Flory, P. J. *Principles of Polymer Chemistry.* Cornell University Press, Ithaca, NY, 1953.
12. Small, P. A. *J. Apl. Chem* 1953, 3, 71.

# CHAPTER 13

# Photocrosslinked Polyanhydrides with Controlled Hydrolytic Degradation

AMY K. BURKOTH[1]
KRISTI S. ANSETH[1]

## INTRODUCTION

DEGRADABLE polymers have found diverse applications in medicine ranging from resorbable vascular grafts to tissue engineering scaffolds. However, polymeric devices with controlled degradation mechanisms have made a significant impact on the success of drug delivery implant technology, offering features that are often difficult to obtain in a nonerodible matrix. Specifically, a wide range of drugs may be delivered from erodible polymeric devices because there is no dependence on diffusivities or drug size. Therefore, bioerodible devices do not display the exponentially decaying release profile, as seen typically from nonresorbable implants where diffusion controls release, and can often be designed to maintain a particular profile of drug release depending on the polymer's degradation kinetics.

The potential clinical applications of polymers as vehicles for drug delivery have grown recently, with most research focusing on polyesters. Polyesters, in particular poly(lactic acid), poly(glycolic acid), and their copolymers (PLGA), have been successfully used to deliver steroids [1,2], hormones and growth factors [3,4], vaccines [5,6], and cancer treatments [7–9]. Although this class of degradable polymers is FDA approved for clinical use in a wide range of applications, many analyses have shown an inflammatory reaction to these materials with time [10]. This adverse reaction is thought to coincide with the onset of degradation, in which acidic

[1]Department of Chemical Engineering, University of Colorado, Boulder, CO 80309-0424, USA.

monomers and oligomers are accumulated at an implant site. Furthermore, polyesters undergo homogenous degradation, where hydrolytic cleavage of the ester linkages occurs randomly throughout the polymer. This mode of degradation causes sudden losses in molecular weight without accompanying losses in mass, leading to a majority of mass loss at the end of degradation and a burst of acid products with complete polymer erosion [10,11]. As a result of the random scission of ester bonds throughout the polymer, release profiles from polyesters are often unpredictable and the high concentration of water in the matrix can reduce the stability of the loaded drug.

Over the past 10 years, interest in polyanhydrides, a new class of surface eroding polymers, has grown rapidly. In contrast to the bulk degrading polyesters, polyanhydrides, with hydrophobic backbone substituents and extremely reactive, hydrolytically labile anhydride linkages, degrade from the surface inward. By hindering solvent penetration into the hydrophobic polymer matrix, the surface erosion mechanism results in well-defined and predictable degradation rates that depend on the polymer composition and the size and shape of the implant. Polyanhydrides and their degradation products have been shown by several researchers to be neither mutagenic nor cytotoxic [12,13] and were FDA approved in 1996 for local delivery of chemotherapy agents to brain tumors [14]. For these reasons, polyanhydrides are advantageous as drug delivery vehicles and have been researched for the delivery of several agents, including osteogenic proteins [15], cancer chemotherapeutic agents [16–18] and insulin [19].

Recognizing that polyanhydrides have desirable degradation kinetics for drug delivery and other medical applications, we chose to modify the anhydride monomers to fit applications where strength may be important, for example, orthopedic applications. By functionalizing the anhydride monomers with two reactive end groups, we designed a new class of photocrosslinkable polyanhydrides that still maintains a surface erosion mechanism (for controlled delivery of therapeutic agents) but with additional material properties that are advantageous for high strength applications, especially where maintenance of strength with degradation is important (e.g., orthopedic applications). Specifically, we have designed a new biomaterial that has good handling and manageability properties for application in a clinical setting (e.g., ease of fashioning and contourability) and high strengths for load-bearing applications.

We chose to modify the anhydride monomers with photopolymerizable methacrylate functionalities. Methacrylate-based polymers have a long history in biomedical applications, ranging from photocured dental composites [20] to thermally cured bone cements [21]. Furthermore, photopolymerizations provide many advantages for material handling and processing, including spatial and temporal control of the polymerization and rapid rates at ambient temperatures. Liquid or putty-like monomer/initiator solutions can be molded to complex, three-dimensional geometries and photoreacted to form rigid polymers in seconds to minutes, with minimal heat released. Most importantly, the precise control of photopolymerization of the methacrylated anhydride monomers brings in the powerful potential to polymerize *in vivo* and, therefore, minimize the invasiveness of certain orthopedic surgical procedures.

To produce high-strength polyanhydrides, we functionalized *both* ends of the anhydride monomer with methacrylate groups in order to facilitate crosslinking. A high degree of crosslinking not only increased the mechanical strength of the polymer but also promoted the surface degradation mechanism by further minimizing uptake of water by the network. Furthermore, controlled degradation in combination with the high material strength allow for the polymer to re-establish the mechanical integrity of the injured bone and subsequently degrade to allow new bone ingrowth and remodeling. A degradation rate that is dependent only on the polymer composition and the ratio of implant surface area to volume also allows for predictable, controlled release of bone morphogenetic proteins for enhanced bone growth or antibiotics to prevent infection. However, as the polymer degrades, the presence of methacrylate groups produces a new degradation product, poly(methacrylic acid) (PMAA), a linear, hydrophilic polymer. PMAA is the only degradation product that differs from the FDA-approved linear polyanhydrides. However, characterization of the distribution of PMAA chains is important because there is a threshold molecular weight under which the body can resorb and excrete linear polymers [22].

Although the polymerization kinetics [23] as well as mechanics and degradation behavior [24] for these materials have been published in detail elsewhere, the focus of this chapter is to present a broader perspective, characterizing the wide range of possible monomers and the advantages of photopolymerization processing in biomedical applications. Polymerization behavior and processing, degradation kinetics and release profiles as well as the degradation product molecular weight distribution are presented.

## EXPERIMENTAL

### MATERIALS

Methacrylated sebacic acid monomer (MSA) was synthesized from sebacic acid (Aldrich) and methacrylic acid (Aldrich). Sebacic acid was stirred in 1.2 equivalents of methacrylic anhydride at 80°C until dissolution (approximately 1 hour). The excess methacrylic anhydride and methacrylic acid by-product were removed by rotary evaporation at 65°C. The MSA was subsequently precipitated in anhydrous petroleum ether from a methylene chloride solution in 75% yield. A more hydrophobic monomer, methacrylated 1,6-bis(carboxyphenoxy) hexane (MCPH), was synthesized as described elsewhere [23]. A melt condensation at 180°C of equal molar amounts of acetylated 1,6-bis(carboxyphenoxy) hexane and 1,6-bis (carboxyphenoxy) propane produced a linear polyanhydride chain [poly(CPH:CPP)].

The methacrylated monomers were characterized by $^1$H-NMR to confirm the success of methacrylation and, therefore, measure the degree of crosslinking. Methacrylate $= CH_2$ protons were located at $\delta = 5.8$ and $\delta = 6.4$ ppm for MSA and compared with the aliphatic protons to measure the degree of methacrylation. Similarly, the methacrylate protons located at $\delta = 6.0$ and $\delta = 6.5$ ppm for MCPH and MCPP were compared with the aromatic protons.

## METHODS

Photopolymerizations were initiated by 2,2-dimethoxy-2-phenyl acetophenone (DMPA, Ciba Geigy), an ultraviolet initiator, or camphorquinone (CQ, Aldrich) and triethanolamine (TEA, Aldrich), a common visible lightinitiating system used in dental applications. A full-beam ultraviolet light source (EFOS, 100SS) and blue light source (DenMat Marathon Two) were used at light intensities ranging from 20–100 mW/cm². Photocurable solutions were prepared by dissolving the initiating system into dimethacrylated monomer at elevated temperatures, spreading it into a desired mold and bulk polymerizing in the presence of oxygen at room temperature.

Differential scanning photocalorimetry (Perkin-Elmer, DSC7) was used as a tool for determining the rate of polymerization for various initiating schemes. The photopolymerization rate is dependent on the incident light intensity, initiator concentration and efficiency as well as the kinetics governing the radical polymerization. To this extent, the purpose of the DSC studies was to determine initiating conditions that are clinically feasible for *in vivo* polymerizations while minimizing the rate at which heat is released to the local tissue.

DSC studies have shown that multifunctional monomers react quickly to form densely crosslinked networks from liquid monomer solutions. However, even a small amount of unreacted monomer can effectively plasticize a crosslinked network, rendering it more pliable. For this reason, mechanical analysis was combined with DSC studies to characterize the physical changes occurring in the proposed dimethacrylate system as polymerization proceeds. Static compression tests (Perkin-Elmer, DMA7e) were completed on disks ($d = 11.5$ mm, $t = 1.7$) immediately after they were irradiated for varied times.

*In vitro* degradation studies were performed to characterize the degradation mode and kinetics in simulated *in vivo* conditions: phosphate-buffered saline at 37°C with constant orbital shaking (80 rpm). Photopolymerized disks ($d$ = 12 mm and $t$ = 1.4) were degraded, and changes in mass and disk dimensions were measured with degradation time. Release studies were conducted with disks of similar dimensions. Polymer disks were loaded with dextran ($M_w$ ~ 11,000 Da) and rhodamine B base ($M_w$ ~ 443 Da), and the dye release was measured with a fluorimeter.

Matrix-Assisted Laser Desorption/Ionization Time-of-flight Mass Spectrometry (MALDI-TOF MS, PerSeptive Biosystems Voyager-DE STR) was used to characterize the molecular weight distribution and the number average molecular weight ($\overline{M_n}$) of the poly(methacrylic acid) (PMAA) degradation products as described elsewhere [25].

## RESULTS

Methacrylated monomers of varied hydrophobicity and functionality have been synthesized from multi-acids with methacrylic anhydride [23,26]. The adaptability of this synthesis allows for a wide range of possible monomers, as shown in Figure 1, and resulting polymer properties. For example, a dimethacrylated monomer, such as methacrylated sebacic acid, can produce networks with different degradation, mechanics, and polymerization properties compared to a trimethacrylated monomer such as methacrylated citric acid. An increase in the functionality of the monomer can lead to significant increases in the crosslinking density and, therefore, a higher modulus network. In addition, increased functionality has been shown to have a significant effect on the rate of polymerization by increasing the concentration of double bonds [26].

The backbone chemistry can easily be modified to incorporate a wide range of substituents, from hydrophobic aromatic groups to rigid imide units, as seen with methacrylated 1,6-bis(carboxyphenoxy) hexane and methacrylated pyromelltylimidoalanine, respectively. As in the case with varied functionality, the backbone chemistry will have a significant impact on the resulting crosslinked network's mechanical and degradative properties. The incorporation of stiff, imide groups into the backbone of polyanhydride chains by functionalized amino acids increases mechanical and thermal stability without significantly altering the rate and mode of degradation [27]. Although the number of different amino acids (naturally occurring or otherwise) is immense, the resulting combination of polymer structures with imide units incorporated is equally as diverse and, furthermore, functionalized biosequences may be added to promote targeted biological responses. Of further significance is the ability of the hydrophobicity of the

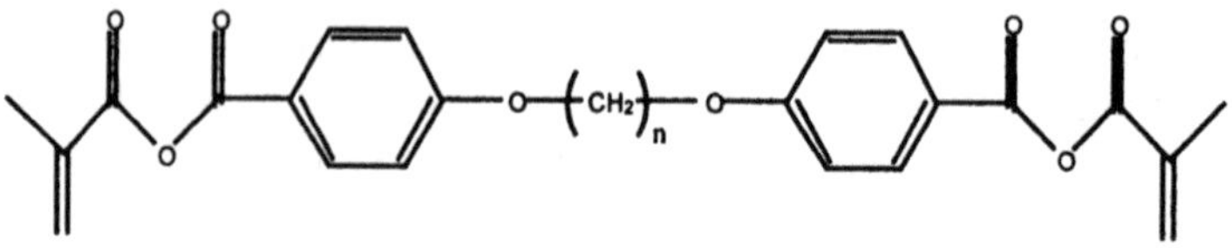

n=1, methacrylated 1,6-bis(carboxyphenoxy) methane (MCPM)
n=3, methacrylated 1,6-bis(carboxyphenoxy) propane (MCPP)
n=6, methacrylated 1,6-bis(carboxyphenoxy) hexane (MCPH)

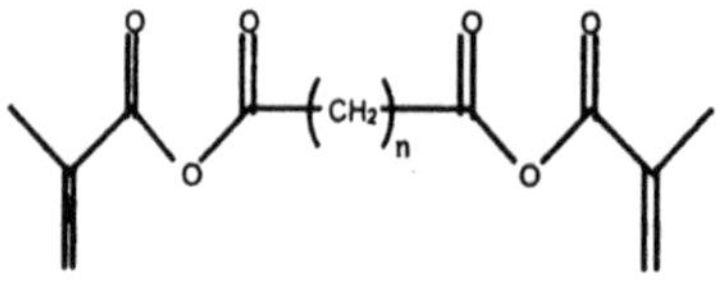

n=8, methacrylated sebacic acid (MSA)
n=10, methacrylated dodecaneoic acid (MDA)

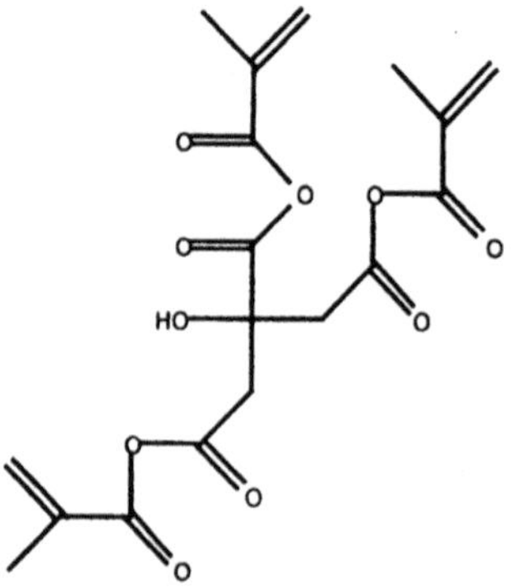

methacrylated citric acid (MCA)

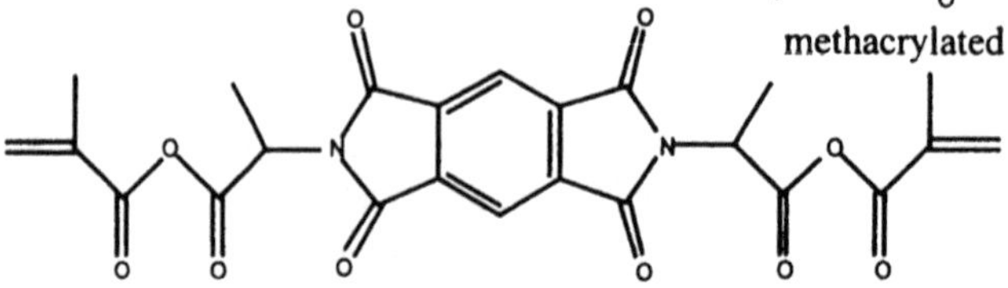

methacrylated pyromellitylimidoalanine (MPMA-ala)

**Figure 1** Methacrylated monomers with varying hydrophobicity, functionality, and stiffness.

backbone chemistry to control the rate of degradation, as will be discussed in the latter portion of this chapter.

Multifunctional monomers react through complex, diffusion limited polymerization mechanisms. Figure 2 shows the complex rate behavior for the photopolymerization of MSA with three different initiation conditions as measured by differential scanning calorimetry. The polymerization undergoes immediate autoacceleration until viscosity limitations restrict the propagation reaction, and autodeceleration leads to diminishingly small rates of polymerization. The restrictions on mobility in crosslinked systems often results in a limiting double bond conversion, which is of significant importance in biomedical applications since unreacted monomer may leach out of the implant and be toxic to nearby cellular material. Also, for reaction of these systems *in vivo*, the polymerization time must be sufficiently fast to remain clinically safe and surgically feasible.

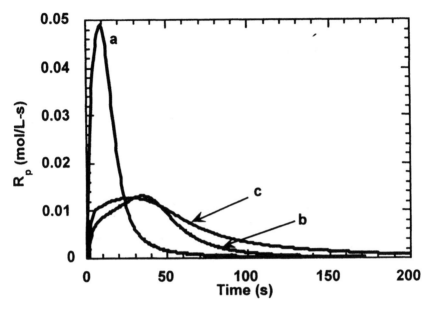

**Figure 2** The rate of polymerization of MSA as a function of time initiated with (a) 1.0 wt% DMPA and 60 mW/cm² ultraviolet light, (b) 0.05 wt% DMPA and 60 mW/cm² ultraviolet light, and (c) 1.0 wt% camphorquinone, 1.0 wt% triethanolamine and 25 mW/cm² visible light.

Illustrated in Figure 2 is that the polymerization of a thin film of MSA is complete in a clinically acceptable timeframe, ranging from seconds to several minutes, depending on the initiating scheme. MSA photopolymerized with 1.0 wt% DMPA and 60 mW/cm² ultraviolet light is nearly completely reacted in less than 50 seconds. Furthermore, a 20-fold decrease in initiator concentration results in a slower polymerization, but the reaction is still complete in approximately 100 seconds. Visible light initiating systems are commonly used in dental applications for curing dental composites and provide a safer, more desirable alternative to using ultraviolet radiation in photopolymerizations. The rate of polymerization of MSA initiated with 1.0 wt% CQ and 1.0 wt% TEA and 25 mW/cm² blue light is also shown in Figure 2. Although the polymerization proceeds slower, due to the decreased initiator efficiency and lower light intensity, the polymerization is still complete in 150 seconds, demonstrating the possible success of using visible light to cure *in vivo*.

Furthermore, the wide range of polymerization rates is controlled by the photoinitiation conditions. Specifically, the initiator concentration and incident light intensity control the rate of polymerization and, therefore, the rate of heat released upon curing. These conditions can be conveniently altered for *in vivo* applications to minimize local tissue necrosis from the

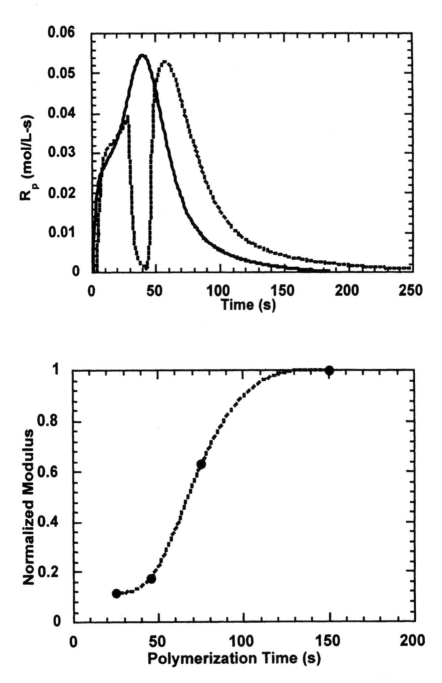

**Figure 3** Polymeriztion of 50/50 w/w methecrylic anhydride initiated with 0.5 wt% DMPA and 80 mW/cm² of ultraviolet light (a) Illustration of the temporal control of the photopolymerization. Initiation is ceased at ~30 s and reinitiated at ~40 s (dashed line) compared to continuous initiation (solid line). (b) Evaluation of the compressive modulus as a function of exposure time.

polymerization exotherm. In addition, the polymerization time can be adapted to minimize the patient's time and money spent in a surgical facility.

Photopolymerizations have further advantages besides fast curing rates at ambient temperatures, including spatial and temporal control of the polymerization process. Specifically, the polymerization can be turned on and off by shuttering the initiating light source. Figure 3(a) illustrates the rapid control of the polymerization rate for a system of MSA (50/50 w/w MSA/methacrylic anhydride) polymerized with 0.5 wt% DMPA and 80 mW/cm² ultraviolet light. The polymerization proceeds until the light is shuttered (after approximately 30 seconds), after which the rate of polymerization immediately drops to zero. The polymerization can be "turned on" again with further irradiation, as shown at approximately 40 seconds, to complete the reaction.

This reliable control of the curing process allows for stopping the polymerization after gelation, when the system has a fixed three-dimensional structure but is still rubbery and contourable. One potential application for employing the temporal control of the photopolymerization process would be in designing contourable fraction fixation plates. Figure 3(b) are the

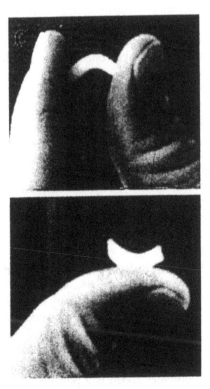

**Figure 3 (continued)** (c) Partially cured, flexible and contourable plate. (d) Fully cured, glassy and rigid plate.

normalized compressive modulus as a function of polymerization time, which can be compared to Figure 3(a) at 30 seconds of irradiation time to demonstrate the rubbery nature of the polymer with incomplete conversion. With a partially cured system, the unreacted monomer plasticizes the network and reduces its strength to only 20% of its final value. At this point, the rubbery device can be contoured [as in Figure 3(c)], for example to a bone defect and, with further irradiation, can form a rigid polymer Figure 3(d) with a compressive modulus more than an order of magnitude higher than the partially cured network.

Although some polymers used in medical applications are designed to be inert and resistant to degradation (e.g., artificial prostheses), the proposed multimethacrylate anhydride monomers react readily to form high-strength networks that are designed to degrade hydrolytically with predictable kinetics. In particular, the high crosslinking density and hydrophobic backbone substituents act to resist solvent uptake from the environment, whereas the anhydride linkages are readily hydrolyzed. The result is a dimensionally stable polymer in which the anhydride linkages are cleaved at the surface only, producing a controlled surface erosion mechanism. The rate of hydrolytic degradation is controlled by the hydrophobicity of the backbone chemistry and, therefore, can be conveniently controlled from one application to another.

Disks were fabricated to simulate one-dimensional degradation for several polyanhydride compositions and the linear profiles for mass loss shown in Figure 4 that confirm the surface erosion mechanism. In addition, Figure 4 illustrates how increased hydrophobicity in the backbone chemistry can dramatically decrease the rate of degradation. A disk polymerized from an aliphatic monomer, MSA, was 100% degraded in approximately 50 hours, whereas the aromaticity of MCPH increased the hydrophobicity of the network and extended the lifetime of the disk by two orders of magnitude. A disk of poly(MCPH) was only 30% degraded after 90 days and would completely erode in approximately 1 year, assuming surface erosion conditions are maintained. Furthermore, incorporation of hydrophobic monomers into the network has been used to tailor the degradation rate of an implant to a particular application [24].

Although copolymerizing hydrophobic monomers into the network can effectively slow the degradation rate, the addition of hydrophobic, inert linear polymers is another mode for controlling the rate of network degradation. Semi-IPNs (semi-interpenetrating networks) were fabricated by polymerizing multifunctional monomers in the presence of linear polymer chains. The result was a crosslinked network with linear chains tangled within, but not chemically bound. Furthermore, the system remains surface eroding if the linear polymer added is a polyanhydride. The degradation profile for a semi-IPN formulated from 50/50 w/w MSA/poly(CPP:CPH)

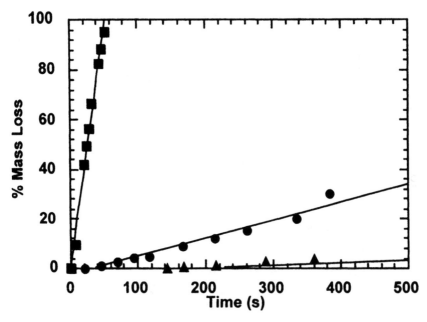

**Figure 4** One-dimensional degradation profiles for disks ($d = 12$ mm, $t = 1.4$ mm) of poly(MSA) (■), poly(MCPH) (▲), and a semi-IPN fabricated from 50/50 w/w MSA/poly(CPP:CPH) (●).

is plotted in Figure 4. Whereas a disk polymerized of only MSA is completely eroded in 2 days, a disk of similar dimensions but with a linear, hydrophobic polyanhydride added is less than 2% degraded at this time. The semi-IPN shown would be 100% degraded in approximately 56 days assuming surface erosion is maintained. In addition to altering the degradation rate, the linear polymers reduce the exothermic temperature rise and minimize the volume shrinkage upon polymerization. Semi-IPNs could also be used to multimode degradation devices. For example, a very hydrolytically unstable polymer that quickly degrades within a slower eroding system could leave pores for new bone ingrowth and cell seeding.

Clearly the degradation rate of the proposed networks can be easily controlled by the addition of hydrophobic monomers or polymers, and the kinetics of surface degradation can be reliably characterized and predicted for more complex geometries based on the one-dimensional kinetics. This is of considerable consequence for predicting the release of bone morphogenetic proteins or antibiotics from a fabricated orthopedic screw, pin, or plate to an injured bone. The one-dimensional release of dextran ($M_w \sim 11{,}000$ Da) and rhodamine ($M_w \sim 443$ Da) from a homopolymer disk of poly(MSA) is shown in Figure 5. The release profiles are consistent with the rate of degradation for poly(MSA), illustrating that the release molecules are not

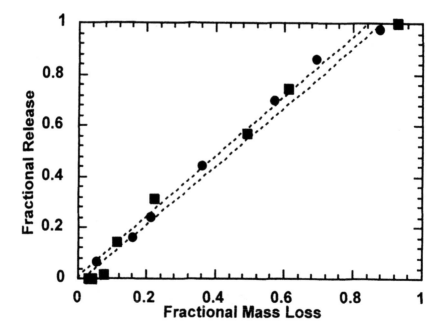

**Figure 5** Fractional release of dextran (■) and rhodamine (●) from a disk of poly(MSA) as a function of the polymer mass loss.

diffusing out of the bulk of the implant, but rather released from the surface as the polymer degrades. Furthermore, the molecular weight of dextran used is almost 25 times as big as that of rhodamine. Therefore, in this molecular weight range, there is no dependence of the size of the release molecule on the release rate. Thus, a wide range of large biomolecules or small antibiotics can be delivered predictably from polyanhydride implants.

As the anhydride crosslinks are broken at the surface and drug is released, the degradation products must also diffuse from the implant site and may compromise the biocompatibility of the material or the effectiveness of the drug. As with the FDA approved linear polyanhydrides, one degradation product is the multi-acid of the backbone substituent. However, crosslinked networks formed from methacrylated monomers also produce linear, hydrophilic poly(methacrylic acid) (PMAA) chains as a degradation product. Since most hydrophilic linear polymers can be resorbed and excreted if under 200,000 Da [22], MALDI-TOF mass spectrometry was used to characterize the molecular weight distribution of the PMAA chains. Figure 6 shows the molecular weight distribution of PMAA degradation products for a disk of poly(MSA) that was cured to 80% conversion with 100 mW/cm2 ultraviolet light and subsequently degraded. Although we have previously shown that the distribution of PMAA chains is dependent on the photopoly-

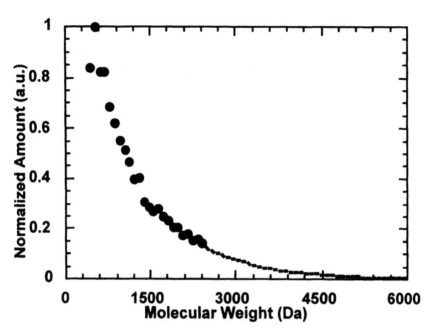

**Figure 6** MALDI-TOF distribution of PMAA chains as a function of molecular weight for the degradation products of a disk of poly(MSA) polymerized with 100 mW/cm² ultraviolet light.

merizations conditions (e.g., rate of initiation and polymerization, conversion of functional groups) [25], Figure 6 is a representative MALDI-TOF spectrum that confirms that the molecular weights of these chains are sufficiently small enough to be processed in the body. For example, the number average molecular weight ($M_n$) as measured by MALDI-TOF for the relatively narrow distribution of degradation products (Q = 1.5) was approximately 1,460 Da. Furthermore, proton NMR was used to confirm the MALDI-TOF results and measured $\overline{M_n}$ to be between 1,120 Da and 2,320 Da for termination by disproportionation and combination, respectively.

## CONCLUSIONS

Multifunctional methacrylate anhydride monomers were investigated that photopolymerize to high-strength and hydrolytically degradable polymers. The rationale for the developed polyanhydrides was driven by several, critical factors. First, the high concentration of methacrylate double bonds (two or more per monomer) leads to very fast reaction rates at room temperature and high-strength, dimensionally stable polymers. In addition, photopolymerizations are temporally and spatially controllable, benefiting the clinical application of these materials by offering advantages such as *in vivo* curing and contourable orthopedic devices from partially cured monomers. Finally, the hydrophobic backbone substituents combined with the hydrolytically labile anhydride linkages leads to a surface erosion mechanism in which the degradation rate can be reliably controlled by the incorporation of hydrophobic monomers or polymers into the network. Release of drugs or agents from these polyanhydride networks depends only on the rate of degradation and not on the size of the agent released. Furthermore, MALDI-TOF confirms that the distribution of PMAA degradation products was narrow and at sufficiently low molecular weights so that the body could process them.

## REFERENCES

1. Singh, M., Saxena, B. B., Graver, R., and Ledger, W. J. *Fertil. Steril.,* 1989, 52, 973.
2. Schakenraad, J. M., Oosterbaan, J. A., Nieuwenhuis, P. and Molenaar, I.; Olijslager, J., Potman, W., Eenink, M. J., Feijen, J. *Biomaterials,* 1988, 9, 116.
3. Miyamoto, S., Takaoka, K., Okada, T., Yoshikawa, H., Hashimoto, J., Suzuki, S. and Ono, K. *Clin. Orthop.,* 1992, 278, 274.
4. Camarata, P. J., Suryanarayanan, R., Turner, D. A., Parker, R. G. and Ebner, T. J. *Neurosurgery,* 1992, 30, 313.
5. Eldridge, J. H., Hammond, C. J., Meulbroek, J. A., Staas, J. K., Gilley, R. M. and Tice, T. R. *J. Control. Rel.,* 1990, 11, 205.
6. Singh, M., Singh, A. and Talwar, G. P. *Pharm. Res.,* 1991, 8, 958.
7. Duysen, E. G., Whiman, S. L., Krinick, N. L., Fujita, S. M. and Yewey, G. L. *Pharm. Res.,* 1994, 11, 88.
8. Furr, B. J. A. and Hutchinson, F. G. *J. Control. Rel.,* 1992, 21, 117.
9. Tice, T. R. and Tabibi, S. E. *Treatise on Controlled Drug Delivery,* Marcel Dekker, New York, 1992.
10. Bos, R. R. M., Rozema, F. R., Boering, G., Nijenhuis, A. J., Pennings, A. J., Verwey, A. B., Nieuwenhuis, P. and Jansen, H. W. B. *Biomaterials,* 1991, 12, 32–36.
11. Migliaresi, C., Fambri, L. and Cohn, D. *J. Biomater. Sci. Polymer Edn.,* 1994, 5, 591–606.
12. Braun, A. G., Buckner, C. A., Emerson, D. J. and Nichinson, B. B. Quantitative correspondence between the *in vivo* and *in vitro* activity of teratogenic agents, *USA,* 1982, 79, 2056.
13. Leong, K. W., D'Amore, P., Marletta, M. and Langer, R. *J. Biomed. Mat. Res.,* 1986, 20, 51–64.
14. Brem, H., Piantadosi, S., Burger, P. C., Walker, M., Selker, R., Vick, N. A., Black, K., Sisti, M., Brem, S., Mohr, G., Muller, P., Morawetz, R. and Schold, S. C. *Lancet* 1995, 345, 1008–1012.
15. Lucas, P. A., Laurencin, C., Syftestad, G. T., Domb, A., Goldberg, V. M., Caplan, A. I. and Langer, R. *J. Biomed. Mat. Res.,* 1990, 24, 901–911.

16. Brem, H., Mahaley, M. S. J., Vick, N. A., Black, K. L., Schold, S. C. J., Burger, P. C., Friedman, A. H., Ciric, I. S., Eller, T. W. and Cozzens, J. W. *J. Neurosurg.*, 1991, 74, 441.
17. Chasin, M., Lewis, D. and Langer, R. *Biopharm. Manufa.*, 1988, 1, 33.
18. Grossman, S. A., Reinhard, C., Colvin, O. M., Chasin, M., Brundrett, R., Tamargo, R. J. and Brem, H. *J. Neurosurg.*, 1992, 76, 640.
19. Mathiowitz, E. and Langer, R. *J. Controlled Release* 1987, 5, 13.
20. Anseth, K. S., Newman, S. M. and Bowman, C. N. Polymeric dental composites: Properties and reaction behavior of multimethacrylate dental restorations. In Peppas, N. A. Langer, R. S. (Eds.), *Biopolymers III, Advances in Polymer Science*, 1995, 122, 177–217.
21. Kohn, D. H. and Ducheyne, P. *Materials Science and Technology;* VCH, New York, 1992, Vol. 14.
22. Murakami, Y., Tabata, Y. and Ikada, Y. *Drug Delivery,* 1996, 3, 231–238.
23. Svaldi-Muggli, D., Burkoth, A. K., Keyser, S. A., Lee, H. R. and Anseth, K. S. *Macromolecules*, 1998, 31, 4120–4125.
24. Svaldi Muggli, D., Burkoth, A. K. and Anseth, K. S. *J. Biomed Mat. Res.*, 1998, 46, 271–278.
25. Burkoth, A. K. and Anseth, K. S. *Macromolecules*, 1998, 32(5), 1438–1444.
26. Young, J. *Fundamentals of Crosslinking Photopolymerizations and Applications to Biomedical Systems*, University of Colorado, Boulder, 1998, 209.
27. Uhrich, E. K., Gupta, A., Thomas, T. T., Laurencin, C. T. and Langer, R. *Macromolecules*, 1995, 28, 2184–2193.

# CHAPTER 14

# Novel Cytokine-Inducing Macromolecular Glycolipids from Gram-Positive Bacteria

MASAHITO HASHIMOTO[1], YASUO SUDA[1], YOSHIMASA IMAMURA[1]
JUN-ICHI YASUOKA[1], KAZUE AOYAMA[2], TOSHIHIDE TAMURA[2]
SHOZO KOTANI[2], SHOICHI KUSUMOTO[1]

## INTRODUCTION

L IPOTEICHOIC acid (LTA) is a macroamphiphile widely distributed on the cell surface of Gram-positive bacteria. The structures of LTAs from various enterococcal and streptococcal species were studied by Fischer [1]. They proposed that LTA is composed of a glycolipid anchor and a 1,3-linked poly(glycerophosphate) chain [Figure 1 (1a and 2a)] and the latter is substituted with oligo-glucose or D-alanine at the 2-positions of the glycerols. In the 1980s, immunostimulating activities (cytokine-inducing and antitumor activities) were reported for several LTAs [2–4]. To find the chemical entity responsible for the observed activity of LTA, Fukase et al. [5,6] synthesized the fundamental structures proposed for LTAs of *Enterococcus hirae* [Figure 1 (1b and 2b)]. However, neither cytokine-inducing nor antitumor activity was observed in the synthetic compounds [7]. The structure of the synthetic analogues had several minor differences from those proposed for natural LTA: (1) the glycolipid anchor in natural LTA contains unsaturated fatty acids, whereas the synthetic ones contain only saturated acids; (2) the poly(glycerophosphate) in natural LTA was assumed to have between 9 and 40 repeating unit, whereas the synthetic ones contain only 4 units; (3) the synthetic poly(glycerophosphate) part had no substituent in contrast to the high glucosyl and appreciable alaninyl substitution existent in the natural LTA.

[1]Department of Chemistry, Graduate School of Science, Osaka University, 1-1, Machikaneyama, Toyonaka, Osaka 560-0043, Japan.
[2]Department of Bacteriology, Hyogo College of Medicine, 1-1, Mukogawa, Nishinomiya, Hyogo 663-8501, Japan.

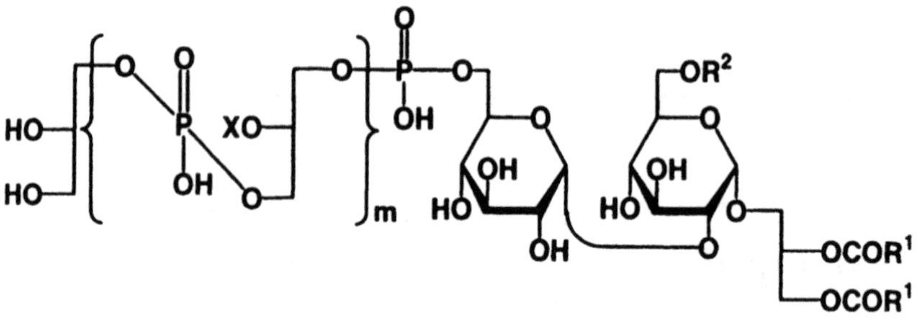

a:   $R^1 = C_{15}H_{31}$, $C_{17}H_{33}$ etc.    b:   $R^1 = C_{15}H_{31}$

n = 8 - 39                                      n = 3

X = H, Ala, or $Glc_n$                          X = H

**1a, 1b:**   $R^2 = H$

**Figure 1** Structures proposed for enterococcal and streptococcal LTAs (1a and 2a) and compounds synthesized according to their fundamental structures (1b and 2b).

The lack of cytokine-inducing activity in the synthetic compounds may thus be attributable to either of the following two possibilities: (1) some or all of the above structural differences are essential for the biological activities of LTA; (2) unknown components present in LTA fraction are responsible for the observed activity, and so-called LTA is not. We have succeeded in separating LTA fraction from *E. hirae* ATCC 9790 into small amounts of cytokine-inducing active macromolecular glycolipids and an inactive major fraction that amounts to over 90 wt% [8]; this favors the latter possibility rather than the former. In this chapter, we review our work on the structural characterization of the major inactive and the minor cytokine-inducing glycolipids.

## MATERIAL AND METHODS

### ISOLATION OF CYTOKINE-INDUCING GLYCOLIPIDS

The cytokine-inducing active fraction (QM-A) was prepared from *E. hirae* ATCC 9790 cells as described [9]. Briefly, the bacterial cells (922 g) were delipidated and extracted with hot phenol-water, and the crude extract was digested with RNase and DNase, to give crude LTA fractions. The crude LTA was subjected to two successive chromatography, hydrophobic interaction chromatography on Octyl-Sepharose CL-4B (Pharmacia, Uppsala, Sweden) by batchwise elution and then ion-exchange membrane chromatography on QMA-Mem Sep 1010 (PerSeptive Biosystems, Framingham, MA, USA) by stepwise elution, to give low-anionic fraction (QM-A, 223 mg) and high-anionic fraction (QM-I, 8 g).

Fractionation of QM-A on Octyl-Sepharose was performed according to the method of Suda et al. [8]. Five cytokine-inducing fractions—OS-1 (19 mg), OS-2 (2.8 mg), OS-3 (8.3 mg), OS-4 (101 mg), and OS-5 (33 mg)— were obtained. Further fractionation of the main fraction OS-4L was performed by QMA-Mem Sep 1010 using 0.01 M acetate buffer (pH 4.5) with a liner gradient of NaCl (0–1 M) to give 11.3 mg of a low-anionic cytokine-inducing glycolipid, named GL4.

## ISOLATION OF CYTOKINE-INDUCING GLYCOLIPIDS BY THE MODIFIED METHOD

The crude extract from 428 g of bacterial cells was digested with RNase and DNase and re-extracted with hot phenol-water, to give a crude glycolipid fraction (20.5 g). This glycolipid fraction was subjected to an ion-exchange column on Macro-Prep High Q Support (Bio-Rad, Hercules, CA, USA) using 0.01 M acetate buffer (pH 4.5) with stepwise NaCl concentration (0, 0.05, 0.1, and 1 M). Two low-anionic cytokine-inducing fractions, HQ-A (0 M NaCl fraction, 1.2 g) and HQ-B (0.05 M NaCl fraction, 1.2 g), were obtained. The fractionation of HQ-A was performed on two successive Octyl-Sepharose columns using 0.1 M formate buffer (pH 4.5) with a liner gradient of 1-propanol; the hydrophobic fraction (28.8 mg) was then subjected to a gel filtration column on Sephacryl HR-200 HR (Pharmacia) using 0.2 M formate buffer (pH 4.5) containing 30% 1-propanol to give 16.5 mg of a cytokine-inducing glycolipid HGL-A. HQ-B was subjected to an Octyl-Sepharose column, the hydrophobic fraction (26 mg), and then subjected to a Sephacryl column, to give two bioactive glycolipids, 17.1 mg of HGL-B1 and 4.3 mg of HGL-B2.

## BIOLOGICAL ASSAYS

IL-6 induction in human peripheral whole-blood cell culture [10,11] and determination of its level using ELISA and *Limulus* activity measured by means of Endospecy Test$^{®}$ (Seikagaku Co., Tokyo, Japan) were performed as described [8]. IL-6 induction in THP-1 cells was performed as reported [12]. In brief, sample solution (25 μL) and 100 μL of THP-1 cells ($1.0 \times 10^6$ cells/mL, preincubation with 50 ng/mL vitamin $D_3$ at 37°C in $CO_2$ for 72 h) were incubated in RPMI-1640 medium with or without 2% fetal calf serum

(FCS) at 37°C in $CO_2$ for 24 h. In these assays, LPS from *Escherichia coli* O111:B4 (Sigma Co., St. Louis, MO, USA) was used as a positive control.

## ANALYTICAL PROCEDURES

Phosphorus, fatty acids, carbohydrates, glycerol, and amino acids were analyzed by the method described in our previous paper [8] and references cited therein. SDS-PAGE [8], TLC [9], HPLC [9], determination of phosphomonoester [8], reducing sugar analysis [13], methylation analysis [14], and hexose analysis [15] were performed as described in the respective literature. Two dimensional TLC was performed on silica-gel plate (Merck Silicagel 60 $F_{254}$ No. 5715) using the solvent systems, chloroform-methanol-acetic acid (65/10/1, v/v/v) for the first development and chloroform-methanol-25% ammonia solution (65/10/1) for the second.

## DEGRADATION PROCEDURES

Deacylation [16], hydrolysis with 47% aqueous hydrofluoric acid (HF) and phase partition of the reaction products [1], and acetolysis [17] were performed as described in the literature.

## SPECTRA

$^1$H-NMR spectra were measured on a JMN-LA500 (JEOL, Tokyo, Japan) at 500 MHz, a UNITY plus (Varian, Palo Alto, CA, USA) at 600 MHz, or a UNITY INOVA (Varian) at 750 MHz. $^{13}$C-NMR spectra were measured on a JMN-LA500 at 126 MHz and a UNITY plus at 151 MHz. FAB-MS was obtained with an SX-102 (JEOL) using *m*-nitrobenzyl alcohol as a matrix. MALDI-TOF-MS was obtained with a Kratos Kompact MALDI IV (Shimadzu, Kyoto, Japan) or a VOYAGER-ELITE-DE (PerSeptive Biosystems Inc., Framingham MA, USA) using 2,5-dihydroxybenzoic acid as a matrix. ESI-MS was obtained with an API III (PE SCIEX, Thornhill, Ontario, Canada) or Mariner® (PerSeptive Biosystems Inc.). ESI-CID-MS/MS was obtained with API-III, and FAB-CID-MS/MS was obtained with JMS-HX110/HX110 (JEOL).

## RESULTS AND DISCUSSION

### FRACTIONATION OF LTA FRACTION

To elucidate the structure of these complicated macromolecular glycolipid molecules, isolation of large amounts of the glycolipids is inevitably required. Separation of bioactive glycolipids was performed according to previous methods [8] with modification (Scheme 1). LTA fraction was prepared by the method of Fischer [1]. Further fractionation of the LTA fraction was performed by ion-exchange membrane chromatography on QMA-Mem Sep 1010. Similar to our previous results, a high anionic major

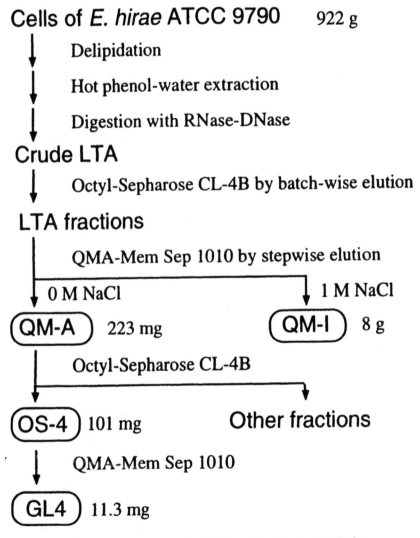

Scheme 1 Separation procedure for the cytokine-inducing glycolipids.

fraction (QM-I) and a low anionic minor fraction (QM-A) were obtained. IL-6 inducing activity from human peripheral whole blood cells was found only in QM-A, but not in QM-I [Figure 2(a)]. The bioactive fraction QM-A was further subjected to Octyl-Sepharose, to give five fractions, OS-1 to OS-5. We focused on the major fraction, OS-4, which was again subjected to QMA Mem Sep, to separate low anionic cytokine-inducing glycolipid (GL4) from high anionic inactive fractions, OS-4H [Figure 2(b)]; 11.3 mg was separated from 922 g of dried bacterial cells. Although the *Limulus* activity GL4 exhibited was similar to that of OS-4H, the IL-6 inducing activity of GL4 was at least 30 times greater than that of OS-4H. It should be emphasized that IL-6 inducing potency of GL4 was more than 150 times as high as the value expected for the identical amount of LPS, which would give the same level of *Limulus* response as observed with GL4. This fact indicated that GL4 itself was responsible for the IL-6 inducing activity.

STRUCTURE OF QM-I [9]

Since bio-inactive glycolipid QM-I was the major (>90 wt%) constituent of the LTA fraction, the structure of QM-I was considered to be close to that of LTA. To clarify that "so-called" LTA is not responsible for cytokine-inducing activity, the structure of QM-I was first elucidated and compared with that of LTA. QM-I was the macromolecular ($1.2 - 1.6 \times 10^4$, estimated by SDS-PAGE) glycolipid, which consisted of glucose (Glc), glycerol (Gro), phosphate, fatty acids (FAs, mainly hexadecanoic and octadecenoic acid), and amino acids (mainly alanine) (Table 1). The direct structural analysis of intact QM-I with NMR was impossible because of the broad signals due to its large molecular features. To reduce the molecule size, QM-I was subjected to HF-hydrolysis to cleave the phosphodiester linkages and the products were partitioned with $CHCl_3$-MeOH-$H_2O$. The hydrophobic products of HF-hydrolysate were isolated by TLC and analyzed by their FAB-MS and NMR, glycolipid anchor, Glc($\alpha$1-2)Glc($\alpha$1-3)acyl$_2$Gro, was mainly found. Acyl groups were mainly composed of hexadecanoate and octadecenoate. In the hydrophilic products of HF-hydrolysate, the ion peaks corresponded to Glc$_{1-5}$Gro and AlaGlc$_{1-4}$Gro, and their phosphorylated derivatives were observed by ESI-MS with API-III in the positive and negative ion mode, respectively. They were dephosphorylated by alkaline phosphatase and then acetylated. The fully acetylated products were separated and characterized on the basis of NMR as [Glc($\alpha$1-2)]$_n$Gro ($n = 0 \sim 4$). The mode of connection among the HF-hydrolysate were elucidated by the $^{13}$C-NMR spectra of deacylated QM-I. The coupling constants corresponding to a value expected for $J_{c,p}$ were observed at the signals of Glc$^{nr}$ C-5,6(5- and 6-positions in the nonreducing side glucose of the glycolipid anchor) and Gro C-1,3 (1- and 3-positions in the glycerol), indicating the phospho-

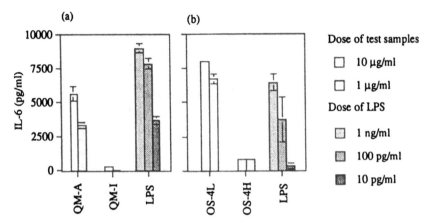

**Figure 2** IL-6-inducing activity of QM fractions (a) and GL4 and OS-4H (b) in human peripheral whole-blood cells. The data present one of several independent experiments with similar results. The IL-6 induction by un-stimulated cells was less than 50 pg/mL. Each bar represents the means ± SD. The blood donor was JY.

diester linkages among them. From the above results, the structure of QM-I was deduced to be as shown in Figure 3, except for the location of the alanyl substitution and the stereochemistry of the glycerols. Consequently, the structure coincided with the one proposed by Fischer's group for a LTA 1a isolated from the same bacterial species. This result clearly proves that the "so-called" LTA is not responsible for the cytokine-inducing activity, but the minor components co-existing in the LTA fraction is essential.

## FUNDAMENTAL STRUCTURE OF GL4

The structure of one of the cytokine-inducing glycolipids, GL4, was then studied [18]. GL4 consisted of phosphate, mannose (Man), Glc, Gro, and FAs (Table 1). MALDI-TOF-MS showed several distinct peaks around

TABLE 1. Chemical Composition of the Glycolipids.

| QM-I | μmol/mg | | | | |
|---|---|---|---|---|---|
| | GL4 | HGL-A | HGL-B1 | HGL-B2 | |
| Phosphate | 1.8 | 0.6 | 0.8 | 1.6 | 1.6 |
| Glucose | 2.6 | 0.5 | 2.2 | 1.8 | 1.8 |
| Mannose | 0 | 2.9 | 0 | 0 | 0 |
| Fatty acids | 0.3 | 0.4 | 0.2 | 0.4 | 0.2 |
| Glycerol | 2.1 | 0.8 | 0.6 | 2.2 | 1.5 |
| Amino acids | 0.3 | 0.3 | 0.04 | 0.3 | — |

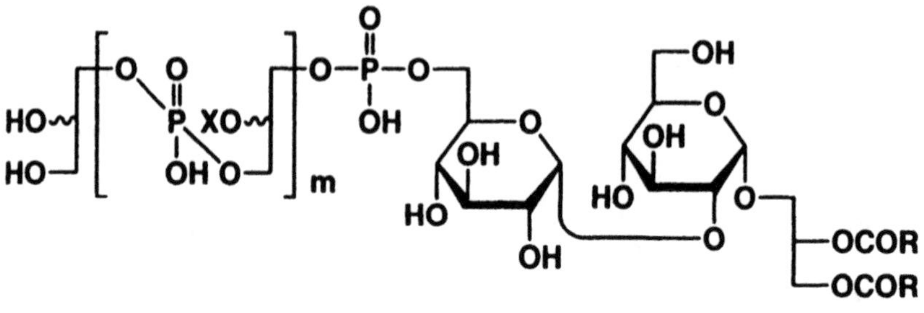

$R = C_{15}H_{31}, C_{17}H_{33}$, etc.

$X = H, Glc_n,$ or $AlaGlc_n$

**Figure 3** The structure of QM-I.

m/z 7,000, indicating that GL4 is a macromolecular glycolipid that contains congeners with different numbers of repeating units and carbohydrate components as well as diversity of fatty acids. Since the presence of phosphodiester was indicated from the composition analysis, GL4 was subjected to HF-hydrolysis. The hydrophilic products in HF-hydrolysate was further separated by gel filtration chromatography on Sephadex G-100 to give low molecular weight products and polysaccharide. The low molecular weight products mainly consisted of Gro and phosphate (molar ratio was ca. 1 : 1) confirmed the presence of a glycerophosphate structure in GL4. Because methylation analysis of the polysaccharide indicated the branched mannan containing 1-2 and 1-6 linkages, the polysaccharide was further subjected to the selective acetolysis to cleave the Man(1-6)Man linkages. The ion peaks corresponded to fully acetylated oligohexose were found in the acetolysis products by MALDI-TOF-MS. They were isolated and characterized on the basis of NMR as Man, Man($\alpha$1-2)Man, Man($\alpha$1-2) Man($\alpha$1-2)Man, and Man($\alpha$1-3)Man($\alpha$1-2)Man($\alpha$1-2)Man. Since these oligomannans were obtained by selective cleavage [17], they were assumed to be connected to each other via Man(1-6)Man linkages. It was impossible to determine the location of the glucose substitution in the polysaccharide because no oligosaccharide containing glucose was detected in the acetolysis products. The molar ratio of each subunit (oligomannan) was calculated from the peak area of anomeric protons in $^1$H NMR of the original polysaccharide. From the above results and further methylation analysis data (data not shown), it was concluded that the hydrophilic region of GL4 consists of a mannose-rich highly branching polysaccharide core linked with glycerophosphate via phosphodiester linkage (Figure 4).

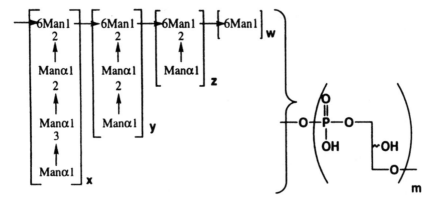

**Figure 4** A fundamental structure proposed for the hydrophilic part of GL4. The sequence of each substructure (oligo-mannan) is interchangeable.

The structure of mannose-rich polysaccharide core in GL4 is close to that of yeast mannan (from *Saccharomyces cerevisiae*), which was inactive for IL-6 induction in a human peripheral whole-blood cells test system. This fact suggests that not the mannose moieties but other components, such as the lipophilic moiety and/or phosphates, are important for the activity. The lipophilic products in HF-hydrolysate of GL4 were then analyzed. In addition to peaks corresponding to the known fatty acids (C16:0, C18:1), two other unknown ion peaks at m/z 330 and 356 were found by FAB-MS (data not shown).

### CYTOKINE-INDUCING OTHER GLYCOLIPIDS OBTAINED BY THE MODIFIED SEPARATION METHOD

To obtain other cytokine-inducing glycolipid in high yield, the separation method was again modified (outlined in Scheme 2). Since it became evident that the structure and the properties of the bioactive glycolipids was totally different from those of inactive LTA as described above, the fractionation of LTA was not necessary and the crude glycolipid fraction from bacteria cells was directly subjected to ion-exchange chromatography to obtain low anionic fractions (HQ-A and -B) and high anionic fractions. IL-6 inducing activity was found only in the low-anionic fraction, HQ-A and -B but not in the high-anionic fractions (data not shown). These results are in good agreement to those of our previous results [8] and one described above. HQ-A and -B were further subjected to Octyl-Sepharose followed by gel filtration chromatography to give three cytokine-inducing glycolipids, HGL-A, -B1, and -B2. The total weight of 38 mg, which was

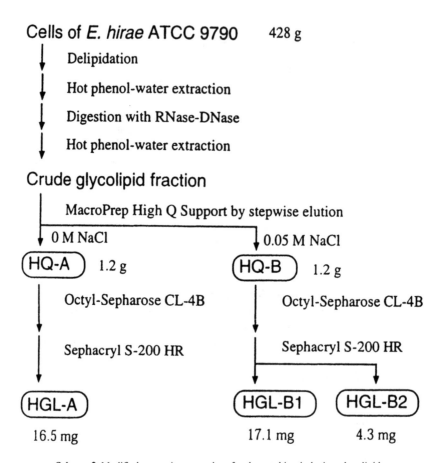

**Scheme 2** Modified separation procedure for the cytokine-inducing glycolipids.

at about five times the yield of the glycolipids separated as above, was separated from 428 g of dried bacterial cells.

Cytokine induction of HGL-A, -B1, and -B2 were performed not only in a human peripheral whole-blood cells test system, but also in human monocytic leukemia cells, THP-1. In THP-1 system, the interleukin-6 (IL-6)-inducing activities of these glycolipids were almost the same as that of LPS, and they were suppressed by the addition of serum (Figure 5). In contrast, the IL-6-inducing activities of LPS was enhanced, probably because of the effect of LPS binding protein that exists in the serum [19,20]. This distinct difference indicates that the activation mechanism of the glycolipids (HGL-A, -B1, and -B2) may be totally different from that of LPS, and the activities of the glycolipids are inherent in themselves: the presence of contaminating LPS is unequivocally excluded.

(a)

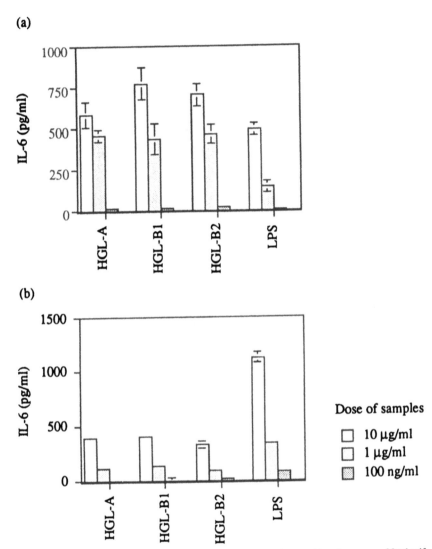

(b)

**Figure 5** IL-6-inducing activity of HGLs in THP-1 cells in the absence (a) and presence of fetal calf serum (FCS) (b). The data present one of several independent experiments with similar results. The IL-6 induction by unstimulated cells was less than 50 pg/mL. Each bar represents the means ± SD.

## PRESENCE OF COMMON LIPOPHILIC BUILDING BLOCKS

HGL-A, -B1, and -B2 consisted of Glc, Gro, phosphate, and FAs, but no Man (Table 1). Although the constituents of the components was quite different from those of GL4, the IL-6-inducing activity of the these glycolipids was almost the same as that of GL4. This suggests that the whole

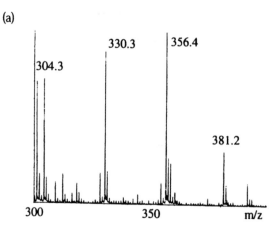

(a)

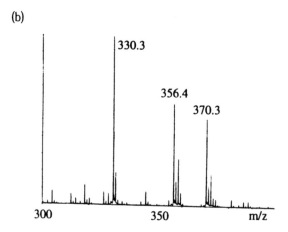

(b)

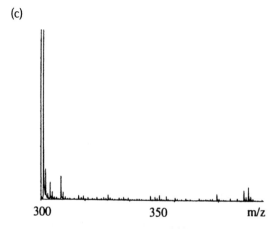

(c)

**Figure 6** ESI-MS of the lipophilic products from the HF-hydrolysis of HGL-A (a), GL4 (b), and QM-I (c) with Mariner® A 1 μg/mL solution (in $CH_2Cl_2$/MeOH 1/1, v/v) was continuously infused at 0.1 mL/h.

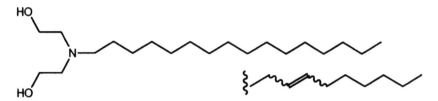

**Figure 7** The structures of the lipophilic amines commonly existent in the cytokine-inducing glycolipids from *E. hirae.*

structure is not necessary for activity, but some common structural essence(s) do exist in all the bioactive components. The HF-hydrolysis products from HGL-A were analyzed. The ion peaks at m/z 330 and 356 were found in the lipophilic products in HF-hydrolysate by ESI-MS (Figure 6). These peaks were also found in HGL-B and GL4, but not in QM-I.

The hydrophobic products were isolated by TLC and characterized by means of NMR, ESI-CID-MS/MS, and FAB-CID-MS/MS as lipophilic long-chain alkyl diethanol amines, which are shown in Figure 7. In the literature these unusual lipophilic amines were not detected as constituents of any bacterial cells. Although the mode of linkage in the whole glycolipid is not clear at this moment, we speculate that these lipophilic amines that commonly existent in the cytokine-inducing glycolipids from *E. hirae* are essential moieties for the cytokine-inducing activity.

## REFERENCES

1. Fischer, W. *Glycolipids, Phosphoglycolipids and Sulfoglycolipids,* Kates, M. ed., 123, Plenum Press, New York, 1990.
2. Yamamoto, A., Usami, H., Nagamuta, M., Sugawara, Y., Hamada, S., Yamamoto, T., Kato, K., Kokeguchi, S. and Kotani, S. *Br. J. Cancer,* 1985, 51, 739.
3. Usami, H., Yamamoto, A., Yamashita, W., Sugawara, Y., Hamada, S., Yamamoto, T., Kato, K., Kokeguchi, S., Ohokuni, H. and Kotani, S. *Br. J. Cancer,* 1988, 57, 70.
4. Bhakdi, S., Klonisch, T., Nuber, P. and Fischer, W. *Infect. Immun.,* 1991, 59, 4614.
5. Fukase, K., Matsumoto, T., Ito, N., Yoshimura, T., Kotani, S. and Kusumoto, S. *Bull.Chem. Soc. Jpn.,* 1992, 65, 2643.
6. Fukase, K., Yoshimura, T., Kotani, S. and Kusumoto, S. *Bull. Chem. Soc. Jpn.,* 1994, 67, 473.
7. Takada, H., Kawabata, Y., Arakaki, R., Kusumoto, S., Fukase, K., Suda, Y., Yoshimura, T., Kokeguchi, S., Kato. K., Komuro, T., Tanaka, N., Saito, M., Yoshida, T., Sato, M. and Kotani, S. *Infect. Immun.,* 1995, 63, 57.
8. Suda, Y., Tochio, H., Kawano, K., Takada, H., Yoshida, T., Kotani, S. and Kusumoto, S. *FEMS Immunol. Med. Microbiol.,* 1995, 12, 97.

9. Hashimoto, M., Yasuoka, J-I., Suda, Y., Takada, H., Yoshida, T., Kotani, S. and Kusumoto, S. *J. Biochem.*, 1997, 121, 779.

10. Hirano, T., Taga, T., Nakano, N., Yasukawa, K., Kashiwamura, S., Shimizu, K., Nakajima, K., Pyun, K. H. and Kishimoto, T. *Proc. Natl. Acad. Sci. USA*, 1985, 82, 5490.

11. Matsuda, T., Hirano, T. and Kishimoto T. *Eur. J. Immunol.*, 1988, 18, 951.

12. Suda, Y., Aoyama, K., Arimoto, K., Tamura T. and Kusumoto S., *Biochem. Biophys. Res. Commun.*, 1999, 257, 327.

13. Momose, T. *Yuuki Teusei Bunseki*, 1972, 226. Hirokawa, Tokyo.

14. Ciucanu, I. and Kerek, F. *Carbohydr. Res.*, 1984, 131, 209.

15. Ashwell, G. *Method Enzymol.*, 1957, 3, 73.

16. Kochanowski, B., Fischer, W., Iida-Tanaka, N. and Ishizuka, I. *Eur. J. Biochem.*, 1993, 214, 747.

17. Shibata, N., Kojima, C., Satoh, Y., Suzuki, A., Kobayashi, H. and Suzuki, S. *Eur J Biochem.*, 1993, 217, 1.

18. Hashimoto, M., Imamura, Y., Yasuoka, J., Kotani, S., Shoichi, K. and Suda, Y. *Glycoconjugates J.*, 1999, 16, 213.

19. Vosbeck, K., Tobias, P., Mueller, H., Allen, R. A., Arfors, K. E., Ulevitch, R. J. and Sklar, L. A. *J. Leukoc. Biol.*, 1990, 47, 97.

20. Suda, Y., Kirikae, T., Shiyama, T., Yasukochi, T., Kirikae, F., Nakano, M., Rietschel, E. T. and Kusumoto, S. *Biochem. Biophys. Res. Commun.*, 1995, 210, 678.

# CHAPTER 15

# Hydrophobe Modified Cationic Polysaccharides for Topical Microbicide Delivery

GEORGE L. BRODE[1]
GUSTAVO F. DONCEL[2]
JOHN E. KEMNITZER[1]

## INTRODUCTION

THE prevalence of sexually transmitted diseases (STDs) has been considerable. The STD problem is steadily growing with increasing numbers of people contracting more severe and resistant infections. Worldwide, 333 million new STD cases were estimated for 1995, including trichomaniasis (170 million), chlamydia (89 million), gonorrhea (62 million), and syphilis (12 million) [1]. Persistent viral infections, including HIV (human immunodeficiency virus), HSV (herpes simplex virus), HBV (hepatitis B virus), and HPV (human papillomavirus), have afflicted millions with no definitive cures at hand. The World Health Organization has recently estimated 30 million people worldwide are infected with HIV [2]. In the U.S., 14 million new STD cases have been reported with 86% occurring among the 15–29 year-old age group [1,3]. In the U.S., AIDS accounts for the largest proportion of STD-related deaths, with 319,849 deaths among 513,486 reported cases in 1995 [4]. A 1992 CDC study found that the most common cause of STD related deaths among U.S. women was cervical cancer (57%), AIDS (29%), hepatitis B and C (10.5%), and other STDs (3.5%) [4].

[1]Integra LifeSciences Corporation, 105 Morgan Lane, Plainsboro, NJ 08536, USA.
[2]CONRAD and Eastern Virginia Medical School, 601 Colley Avenue, Norfolk, VA 23507, USA.

The U.S. 1994 economic burden impact of STDs and associated sequelae has been estimated to be approximately $10 billion for the major STDs and increases to $17 billion when sexually transmitted HIV infections are included [4]. Greater than 10% of new AIDS cases reported to the CDC in 1990 were from heterosexual transmission [5], with double of such heterosexual transmission between 1990 and 1995 [6]. Clearly, effective STD prevention can dramatically reduce death, morbidity, and health care costs.

In spite of these sobering statistics, STD prevention is not currently designed into today's vaginal contraceptive formulations. Contraceptive efficacy of vaginal formulations in itself is a very important aspect of fertility control and health. Infection prevention is the most effective way to reduce the adverse consequences of STDs and is currently addressed by use of barrier methods, such as condoms. Formulations that address this major health need are becoming increasingly critical. Unfortunately, in many cultural and social settings, it is the woman who bears not only the primary responsibility of contraceptive control but also the tragic health consequences associated with partners who harbor STDs including HIV. The risk of HIV transmission appears to be higher when one currently has a STD or in women with an unbalanced vaginal flora [7,8,9].

Appropriate barrier methods and vaginal formulations are the first line of defense against the prevention of pregnancy and STDs. Recognition of this fact and initiation of work to develop truly novel systems should be a major development focus. Indeed, major efforts are under way to develop new and more effective microbiocides, excipient delivery vehicles, and subsequent formulations.

A new class of water-soluble materials [10,11], was developed as a result of such design parameters that will be referred to as double substituted cationic cellulose ethers (DCEs). These materials contain both a cationic substituent and a hydrophobic substituent, attached to a cellulose ether backbone. The use of a double-substituted hydrophobe modified cationic polysaccharide is fundamentally different from current commercial vaginal formulations, which rely exclusively on nonionic or anionic vehicles.

The key objective of our efforts has been to develop a vaginal formulation that optimizes spermicidal and antiviral activity while enhancing spreading and true bioadhesiveness. Utilization of strict design principles for an excipient delivery vehicle, which included substantivity to vaginal mucosa, saline compatibility, compatibility with a wide range of spermicidal and antiviral compounds, low irritation potential, sperm impedance, system stability, and efficacy after stressed storage conditions, resulted in the development of DCE's [11,12,13]. Based on the results from *in vitro* studies, the DCE vehicle was selected for clinical development.

## RATIONALE FOR DESIGN

Traditionally, OTC vaginal contraceptives have several common characteristics. Design of such vaginal formulations for contraceptive indications have relied upon spermicide [N9 is currently the only approved active in U.S.), water-soluble polymer vehicle thickener (for viscosity control and to reduce seepage; bioadhesiveness), preservative (to meet USP microbiological challenge (multi-use), or prevent growth during manufacture (unit dose)], additives (vehicle property enhancement, pH, fragrance, and system compatibility), and water (formulation solvent).

It is obvious that these commonalties can be tailored to achieve successful contraceptive formulations with varying degrees of efficacy, user friendliness, and aesthetics. U.S. marketed contraceptive products all contain N9 as the spermicide at various concentrations. These products rely only on the mechanism of sperm destruction by the nondiscriminating surfactant effects of the spermicide, as effects of the carrier system on sperm motility are generally minimal. No claim of activity against STDs and HIV are made by current contraceptive products, with the exception of full barrier methods such as condoms.

An interesting feature of current commercial products is that the polymer vehicles available for formulation have been limited to nonionic and anionic materials. The delivery vehicles available included off-the-shelf polymers such as carboxymethylcellulose, soluble starch, hydroxyethylcellulose, polyvinyl alcohol, poly(acrylic acid), and polyvinylpyrrolidone, or mixtures thereof. The choice of available polymeric delivery system primarily depends on component compatibility, aesthetics, and efficacy. However, by reliance upon available (off-the-shelf) systems, limitations on bioadhesion, drug bioavailability, contraceptive efficacy, and end-use characteristics has been limited.

A more advantageous approach is the rational design of a delivery vehicle (polymer) to optimize performance. Vehicle design would have to include knowledge of general polymeric structure–property relationships, and especially, prediction of solution properties and compatibility with the active(s). Off-the-shelf excipients do not take full advantage of the knowledge base that exists for the vaginal environment (structure [14,15], pH [16], and rheology [17] of the cervical mucus, characteristics of vaginal mucosa, etc.), since in every case the polymers were designed for other applications.

Water solubility is determined by polymer structure (linear, branched, etc.), concentration and placement of charged species [ionomeric (cationic or anionic) or amphoteric (cationic and anionic)], hydrophilic/hydrophobic substituents, and hydrogen bonding, to name the more commonly encountered factors. In general, polymer water solubility requires polar functional groups and sites for specific interactions to occur. The solution properties, such as viscosity, are a consequence of the interactions of all these factors. Viscosity control in conventional vaginal formulations is addressed by the addition of a low concentration of a high molecular weight water-soluble polymer (natural, synthetically modified natural, or synthetic based) and fall within the families of nonionic and anionic polymers. It is important to note, however, that poor correlation is achieved between vaginal retention time and bulk viscosity measurements.

The only functional purpose of these conventional polymers is to tailor the viscosity. The polymer, as an active, is generally a secondary consideration, with few exceptions. Those polymers that confer strong anti-HIV activity *in vitro* include low molecular weight povidone-iodine [18], the sulfated polysaccharides/glycosaminoglycans (e.g., dextran sulfate, carageenan, heparin) [19], PAVAS (a copolymer of acrylic acid with vinyl alcohol sulfate) [20], and a sulfated polystyrene derivative [21,22] (which is contraceptive but nonspermicidal [23]). Cationic polymers, surprisingly, are not encountered in vaginal preparations, but they are found in cosmetic and dermatological products [24,25] (skin creams, conditioners, mousse, etc.) due to their excellent substantivity to anionic surfaces (e.g., skin and hair). Typically encountered polymer systems in water-soluble applications are listed in Table 1. Of course, many other water-soluble polymers exist that may be incorporated to meet the demands of a particular application [26].

In the mucosal environment, effects of salt, pH, temperature, and lipids need to be taken into consideration for possible effects on viscosity and solubility. A pH range of 4–7 and a relatively constant temperature of 37°C can generally be expected. Observed solution properties as a function of salt and polymer concentration can be referred to as saline compatibility. Polyelectrolyte solution behavior [27] is generally dominated by ionic interactions, such as with other materials of like charge (repulsive), opposite charge (attractive), solvent ionic character (dielectric), and dissolved ions (i.e., salt). In general, at a constant polymer concentration, an increase in the salt concentration decreases the viscosity, due to decreasing the hydrodynamic volume of the polymer; at a critical salt concentration precipitation may occur.

It is a well accepted axiom that the efficacy of any biomaterial is determined at the interface between the active and the targeted substrate.

Many factors were considered at the outset of the program in mucosal delivery systems and more specifically in the contraceptive and antiviral area, including:

- composition of the vaginal mucosa and cervical mucus
- molecular structure of active (originally N9)
- delivery system options; polymer type, vehicle, buffers, preservatives, etc.

TABLE 1. Examples of Commonly Encountered Water Soluble Polymers.

| Water-Soluble Polymer | Charge Type[a] | Chemical Functionality |
|---|---|---|
| Hydroxymethylcellulose | N | $-OH$ |
| Hydroxyethylcellulose | N | $-OH$ |
| Hydroxypropylcellulose | N | $-OH$ |
| Hydroxypropylmethylcellolose | N | $-OH$ |
| Amylopectin | N | $-OH$ |
| Carboxymethylcellulose | – | $-CO_2H, -OH$ |
| Xanthan | – | $-CO_2H, -OH$ |
| Hyaluronic acid | – | $-CO_2H, -OH$ |
| Poly(acrylic acid) | – | $-CO_2H$ |
| Polyacrylamide | N | $\underset{\displaystyle -CNH_2}{\overset{\displaystyle O \atop \displaystyle \|}{}}$ |
| Poly(ethylene oxide) | N | None |
| Poly(vinyl alcohol) | N | $-OH$ |
| Poly(N-vinyl pyrrolidinone) | N | None |
| Poly(diallyldimethyl-ammonium chloride) | + | |
| "Cationic" polyacrylamide | + | |
| Chitosan | + | $-NH_2, -OH$ |
| Polymer JR® | + | $(CH_3)_3N^+-$ |
| Quatrisoft® | + | $CH_3(CH_2)_{11}-\overset{\displaystyle CH_3}{\underset{\displaystyle CH_3}{\overset{|}{\underset{|}{N}}}}\overset{+}{-}$ |
| DCE | + | $(CH_3)_3N^+-$ |
| | | $CH_3(CH_2)_{11}-\overset{\displaystyle CH_3}{\underset{\displaystyle CH_3}{\overset{|}{\underset{|}{N}}}}\overset{+}{-}$ |

[a]N = nonionic, – = anionic, + = cationic.

215

- compatibility of the active and delivery system
- saline compatibility
- biocompatability/toxicology of the active and delivery system
- user-friendly aspects of formulated products
- formulation efficacy and stability
- economics

The chemistry of the substrate, mucosa, was a primary consideration in the design of the delivery system. Mucins, whether gastrointestinal, respiratory, ophthamological, cervical, or vaginal, are glycoproteins whose major function is to protect the epithelial surface [28]. These extremely hydrophilic substances are composed of a polypeptide backbone covalently linked to oligosaccharide and polysaccharide units carrying a negative charge, with sialic acid being present in all mucins and mainly responsible for their anionic nature (pH 4.5) [29]. It is also interesting to note that it has been observed that no significant differences in mucin composition are found in women belonging to different blood groups [30].

Based on earlier drug delivery work in dermatology and opthamology, and the related chemistry of vaginal mucosa, it was our hypothesis that a biocompatible, cationic polymer could complex with the anionic mucosa and reduce irritation of vaginal and cervical epithelia associated with certain actives, for example, nonoxynol 9 (N9) [31]. By complexing the hydrophobe portion of N9, with DCE (without decreasing spermicidal activity), the hypothesis was that irritation, associated with the strong detergent properties of N9, would be reduced. Further, reducing epithelial damage would remove a pathway for entry of the human immunodeficiency virus (HIV) [32]. In the latter phases of this work, formulations with DCEs containing either a spermicide, virucide, or both were evolved. In some of the later cases, new problems in formulation surfaced, the solution to which stemmed from dermatological complexes studied earlier (*vide infra*).

## CHEMISTRY OF VEHICLES

Since sialic acid is dissociated at normal vaginal pH, it contains a large hydration sphere that strongly attracts cationic substances. Thus, cationic polymers were designed, synthesized, and screened to bioadhere to these sites.

Cationic and hydrophobe modified polysaccharides are preferred excipients for personal care products since they are substantive (adherent) to anionic or hydrophobic substrates (skin, hair, mucosa), hydrophilic, film forming, compatible with many therapeutic agents, nonpenetrating, and nonirritating. Polymer JR® and Quatrisoft® (Figure 1), are two such materials

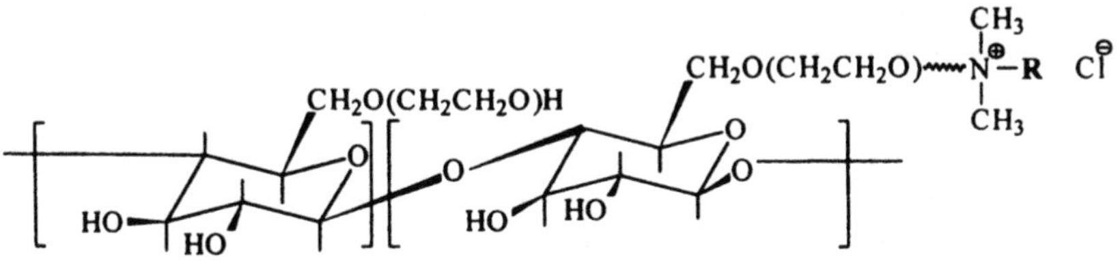

R= -C₁₂H₂₅     QUATRISOFT®
R= -CH₃       Polymer JR®

**Figure 1** "Single substituted" cellulose ethers.

routinely used in topical formulations for cosmetic and personal care products. The objective in this case was to evaluate these polymers and modify the cationic and hydrophobic moieties to improve mucin substantivity and spermicidal effects.

Topical spermicides such as nonoxynol-9 (N9) and benzalkonium chloride act on sperm membranes through a detergent effect, namely, hydrophobe–hydrophobe interaction between the active and substrate (spermatozoa). The idea was to optimize the cationic/hydrophobic polymer in the drug delivery system so epithelial cells were protected without sacrificing the drug's spermicidal activity. One of the questions that needed to be answered in designing an optimum cationic/hydrophobe modified polymer was the effect of the hydrophobe on the drug activity (N9 initially, and other actives subsequently).

Summarized in Table 2 is the sperm penetration in cervical mucus data as a function of molecular structure for various polysaccharides, with the following observations:

- Hydroxyethyl cellulose (A) or a mixture of carboxymethyl cellulose/polyvinyl pyrrolidone (B) did not affect sperm penetration. These polymers do not contain a hydrophobic moiety.
- Cationic polymer C (hydroxyethyl cellulose with a cationic moiety without a hydrophobic group) has a minor effect on sperm penetration.
- Hydrophobe incorporation has an important effect, independent of the polymer charge (D, E, F).
  —Hydrophobe modified carboxymethyl cellulose (D) reduces penetration (compare to B).
  —Cationic/hydrophobe modified hydroxyethyl celluloses, (E) and (F), eliminate sperm penetration, even though they are not spermicidal. These DCEs physically impede sperm penetration, without affecting motility.

TABLE 2. Molecular Structure Effects on Sperm Penetration [12].

| Item | Polymer | Class | Sperm Penetration in Cervical Mucus (% of Control) |
|------|---------|-------|---------------------------------------------------|
| A | Hydroxyethyl cellulose | Nonionic | 100 |
| B | Carboxymethyl cellulose/ polyvinyl pyrrolidone | Anionic | 97 |
| C | Polymer JR® | Cationic | 77 |
| D | Carboxymethyl cellulose | Anionic/hydrophobic | 22 |
| E | Quatrisoft® | Cationic/hydrophobic | 0 |
| F | DCE | Cationic/hydrophobic | 0 |

—The effect in (E) and (F) is primarily hydrophobe related, not cationic concentration related. (C) has a much higher cationic content than (E), but sperm penetration in the presence of (C) is significantly higher than (E).

• Composition (E), (Figure 1, $R=C_{12}H_{25}$) had many of the sought-after attributes; however, border line saline compatibility in topical formulations was viewed as a weak point. By contrast, composition (F) had the same desirable features with the added advantage of saline compatibility in formulated products. Therefore, composition (F) became a leading candidate for topical contraceptive development.

In the course of screening N9/cationic polymer formulations, an important observation was made. Hydrophobe modified cationic polysaccharides [33] displayed unique sperm impedance, but not spermicidal, properties. By contrast, the related non-hydrophobe modified material was devoid of that effect. It is important to state that hydrophobe incorporation into water soluble polymers, at the desired level for optimum efficacy, is complex. Chemical efficiency can be low in the derivatized polymer and solubility characteristics change dramatically. Of the various hydrophobes evaluated in these studies, the $-C_{12}H_{25}$ hydrophobe was preferred.

## DCE VEHICLE CHARACTERIZATION

DCE characterization can be conveniently carried out by $^1$H-NMR, elemental analysis, pH, and viscosity determination. The trimethyl ammonium and dimethyldodecyl ammonium content were evaluated, as well as an indirect percent nitrogen (% N), by $^1$H-NMR (Gemini 200 NMR, $^1$H/$^{13}$C dual probe at 25°C, as 3.3% solutions in $d_6$-DMSO). The %N via $^1$H-NMR correlated extremely well with elemental analysis (Tables 3 and 4).

TABLE 3. DCE Component Information.

| Component | Structure | Formula Weight (g/mol) | Designation | Formula Weight for NMR Calculations | $^1$H-NMR ppm Range |
|---|---|---|---|---|---|
| Polysaccharide base | | 294.3 | HEC | 294.3 | 3.3–4.5 |
| Quab 151 | | 151.6 | TM | 151.6 | 3.0–3.3[a] |
| Quab 342 | | 342.4 | DD | 306.9 | 0.6–1.7<br>3.0–3.3[a] |

[a]Overlap area of TM and DD.

TABLE 4. $^1$H-NMR Relevant Assignment Ranges.

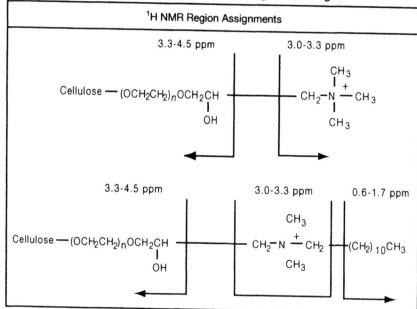

It follows that an indirect % N from $^1$H-NMR (Table 5) can be calculated using the following relationship:

$$\% \ N = \frac{[(DD_{MS}) + (TM_{MS})] \ N_{FW}}{(DD_{MS})(DD_{FW}) + (TM_{MS})(TM_{FW}) + HEC_{FW}} \times 100$$

where:

MS = molar substitution of the appropriate substituent on the HEC, as determined by $^1$H NMR

$N_{FW}$ = nitrogen formula weight

$DD_{FW}$ = Quab 342 formula weight

$TM_{FW}$ = Quab 151 formula weight

$HEC_{FW}$ = polysaccharide base approximate formula weight

## ANTI-VIRAL DESIGN PARAMETERS

Prevention of HIV transmission is a major worldwide concern. A recent recommendation by the International Working Group on Vaginal Microbiocides states [34] that ". . . It is desirable that an agent be evaluated for activity against HIV and other STD regardless of its intended HIV indication since a clinical outcome of HIV prevention may be achieved by the prevention

TABLE 5. Analytical Comparison of DCE Lots.

| DCE Lot | % N | | Cationic Molar Substitution (NMR) | Hydrophobic Molar Substitution (NMR) | pH (2.5% solution) |
|---|---|---|---|---|---|
| | NMR | Elem. Anal. | | | |
| A | 1.91 | 1.93 | 0.331 | 0.240 | 6.2 |
| B | 1.85 | 1.81 | 0.325 | 0.215 | 5.7 |
| C | 1.67 | 1.77 | 0.272 | 0.201 | 5.9 |
| D | 1.80 | 1.75 | 0.326 | 0.191 | 6.1 |
| E | 1.70 | 1.72 | 0.275 | 0.213 | 6.1 |

of other STD." Clearly, formulations that have only contraceptive indications are doing half the job necessary from a health standpoint. Formulations need to be intentionally designed to exhibit antiviral properties [35].

Specific polyanions such as dextran sulfate (DS) appear to exhibit strong anti-HIV activity *in vitro* [36,37]. Human oral administration of DS is poorly absorbed, but intravaneous administration does result in increased plasma lipolytic activity [38]. Polyanions that have been considered for intravaginal anti-HIV activity include DS, carrageenan, heparin, heparan sulfate, dermatan sulfate, pentosan polysulfate, fucoidan chondroitin sulfate, keratan sulfate, and PAVAS [21,22,39,40].

Emphasis in the initial phase of our work was placed on sulfated polysaccharides that are antiviral. Not only were the desired rheological properties and long-term stability achieved in DCE formulations, the activity of the dextran sulfate or N9 were not compromised. DCE formulations containing DS display strong anti-HIV activity *in vitro* in comparison with negative (not shown) and positive controls (Figure 2). This is an important first step in the screening process towards clinical effectiveness.

One path for HIV infection may involve damaged (or healthy) vaginal epithelia [41,42] through cell surface adhesion [40]. Recently it has been reported that the major HIV envelope protein is held intact through hydrophobic residues [43]. For these and other reasons, we sought to combine hydrophobe substituted cationic polymers, particularly DCEs with active anionic polymers, with or without N9 present.

We were successful earlier in achieving anionic/cationic complexes with glycosaminoglycans, e.g., hyaluronic acid/cationic polysaccharide compositions [44], now used in personal care applications. Those compositions are substantive, stable and user-friendly. Unfortunately, despite the highly anionic nature of hyaluronic acid (and other glycosaminoglycans), these materials do not block viral infections [39]. Furthermore, as shown by the *in vivo* rabbit irritation studies, such compositions displayed minimal irritation [13].

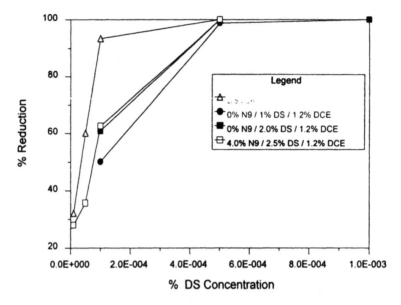

**Figure 2**  Anti-HIV results of vaginal formulations via viral binding assay.

## CONTRACEPTIVE DESIGN PARAMETERS

The most important aspect of a contraceptive vaginal formulation is obviously its effectiveness in the prevention of pregnancy. Initial evaluations prior to any clinical study should include:

(1)  Spermicidal effectiveness (Sander-Cramer Assay [sperm dilution and minimum effective concentration (MEC)] [45]
(2)  Ability to inhibit sperm penetration in cervical mucus [the Modified One End Test (MOET)]
(3)  Biodiffusion in cervical mucus [Double End Test (DET)]

A series of DCE-based formulations containing Igepal CO-630 Special (N9) and dextran sulfate (DS) were evaluated for *in vitro* contraceptive testing. Also included in the screening tests were placebos, N9, two commercial spermicidal products (Conceptrol® and KY® Plus), vaginal moisturizer products (KY® Jelly and Replens®), and saline. Test results are summarized in Table 6. The N9-containing products exhibited similar spermicidal activity, as illustrated by the minimum effective concentration (MEC). Samples without N9 did not have spermicidal activity. DCE placebo vehicle inhibits sperm penetration into the cervical mucus, illustrated by the very low MOET values after 1 : 10 and 1 : 160 dilutions. This activity has not been reported for anionic or nonionic polymer vehicles. There are no striking differences

TABLE 6. Spermicidal/Cervical Mucus Blocking/Cervical Mucus Biodiffusion Activity of Various Formulations.

| Formulation | MEC (mg/mL) | DET | MOET Conc (dilution) | MOET % CTL |
|---|---|---|---|---|
| 0% N9/2% DS/ 1.25% DCE | na | 98.9 ± 0.9 | 1:10 | 5.2 ± 1.7 |
| | | | 1:160 | 25.7 ± 4.5 |
| 1% N9/2% DS/ 1.25% DCE | 0.084 ± 0.012 | — | — | — |
| 2% N9/2% DS/ 1.25% DCE | 0.119 ± 0.016 | 93.7 ± 3.8 | — | — |
| 4% N9/2% DS/ 1.25% DCE | 0.087 ± 0.017 | — | — | — |
| Conceptrol® (4% N9) | 0.119 ± 0.016 | 88.4 ± 4.1 | — | — |
| KY® Plus (2.2% N9) | 0.151 ± 0.016 | 89.5 ± 3.9 | — | — |
| Igepal CO-630 Special | 0.107 ± 0.012 | — | — | — |
| KY Jelly® (0% N9) | na | na | 1:10 | 26.9 ± 5.4 |
| | | | 1:160 | 71.1 ± 5.3 |
| Replens® (0% N9) | na | — | 1:160 | 82.8 ± 6.1 |
| 0.9% NaCl | — | — | — | 100 |

observed in the cervical mucus biodiffusion (DET) between the contraceptive products or the DCE systems. Hence, a contraceptive effect presumably exists for the DCE vehicle by a sperm impedance mechanism, not spermicidal, that does not exist with currently used anionic and nonionic vehicles. This activity was observed in a preliminary *in vivo* model [46] (Figure 3).

Despite the fact that DCE is not spermicidal, a DCE gel composition without N9 showed a pregnancy rate almost equivalent to the commercial control (with 4% N9). Furthermore, the DCE gel with only 2% N9 had the lowest pregnancy rate (and mean number of implantation sites).

N9 is currently the only approved spermicide in the U.S., and formulators must take this into strategic account. However, in the design of new polymer vehicles, the use of an approved spermicide is a logical choice for product introduction. However, compatibility of such excipients is an important consideration in the formulation of new chemical spermicides.

A correlation exists for a surfactant's spermicidal activity and partition coefficient, such that the MEC is in the order of nonionic > cationic > anionic for given structural variables [47]. Such design parameters need to be factored when designing polymer vehicles and formulations with ionic and hydrophobic features.

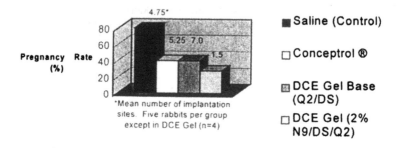

Figure 3 Rabbit contraceptive efficacy trial.

## VEHICLE BIOADHESION

Bioadhesion to the vaginal substrate implies knowledge and demonstration of the mechanism of interaction. In the present case, mucoadhesion, bioadhesion, and substantivity can be used interchangeably when evaluating the mucosa as a substrate. Several forces may be active, such as electrostatic, hydrophobic, and hydrogen bonding, giving rise to surface interactions between the polymer and the mucosal substrate or diffusion into the mucus layer [48,49].

It is well established that the vaginal and cervical mucus glycoproteins have a net negative charge [15]. Mucosal tissue has also been shown to possess appreciable hydrophobicity as determined by contact angle measurements [50]. Detection of material adhesion to a biological surface must be determined by techniques that unequivocally show that the material of interest actually interacts with a substrate, possibly through one of the above interactions. Surface analysis of appropriate *in vitro* models, using techniques such as ATR-IR [51,52,53,54] (attenuated total reflectance-infrared), ESCA [55,56] (electron spectroscopy for chemical analysis), and SIMS [57] (secondary ion mass spectrometry) are techniques that readily allow chemical depth profiling distinctions.

Interaction with the cervical mucus has been anticipated to be highest with cationic species [58], such as benzalkonium chloride, chlorhexidene, and vantocil (polyhexamethylene biguanide). A clear exception is the water-soluble sulfated polystyrene derivative (ORF 13904) [23]. In general, sperm penetration is lower for water-soluble cationic polymers than for anionic or nonionic polymers [59].

Substantivity [24] of a series of hydrophobe modified cationic polysaccharides to skin, hair, and an anionic vinyl substrate (skin substitute) was shown using ESCA, even after repeated washings with distilled water. ESCA, being an expensive technique not suitable for screening purposes, can be corroborated qualitatively using a "skin substitute" followed by exposure to a fluorescent dye such as fluorescein.

Using this technique, substantivity (Table 7) was assessed using a negatively charged vinyl substrate that is often used to simulate skin (the anionic UCARMAG Binder 527 resin). Fluorescein, sodium salt, at 0.5% in water was added before and/or after a 0.9% saline wash (2 × 25 mL) step. If added later, then samples were rewashed to determine whether any substantive coating remained.

Greater substantivity of DCE formulations was observed against a bioadhesive-claiming commercial product. This substantivity was maintained after rinsing with saline. Substantivity to the vaginal mucosa and epithelia over a prolonged period is anticipated, and will be clinically evaluated.

## SUMMARY

(1) Hydrophobe, cationic modified cellulose ethers (DCEs) are drug delivery vehicles specifically designed to interact with mucosae. DCE's are closely related to cationic polysaccharides used safely and effectively for many years in topical personal care products.
(2) DCEs are not spermicidal. However, DCEs effectively impede sperm penetration (impedance). Incorporation of N9 facilitates impedance and confers cidal properties.
(3) DCE gel formulations displayed minimal irritation in two separate *in vivo* rabbit studies.
(4) In a preliminary *in vivo* animal study, DCE gel, without pharmaceutical active, reduced pregnancy rates to the same degree as a leading commercial product containing 4% N9. The addition of 2% N9 to DCE gel induced a significantly lower embryo implantation rate than the commercial product containing 4% N9.
(5) DCE formulations with nonionic, cationic, and anionic actives have been formulated successfully.
(6) The addition of sulfated polysaccharides to DCE gels have yielded effective microbicides *in vitro*, including activity against human immunodeficiency virus (HIV).

## FUTURE CONSIDERATIONS

Contraceptive formulations based on available water-soluble polymers are used quite extensively and successfully. However, to meet the challenge of today's emerging health crisis, rationally designed water-soluble polymer vehicles are poised to play an important role in the arsenal against

TABLE 7. Formulation Substantivity to a "Simulated" Mucosal Substrate.

| Test System | Remarks | Results |
|---|---|---|
| UCARMAG 527 Control | Fluorescein before saline wash | Not substantive (very light violet) |
| DCE Sol'n | Fluorescein before saline wash | Substantive (strong green color) |
| 2% DS/4% N9/1.25% DCE | Fluorescein before saline wash | Substantive (strong green color) |
| 2% DS/4% N9/1.25% DCE | Fluorescein after saline wash; repeat wash | Substantive (strong green color) |
| 2% DS/0% N9/1.25% DCE | Fluorescein before saline wash | Substantive (strong green color) |
| 2% DS/0% N9/1.25% DCE | Fluorescein after saline wash; repeat wash | Substantive (strong green color) |
| Advantage 24® | Fluorescein before saline wash | Partially substantive (faint violet with streaks of green) |
| Advantage 24® | Fluorescein after saline wash; repeat wash | Partially substantive (faint violet with streaks of green) |

pregnancy and STDs. Double-substituted hydrophobe modified cationic polysaccharides, designed specifically for mucosal applications, begin to fulfill this role by displaying sperm impedance, bactericidal activity, and excellent formulation characteristics with a broad range of actives and additional excipients as well as being nonirritating.

Incorporation of an antiviral component brings an additional element of protection to the user. Dextran sulfate in current DCE formulations demonstrates excellent antiviral activity *in vitro* and demonstrates the potential usefulness of a highly discussed strategy.

Exploitation of the cationic/hydrophobic effect for mucosal substrate use is a design principle that lends itself toward the investigation of a variety of structural analogues. However, the measure of success for any vehicle, combination of activities, and a final formulated product will be the demonstration of unequivocal efficacy *in vivo*.

## REFERENCES

1. World Health Organization/Global Programme on AIDS, WHO/GPA/STD/95.1, 1995, An overview of selected curable sexually transmitted diseases.
2. April, K., Koster R., and Kost, M. G. Worldwide HIV incidence-aspects and dynamics of a tardive epidemic. *Schweiz. Med. Wochenschr.,* 1997, 127(45), 1853–61.
3. National Adolescent Reproductive Health Partnership Update, Washington, DC: Association of Reproductive Health Professionals, 1995.
4. Eng, T. R. and Butler, W. T. (Ed). *The Hidden Epidemic: Confronting Sexually Transmitted Diseases,* Institute of Medicine, National Academy Press, Washington, DC, 1997.
5. Centers for Disease Control. Update: acquired immune deficiency syndrome— United States, 1981–1990. *MMWR* 1991, 40, 358–369.
6. Brookmyer, R. Reconstruction and future trends of the AIDS epidemic in the United States. *Science* 1991, 235, 37–42.
7. Laga, M., Manoka, A., Kivuvu, M. et al. Non-ulcerative sexually transmitted diseases as risk factors for HIV-1 transmission in women: Results from a cohort study. *AIDS,* 1993, 7, 9–102.
8. Plummer, F. A., Simonsen, J. N., Cameron, D. W. et al. Cofactors in male-female transmission of human immunodeficiency virus type 1. *J. Infect. Dis.,* 1991, 163, 233–239.
9. Cameron, D. W., Simonsen, J. N., D'Costa, L. J. et al. Female to male transmission of human immunodeficiency virus type 1: risk factors for seroconversion in men. *Lancet,* 1989, 2, 403–407.
10. Brode, G. L. and Salensky, G. A. U.S. Patent # 5,300,494 (1994), Delivery systems for quaternary and related compounds.
11. Brode, G. L., Kreeger, R. L., and Salensky, G. A. U.S. Patent # 5,407,919 (1995), Double-substituted cationic cellulose ethers.
12. Brode, G. L., Doncel, G. F., Kreeger, R. L., and Salensky, G. A. U.S. Patent # 5,595,980 (1997), Contraceptive compositions.

13. Brode, G. L. in *Vaginal Microbicide Formulations Workshop*, Rencher, W. F. (Ed), Lippincott-Raven (Philadelphia, PA), Chemical/physical principles in microbiocide formulations with emphasis on hydrophobe modified cationic polysaccharides-a new excipient class, 1998, Chap. 5, pp. 38–50.

14. Volochine, B., Cazabat, A. M., Chretien, F. C., and Kuntsmann, J. M. Structure of human cervical mucus from light scattering measurements. *Human Reproduction* 1988, 3(5), 577–582.

15. Carlstedt, I. and Sheehan, J. K. Structure and macromolecular properties of cervical mucus glycoproteins. *Symposia of the Society for Experimental Biology* 1989, 43, 289–316.

16. Eggert-Kruse, W., Kohler, A., Rohr, G., and Runnebaum, B. The pH as an important determinant of sperm-mucus interaction. *Fertility and Sterility* 1993, 59(3), 617–628.

17. Hassan, E. E. and Gallo, J. M. A simple rheological method for the in vitro assessment of mucin-polymer bioadhesive bond strength. *Pharmaceutical Research* 1993, 7(5), 491–495.

18. Shanbrom, E. U.S. Patent # 5,545,401 (1996), Antiviral, spermicidal vaginal gel and foam containing low molecular weight povidone-iodine.

19. Pearce-Pratt R. and Phillips, D. M. Sulfated polysaccharides inhibit lymphocyte-to-epithelial transmission of human immunodeficiency virus-1. *Biology of Reproduction* 1996, 54(1), 173–82.

20. Neyts, J. and De Clercq, E. Effect of polyanionic compounds on intracutaneous and intravaginal herpesvirus infection in mice: impact on the search for vaginal microbiocides with anti-HIV activity. *Journal of Acquired Immune Deficiency Syndromes and Human Retrovirology* 1995, 10, 8–12.

21. Szymanski, C. D. U.S. Patent # 4,432,967 (1982), Contraceptive composition.

22. Lee, J. J. U.S. Patent # 5,308,612 (1992), Uses of polystyrenesulfonate and related compounds as inhibitors of transactivating transcription factor (tat) and as therapeutics for HIV Infection and AIDS.

23. Homm, R. E., Foldsey R. G., and Hahn, D. W. ORF 13904, A new long-acting vaginal contraceptive. *Contraception* 1985, 32(3), 267–274.

24. Brode, G. L., Goddard, E. D., Harris, W. C., and Salensky, G. A. Cationic polysaccharides for cosmetics and therapeutics. *Cosmetic and Pharmaceutical Applications of Polymers*, 1991, pp. 117–128.

25. Pugliese, P., Himes G., and Wielinga, W. Skin protection properties of a cationic guar gum. *Cosmetics and Toiltries* 1990, 105, 105–111.

26. McCormick, C. L., Bock, J., and Schulz, D. N. Water-soluble polymers, in the *Encyclopedia of Polymer Science and Engineering*, Vol. 17, 1984, pp. 731–785.

27. For a review, see: Kotz, J. and Beitz, T. The phase behaviour of polyanion-polycation systems. *Trends in Polymer Science* 1997, 5(3), 86–90.

28. Strous, G. J. and Dekker, J. Mucin-Type Glycoproteins. *Critical Reviews in Biochemistry and Molecular Biology*, 1992, 27(2), 57–92.

29. Carlstedt, I., Lindgren, H., Sheehan, J. K., Ulmsten, V., and Wingerup, J. Isolation and characterization of human cervical-mucus glycoproteins. *Biochem J.* 1983, 211, 13–22.

30. Van Kooj, R. J., Kathmann, H. J. M., and Kramer, M. F. Human cervical mucus and its mucous glycoprotein during the menstrual cycle. *Fertility and Sterility* 1980, 34, 226–233.

31. Roddy, R. E., Cordero, M., Cordero, C., and Fortney, J. A. *Int. J. Studies on AIDS,* 1993, 4, 165–170.

32. Niruthisard, S., Roddy, R. E., and Chutivongse, S. The effects of frequent nonoxynol-9 use on the vaginal and cervical mucosa. *Sexually Transmitted Diseases* 1991, 18, 176–9.

33. Brode, G. L., Kreeger, R. L., Goddard, E. D., Merritt, F. M., and Braun, D. B. U.S. Patent # 4,663,159 (1987), Hydrophobe substituted water soluble cationic polysaccharides.

34. The International Working Group on Vaginal Microbiocides (Geneva, Switzerland), Recommendations for the development of vaginal microbiocides. *AIDS* 1996, 10, UNAIDS1–UNAIDS6.

35. Doncel, G. F. Chemical vaginal contraceptives: Preclinical evaluation, in *Barrier Contraceptives: Current Status and Future Prospects,* Mauck, C., Cordero, M., Gabelnick, J. L., Spieler, J. M., and Rivera, R. (Eds), pp. 147–162, Wiley-Liss, New York, 1994.

36. Ueno, R. and Kuno, S. Dextran sulfate, a potent anti-HIV agent in vitro having synergism with zidovudine. *Lancet,* 1987, 1379.

37. Bugelski, P. J., Ellens, H., Hart, T. K., and Kirsh, R. L. Soluble CD4 and dextran sulfate mediate release of gp 120 from HIV-1: Implications for clinical trials. *Journal of Acquired Immune Deficiency Syndromes* 1991, 4(9), 923–924.

38. Lorentsen, K. J., Hendrix, C. W., Collins, J. M., Kornhauser, D. M., Petty, B. G., Klecker, R. W., Flexner, C., Eckel, R. H., and Lietman, P. S. Dextran sulfate is poorly absorbed after oral administration. *Annals of Internal Medicine* 1989, 111(7), 561–566.

39. Zaretzky, F. R., Pearce-Pratt, R., and Phillips, D. M. Sulfated polyanions block *Chlamydia trachomatis* infection of cervix-derived human epithelia. *Infection and Immunity* 1995, 63(9), 3520–3526.

40. Phillips, D. M. Intravaginal formulations to prevent HIV infection. *Perspectives in Drug Discovery and Design* 1996, 5, 213–224.

41. Phillips, D. M., Zacharopoulos, V. R., Tan, X., and Pearce-Pratt, R. Mechanism of sexual transmission of HIV; does HIV infect intact epithelia. *Trends in Microbiology* 1994, 2(11), 454–8.

42. Pearce-Pratt, R. and Phillips, D. M. Studies of adhesion of lymphocytic cells; implications for sexual transmission of human immuno deficiency virus. *Biology of Reproduction* 1993, 48, 431–45.

43. Kim, P. S., Chan, D. C., Fass, D., and Berger, J. M. Core structure of gp41 from the HIV envelope glycoprotein. *Cell* 1997, 89, 263.

44. Brode, G. L., Goddard, E. D., Partian, E. M., and Leung, P. S. U.S. Patent # 4,767,463 (1988), Gylcosaminoglycan and cationic polymer combinations.

45. For a review of such tests: C-H. Lee. Review: in vitro spermicidal tests. *Contraception* 1996, 54, 131–147.

46. Doncel, G. F. (1997), Private Communication: Eastern Virginia Medical School.

47. Furuse, K., Ishizeki, C., and Iwahara, S. Studies on spermicidal activity of surfactants. I. Correlation between spermicidal effect and physicochemical properties of *p*-menthanylphenyl polyoxyethylene (8.8) ether and related surfactants. *J. Pharm. Dyn.* 1983, 6, 359–372.

48. Thermes, F., Grove, J., Rozier, A., Plazonnet, B., Constancis, A., Bunel, C., and Vairon, J.-P. Mucoadhesion of copolymers and mixtures containing polyacrylic acid. *Pharmaceutical Research* 1992, 9(12), 1563–1567.

49. Gu, J. M., Robinson, J. R., and Leung, S. H. Binding of acrylic polymers to mucin/epithelial surfaces: Structure-property relationships. *Crit. Rev. Ther. Drug Carrier Syst.* 1988, 5(1), 21–67.

50. Lehr, C. M., Bouwstra, J. A., Bodde, H. E., and Junginger, H. E. A surface energy analysis of mucoadhesion: Contact angle measurements on polycarbophil and pig intestinal mucosa in physiologically relevant fluids. *Pharmaceutical Research* 1992, 9(1), 70–75.

51. Durrer, C., Irache, J. M., Puisieux, F., Duchene, D., and Ponchel, G. Mucoadhesion of latexes. I. Analystical methods and kinetic studies. *Pharmaceutical Research* 1994, 11(5), 674–679.

52. Barbucci, R., Magnani, A., Baszkin, A., DaCosta, M. L., Bauser, H., Hellwig, G., Martuscelli, E., and Cimmino, S. Physico-chemical surface characterization of hyaluronic acid derivatives as a new class of biomaterials. *J. Biomater. Sci. Polym. Ed.* 1993, 4(3), 245–173.

53. Castillo, E. J., Koenig, J. L., Anderson, J. M., Kliment, C. K., and Lo, J. Surface analysis of biomedical polymers by attenuated total reflectance-fourier transform infra-red. *Biomaterials* 1984, 5(4), 186–193.

54. Hofer, P. and Fringeli, U. P. Stuctural investigation o biological material in aqueous environment by means of infrared-ATR spectroscopy. *Biophys. Struct. Mech.* 1979, 6(1), 67–80.

55. Ratner, B. D. Surface characterization of biomaterials by electron spectroscopy for chemical analysis. *Ann. Biomed. Eng.* 1983, 11(3–4), 313–336.

56. Ruckenstein, E. and Gourisankar, S. V. Preparation and characterization of thin film surface coatings for biological environments. *Biomaterials* 1986, 7(6), 403–422.

57. Linton, R. W., and Goldsmith, J. G. The role of secondary ion mass spectrometry (SIMS) in biological microanalysis: technique comparisons and prospects. *Biol. Cell.* 1992, 74(1), 147–160.

58. Chantler, E., Sharma, R., and Sharman, D. Changes in cervical mucus that prevent penetration by spermatozoa. *Symposia of the Socitey for Experimental Biology* 1989, 43, 325–336.

59. Sharman, D. A. Ph.D. Thesis, University of Manchester, U.K. (1987), The interaction of polymers with cervical mucus.

# CHAPTER 16

# Novel Antimicrobial *N*-Halamine Polymer Coatings

S. D. WORLEY[1], M. EKNOIAN[1]
J. BICKERT[2], J. F. WILLIAMS[2]

## INTRODUCTION

THERE are many applications in the medical and food sanitization fields for which antimicrobial surface coatings are needed. There are commercially available products making antibacterial claims, but they are limited in their applications by such factors as a poor spectra of activity, high cost, and toxicity. Work in the laboratories at Auburn University over the past two decades has established a novel class of heterocyclic organic compounds termed *N*-halamines that have been demonstrated to be excellent antimicrobial materials for a broad variety of applications.

First, a series of *N*-halamine compounds in the classes oxazolidinones and imidazolidinones (Figure 1) were synthesized and tested [1,2]. These compounds, some of which are being commercially developed, are state-of-the-art disinfectants for applications requiring water solubility. Second, a new insoluble *N*-halamine polymer, which is a derivative of poly(styrene-hydantoin), was prepared (Figure 2). This polymer inactivates pathogenic microorganisms upon contact of the solid particles with flowing suspensions containing the pathogens [3–5] and is being commercially developed as an antimicrobial filter for potable water. All of the compounds investigated in these laboratories contain nitrogen-chlorine or nitrogen-bromine bonds stabilized by neighboring alkyl groups, such that the compounds are remarkably stable toward hydrolyses, i.e., very little free halogen is eluted into the aqueous solution.

[1]Department of Chemistry, Auburn University, Auburn, AL 36849, USA.
[2]Department of Microbiology, Michigan State University, East Lansing, MI 48824, USA.

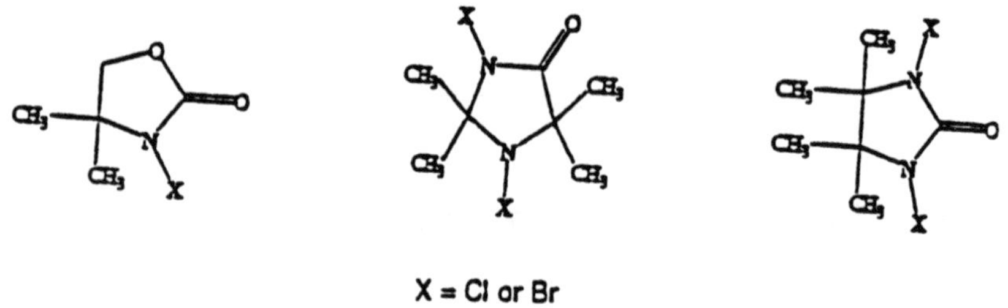

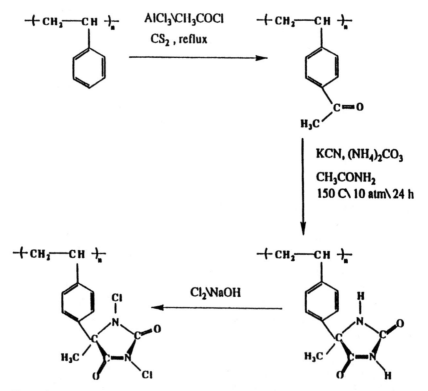

X = Cl or Br

**Figure 1** Water-soluble *N*-halamine disinfectant compounds.

The probable mechanism of action for all of the stable *N*-halamines involves direct contact of the microorganism with the combined *N*-halamine, followed by donation of a $Cl^+$ or $Br^+$ species to a receptor site within the cell, causing oxidation, and subsequent cell inactivation. Once the halogen content is exhausted through reaction with reducing agents, organic impurities,

**Figure 2** Synthetic scheme for a water-insoluble *N*-halamine polymer disinfectant for use in potable water sanitization.

and microbial species, it can be replenished by subsequent reaction *in situ* with free halogen.

Recently the work in these laboratories has been extended to the production of *N*-halamine polymer coating materials utilizing new oxazolidinone monomers that should be useful for a variety of applications [6–8]. The structures of the new monomers are shown in Figure 3. This work will be summarized herein.

## EXPERIMENTAL

The synthetic methods and chemical characterization data for the various polymeric materials to be discussed in this work have been reported elsewhere [6–8]. In some cases copolymerization of the unchlorinated oxazolidinone monomer with other common monomers such as acrylonitrile, vinyl chloride, styrene, and vinyl acetate, using potassium persulfate as an initiator, was performed. In other cases the unchlorinated oxazolidinone monomer was grafted onto polymers such as poly(acrylonitrile), poly(vinyl chloride), poly(styrene), poly(vinyl acetate), and poly(vinyl alcohol), again using potassium persulfate as an initiator.

The unchlorinated polymer latex compounds were coated onto various substrate surfaces [glass, textile fabric, and poly(urethane)]. This was accomplished by adding an aqueous solution of the polymer, in each case containing 10 wt% oxazolidinone, to the surface of the substrate and then evaporating the water from the coating at 80–100°C. The coatings that were produced were clear, resistant to abrasion, and possessed good adherence to the substrates. The coated surface was then soaked in an aqueous solution of sodium hypochlorite (3,000 mg/L free chlorine) for 10–30 min to produce an *N*-chloramine polymeric coating. Finally, the coated material was washed with chlorine-demand-free water and dried at ambient temperature overnight to remove any occluded free chlorine. Control samples were treated in the same manner except that they were never exposed to free chlorine.

Several types of testing were employed to evaluate the bactericidal efficacies of the coated substrates. Five of the coatings on circular glass coverslips (12 mm diameter) were challenged with the bacterium *Staphylococcus aureus* (ATCC 6538). This was accomplished by adding a 50-μL suspension of $10^6$ CFU *S. aureus* to the surface of each sample. At predetermined contact times a 25-μL aliquot was removed from the surface, quenched with sterile 0.02 N sodium thiosulfate, and plated on nutrient agar. The viable bacterial colonies were then counted after 48 h incubation at 37°C. Fabric samples were tested by two methods. In one, small squares (1.0–1.5 cm) were placed on a Tryptic Soy agar plate that was inoculated

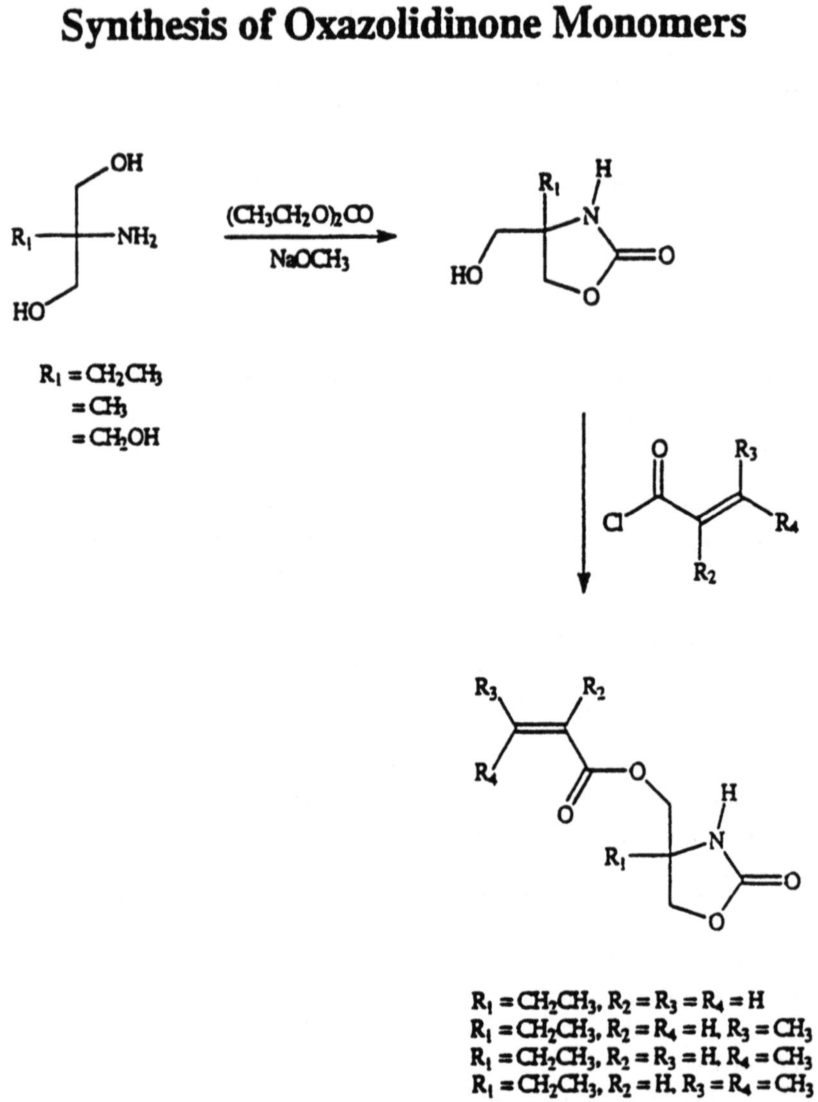
Synthesis of Oxazolidinone Monomers
(CH3CH2O)2CO
NaOCH3
R1 = CH2CH3
= CH3
= CH2OH
R1 = CH2CH3, R2 = R3 = R4 = H
R1 = CH2CH3, R2 = R4 = H, R3 = CH3
R1 = CH2CH3, R2 = R3 = H, R4 = CH3
R1 = CH2CH3, R2 = H, R3 = R4 = CH3

**Figure 3** Novel oxazolidinone monomers that can be copolymerized with other monomers or grafted to polymers.

with *S. aureus* (ATCC 6538). After 24 h incubation at 37°C, zones of inhibition about the pieces of fabric were measured. In the other method, small pieces of fabric were challenged with the bacterium *Salmonella enteritidis* (ATCC 13076) according to Method #100 of the American Association of Textile Chemists and Colorants (AATCC). In this procedure each fabric sample was prepared as a disk of diameter 1.0 cm and challenged with 100 μL of a suspension containing $10^6$ CFU of the bacteria for a contact time of 10 min at ambient temperature. Samples were then immersed in 10 mL of 0.02 N sodium thiosulfate and agitated for 60 sec. Aliquots of 100 μL were serially diluted in sterile water and plated on Trypticase Soy agar plates that were allowed to incubate for 24 h at 37°C before colony enumeration. Finally, two of the chlorinated polymers were coated onto small pieces of poly(urethane) medical catheters as substrates. Unchlorinated coatings served as controls. The samples, which were about 2–3 mm$^2$ in surface area and 150 μm thick, were placed in mesh histological specimen bags in a 15-mL chamber through which a $10^5$ CFU/mL suspension of *Pseudomonas aeruginosa* (field isolate from an "antibacterial" soap) was flowed constantly at a rate of about 200 mL/day for 3 days. Samples of the catheters were removed at 24-h intervals, fixed in 4% glutaraldehyde for 2 h, dehydrated with ethanol, coated with gold (20 nm), and subjected to analysis with a JEOL Scanning Electron Microscope for comparison of adherence of the *Pseudomonas* organisms to the biocidal and control surfaces.

## RESULTS AND DISCUSSION

Results for the challenge of coated glass with *S. aureus* are shown in Table 1. The time of chlorination was a function of the hydrophilicity of the surface coating; the poly(vinyl acetate oxazolidinone) polymers required the least amount of exposure to free chlorine. All of the surface coatings were effective against *S. aureus* in brief contact times. The grafted poly(acrylonitrile oxazolidinone) sample was not tested until 30 days after chlorination, but it still provided a 6-log inactivation of *S. aureus* in less than 10 min of contact.

The zone-of-inhibition studies (Table 2) established that all of the coatings on print cloth or cotton tested prevented the colonization of *S. aureus* on the surface of the material, and all provided small zones of inhibition around the pieces of cloth. The contact times were the incubation periods. In the other type of experiment, in which 10-min contact times were employed, the three polymer coatings tested did provide significant reductions in the numbers of *S. aureus* bacteria in the suspension (Table 3) even in the short period of contact.

Perhaps the most striking demonstration of the biocidal nature of the surface coatings was provided by the flow-chamber experiment utilizing the poly(vinyl acetate oxazolidinone) copolymer coated on the poly(urethane) medical catheter material. The test was designed to determine the efficacies of the coatings to prevent biofouling over a 72-h period. The results are shown in Figure 4. The top electron micrograph shows the surface of an

TABLE 1. Efficacies of *N*-Chloramine Polymeric Biocidal Coatings on Glass against *Staphylococcus aureus*.

| Compound[a] | Time of Chlorination (min)[b] | Age of Sample (days)[c] | Contact Time Required for 6-Log Inactivation (min) |
|---|---|---|---|
| 1 | 20 | 5 | 10–20 |
| 2 | 10 | 10 | 5–10 |
| 3 | 20 | 30 | 5–10 |
| 4 | 10 | 15 | 5–10 |
| 5 | 30 | 5 | 30–60 |

[a]1 = Poly(acrylonitrile-*co*-oxazolidinone), 2 = poly(vinyl acetate-*co*-oxazolidinone), 3 = poly(acrylonitrile-*g*-oxazolidinone), 4 = poly(vinyl acetate-*g*-oxazolidinone), 5 = poly(vinyl alcohol-*g*-oxazolidinone).
[b]Coatings were soaked in 3,000 mg/L free chlorine.
[c]Time after chlorination when biocidal efficacy was measured.

TABLE 2. Zones of Inhibition of *N*-Chloramine Polymeric Biocidal Coatings on Fabric against *Staphylococcus aureus*.

| Compound[a] | Material[b] | Weight % Increase | Zone of Inhibition (mm)[c] |
|---|---|---|---|
| 1 | Printcloth | 16.5 | 0.5 |
| 1 | Cotton | 23.2 | 1.0 |
| 2 | Printcloth | 21.6 | 0.1 |
| 2 | Cotton | 30.5 | 0.2 |
| 4 | Printcloth | 13.5 | 0.8 |
| 4 | Cotton | 26.8 | 1.2 |
| 5 | Printcloth | 29.6 | 0.3 |
| 5 | Cotton | 32.3 | 0.5 |
| 6 | Printcloth | 29.6 | 0.5 |
| 6 | Cotton | 22.1 | 0.5 |

[a]1 = Poly(acrylonitrile-*co*-oxazolidinone), 2 = poly(vinyl acetate-*co*-oxazolidinone), 4 = poly(vinyl acetate-*g*-oxazolidinone), 5 = poly(vinyl alcohol-*g*-oxazolidinone), 6 = poly(vinyl chloride-*g*-oxazolidinone).
[b]Printcloth consists of a 54/46 cotton/polyester blend.
[c]Length in mm from the edge of the fabric to the viable bacteria on an agar plate.

TABLE 3. Efficacies of *N*-Chloramine Polymeric Biocidal Coatings on Fabric against *Salmonella enteritidis*.

| Compound[a] | Material[b] | Weight % Increase | % Reduction of Bacteria in 10 Min Contact |
|---|---|---|---|
| 1 | Printcloth | 20.3 | 97.0 |
| 2 | Printcloth | 18.5 | 99.99 |
| 5 | Printcloth | 22.3 | 99.9 |

[a]1 = Poly(acrylonitrile-*co*-oxazolidinone, 2 = poly(vinyl acetate-*co*-oxazolidinone), 5 = poly(vinyl alcohol-*g*-oxazolidinone.
[b]Printcloth consists of a 54/46 cotton/polyester blend.

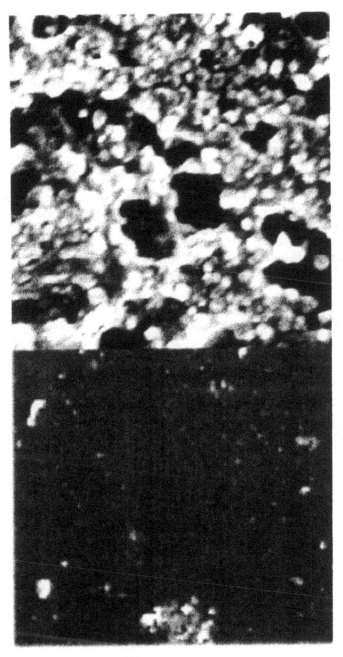

**Figure 4** Electron micrographs of unchlorinated poly(vinyl acetate-*co*-oxazolidinone) (top) and chlorinated poly(vinyl acetate-*co*-oxazolidinone) (bottom) coated medical catheters exposed for 72 h to a flowing aqueous suspension of *Pseudomonas aeruginosa* ($10^5$ CFU/mL).

unchlorinated control catheter sample, and the bottom micrograph refers to the chlorinated poly(vinyl acetate oxazolidinone) copolymer surface coating. Both electron micrographs were taken after 72 h of flow of the *Pseudomonas* suspension over the surfaces. Clearly, a very significant reduction in biofouling is observed for the surface coated with *N*-chloramine biocidal polymer.

## CONCLUSION

From the results presented in this work is should be evident that *N*-halamine biocidal polymer coatings are effective at inactivation of bacteria and the prevention of biofouling on surfaces. Once exhausted of halogen, the biocidal surface films can be regenerated by exposure to aqueous-free halogen. Thus, such coatings should have considerable commercial potential.

## REFERENCES

1. Worley, S. D. and Williams, D. E. *CRC Crit. Rev. Environ. Cntrl.* 1988, 18, 133.
2. Tsao, T. C., Williams, D. E., Worley, C. G., and Worley, S. D. *Biotech. Prog.* 1991, 7, 60.
3. Sun, G., Wheatley, W. B., and Worley, S. D. *Ind. Eng. Chem. Res.* 1994, 33, 168.
4. Sun, G., Allen, L. C., Luckie, E. P., Wheatley, W. B., and Worley, S. D. *Ind. Eng. Chem. Res.* 1995, 34, 4106.
5. Sun, G., Chen, T. Y., Habercom, M. S., Wheatley, W. B., and Worley, S. D. *J. Amer. Water Res. Assoc.* 1996, 32, 793.
6. Eknoian, M. W., Putnam, J. H., and Worley, S. D. *Ind. Eng. Chem. Res.* 1998, 37, 2873.
7. Eknoian, M. W. and Worley, S. D. *J. Bioact. Compat. Polym.* 1998, 13, 303.
8. Eknoian, M. W., Worley, S. D., Bickert, J. R., and Williams, J. F. *Polym.* 1998, 40, 1367.

# CHAPTER 17

# Design of Macromolecular Prodrug of Cisplatin Attached to Dextran through Coordinate Bond

TATSURO OUCHI[1], MITSUO MATSUMOTO[1]
TATSUNORI MASUNAGA[1], YICHI OHYA[1]
KATSUROU ICHINOSE[2], MIKIROU NAKASHIMA[2]
MASATAKA ICHIKAWA[2], TAKASHI KANEMATSU[2]

## INTRODUCTION

Cis-DICHLORODIAMMINEPLATINUM (II) (cisplatin, CDDP) has been widely used for clinical cancer therapy in spite of its severe renal toxicity [1] and low water-solubility.

Recently, drug delivery systems (DDS) using polymers as drug carriers have been investigated to achieve the efficient delivery of anticancer agents to tumor cells. In comparison with a low molecular-weight prodrug, a macromolecular prodrug is expected to overcome the problem of side effects by improving drug distribution in the body and prolongation of its activity. Since the growth rate of tumor cells is very high, the formation of solid tumor tissues is incomplete because of undernourishment. So, solid tumor tissues are hypervascular and lack a lymphatic capillary system compared with normal tissues. Since the polymer/drug conjugate shows the "enhanced perameability and retention" (EPR) effect in solid tumor tissues, the appearance of passive targeting to tumor tissues by polymer/drug conjugation technique can be expected. The "EPR" effect was proposed by Prof. Maeda [2].

We have used of water-soluble polymers, such as polysaccharides, as carriers of antitumor agents to reduce their side effects [3]. In this study,

[1]Faculty of Engineering, Kansai University, Suita, Osaka, 564-8680, Japan.
[2]Nagasaki University, School of Medicine, Sakamoto, Nagasaki 852-8102, Japan.

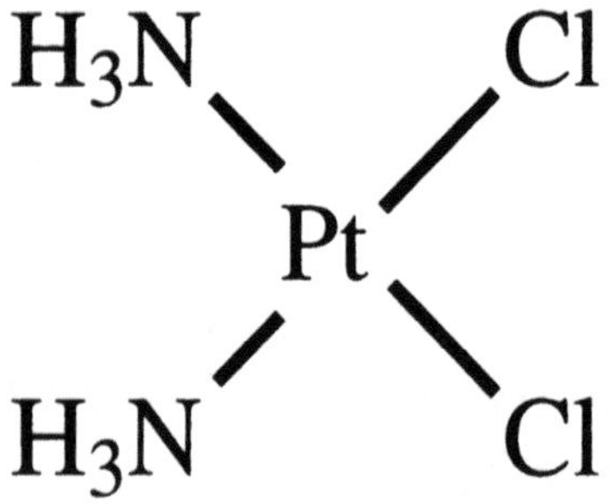

$$H_3N \diagdown \phantom{Pt} \diagup Cl$$
$$Pt$$
$$H_3N \diagup \phantom{Pt} \diagdown Cl$$

# Cisplatin (CDDP)

to provide a macromolecular prodrug of CDDP that has reduced side-effects and good water solubility, we designed CDDP polymer/dextran derivatives complexes with carboxylic acid groups, oxidized dextran (OX-Dex)/CDDP conjugates, and dicarboxymethyl dextran (DCM-Dex)/CDDP conjugates.

The release behavior of free aquo-CDDP from the conjugates was investigated in phosphate buffer solution (PBS; pH = 7.4). The cytotoxic activity of platinum complex decreased in the bloodstream because of ligand exchange with substances containing amino groups such as proteins, and amino acids, in the serum. The polymer/CDDP conjugate is expected to keep the platinum complex away from these deactivating factors and to maintain its cytotoxic activity during circulation in blood. The residual cytotoxic activity of the conjugates against tumor cells after preincubation in medium containing fetal calf serum (FCS) was evaluated and compared with that of free CDDP. The pharmacokinetics of free CDDP, OX-Dex/CDDP, and DCM-Dex/CDDP conjugates after intravenous injection in rat and their growth inhibitory effects against colon 26 tumor bearing mice by intravenous (i.v.) injection were investigated.

**EXPERIMENTAL**

MATERIALS

CDDP was obtained from Nihon Kayaku Co. Ltd. Dextran ($M_w = 6.0 \times 10^4$) was purchased from Wako Pure Chemical Industry and converted to two kinds of dextran derivatives, OX-Dex and DCM-Dex. The organic solvents were of commercial grades and used without further purification.

## OX-Dex

OX-Dex was synthesized according to Scheme 1. Dextran (500 mg) was dissolved in 20 mL of 0.06 M sodium periodate aqueous solution and stirred at 0°C for 6 h in the dark. After the addition of 5 mL of ethylene glycol, the solution was dialyzed in distilled water using a cellulose tube (Seamless Cellulose Tubing, Small Size 30, Viskase Sales Co.) (cut off: $M_w = 1.0 \times 10^4$) for 3 days and was freeze-dried to yield 320 mg of dextran–dialdehyde as a white powder. The dextran–dialdehyde (320 mg) obtained was dissolved in 20 mL of 0.6 M sodium chlorite aqueous solution. The pH of the solution was adjusted to 4 by the addition of acetic acid and stirred at room temperature for 24 h. Nitrogen gas was then passed through the solution until a colorless solution was obtained. The pH of the solution was raised to 9 by the addition of 1 M NaOH solution, which was dialyzed in distilled water for 7 days. The solution was then passed through a column packed with an cation exchange resin (Amberlite 120B, Organo Co.) and freeze-dried to yield 292 mg of OX-Dex as a white powder. The degree of carboxylation per sugar unit (DCA) for OX-Dex was estimated by a titration. The molecular weight obtained for OX-Dex was determined by gel-permeation chromatography (GPC).

## DCM-Dex

DCM-Dex was synthesized according to Scheme 2. Dextran (500 mg) was dissolved in 8.8 M NaOH (10 mL). Bromodiethylmalonate (10 g) dissolved in 10 mL of tetrahydrofuran was added to the solution at 0°C. The solution was then stirred at room temperature for 20 h. The reaction mixture was dialyzed in distilled water using cellulose tubing for 7 days.

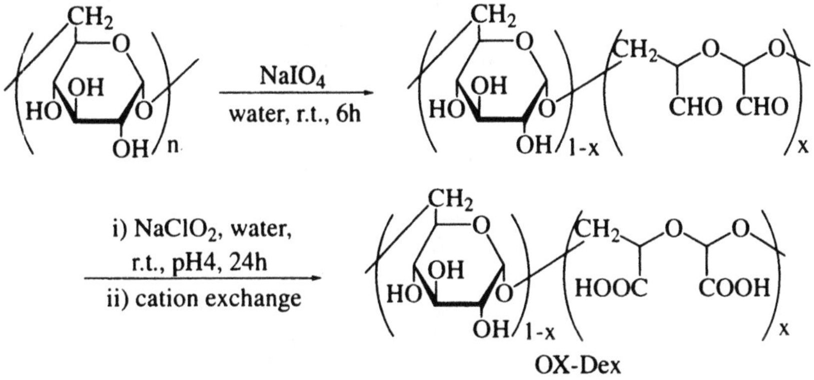

**Scheme 1** Synthetic reaction of OX-Dex.

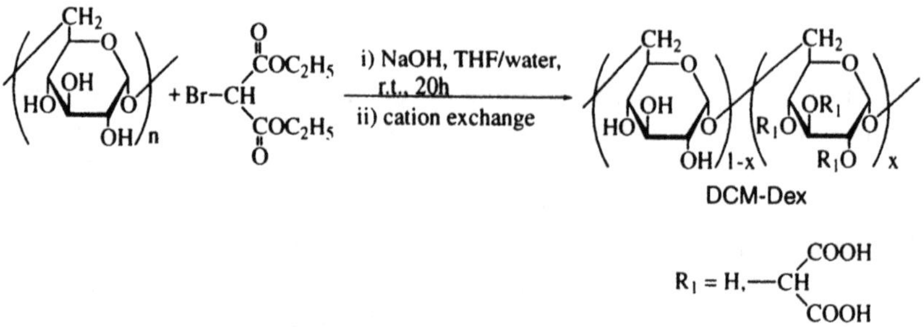

**Scheme 2** Synthetic reaction of DCM-Dex.

The solution obtained was passed through a column packed with a cation exchange resin (Ambelite 120B, Organo Co.) and freeze-dried to give DCM-Dex in 480 mg yield as a white powder. The DCA value and molecular weight of the DCM-Dex obtained were determined by the methods described above.

## PREPARATION OF THE CONJUGATES

The syntheses of OX-Dex/CDDP and DCM-Dex/CDDP conjugates were carried out according to the methods shown in Scheme 3 [4,5].

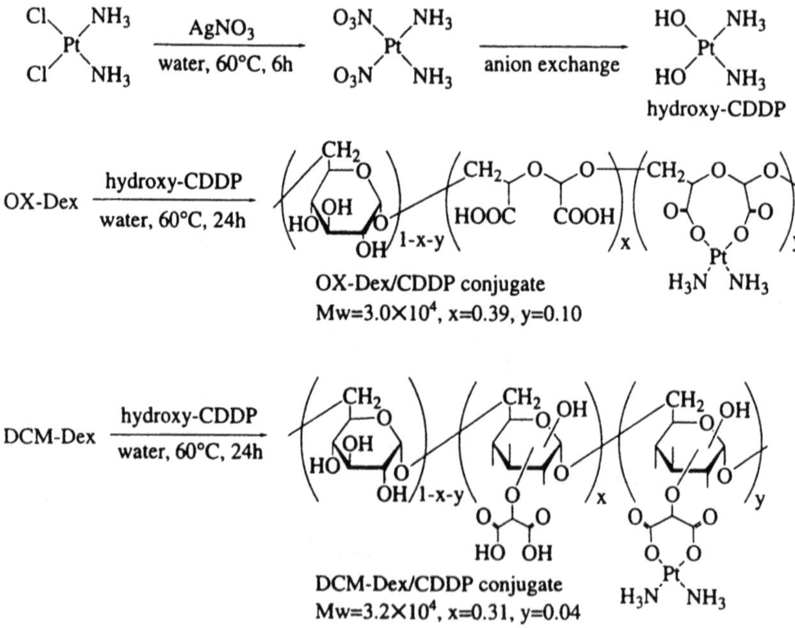

**Scheme 3** Synthetic routes of OX-Dex/CDDP and DCM-Dex/CDDP conjugates.

CDDP (50 mg) was dissolved in 30 mL of water and stirred at 60°C for 3 h, and then aqueous silver nitrate solution (0.1 M, 0.22 mL) was added and the mixture stirred at 60°C for 6 h. The precipitate silver chloride were removed by filtration. The filtrate containing CDDP(nitrato) was passed through a column packed with an anion exchange resin (Diaion SA-10A OH-type) to convert it to CDDP(hydroxo). The solution obtained was added to OX-Dex ($M_w = 3.0 \times 10^4$) or DCM-Dex ($Mw = 3.2 \times 10^4$) (226 mg) in water and stirred at 60°C for 24 h. The product was purified by gel-filtration chromatography (Sephadex G-25, eluent: water). The high molecular-weight fraction was collected and freeze-dried to give Dex/CDDP conjugates. The degree of CDDP introduction per sugar unit was determined by atomic absorption spectrometry (HITACHI Z-8000).

## DETERMINATION OF RELEASE OF AQUO-CDDP FROM THE CONJUGATES

The release behavior of aquo-CDDP from the conjugates was investigated in PBS at 37°C *in vitro*. The conjugates (10 mg) were dialyzed in PBS (pH = 7.4, 20 mL) containing 1M NaCl using a cellulose tube (cut off: $M_w = 1.0 \times 10^4$). The solution outside of the cellulose tube was periodically replaced with fresh PBS. The amount of aquo-CDDP released to the medium from the carrier polymer was estimated using atomic absorption spectrometry.

## MEASUREMENT OF RESIDUAL CYTOTOXIC ACTIVITY

After preincubation in RPMI-1640 medium containing 10% fetal calf serum (FCS), the residual cytotoxic activity of the conjugates against colon 26 tumor cells was measured by MTT assay method at 37°C *in vitro*.

## PHARMACOKINETICS

The blood clearance of the conjugates and free CDDP after intravenous injection was investigated. The conjugates and free CDDP were injected into male rat (6 weeks) intravenously. After specific periods, the amount of total Pt and free Pt in the blood was estimated by atomic absorption spectrometry before and after ultrafiltration (cut-off: $M_w = 1.2 \times 10^4$), respectively.

## ANTITUMOR ACTIVITY *IN VIVO*

The tumor growth inhibitory effect of DCM-Dex/CDDP conjugate was evaluated against colon 26 tumor-bearing mice. Colon 26 tumor cells ($2 \times 10^5$), which were maintained in RPMI-1640 medium containing 10% FCS, were inoculated subcutaneously in the dorsum of the six Balb/c mice. The PBS test solution of DCM-Dex/CDDP (3 mg/kg as CDDP conversion), free CDDP (3 mg/kg), or saline as a control was administrated intravenously 3 times into the tumor mice on days 14, 17, and 20. To estimate the tumor growth inhibitory effects of DCM-Dex/CDDP and free CDDP, the sizes of tumor were recorded. The mice were killed on 30 days, and the tumors were excised to compare the actual tumor weight among the 3 groups.

## RESULTS AND DISCUSSION

The water-soluble OX-Dex/CDDP and DCM-Dex/CDDP conjugates were easily obtained by the ligand-exchange reaction of hydroxo-CDDP with the corresponding dextran derivatize with carboxylic acid groups.

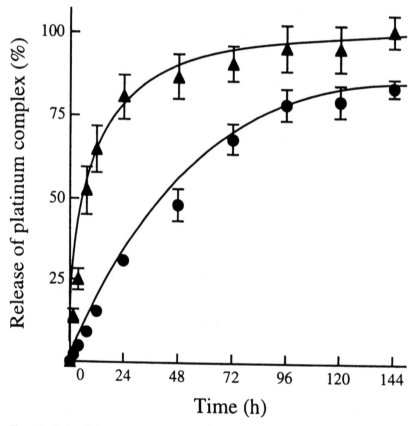

**Figure 1**   Release behavior of platinum complex from OX-Dex/CDDP and DCM-Dex/CDDP conjugates in PBS (pH = 7.4) at 37°C. ▲: OX-Dex/CDDP conjugate. ●: DCM-Dex/CDDP conjugate.

The release behavior of free aquo-CDDP from two Dex/CDDP conjugates is shown in Figure 1. The release rate of free aquo-CDDP from DCM-Dex/CDDP conjugate was found to be slower than that from OX-Dex/CDDP conjugate because former has a more stable six-membered chelation complex.

Shown in Figure 2 are stability data for the conjugates in the in vitro serum medium. Although the cytotoxic activity of free CDDP and OX-Dex/CDDP conjugate decreased gradually with preincubation in the serum, DCM-Dex/CDDP conjugates maintained a higher cytotoxic activity level for long period.

The results of the *in vivo* pharmacokinetics are shown in Table 1. DCM-Dex/CDDP conjugate had a significantly longer Pt half-life in the blood after intravenous injection than free CDDP or the OX-Dex/CDDP conjugate.

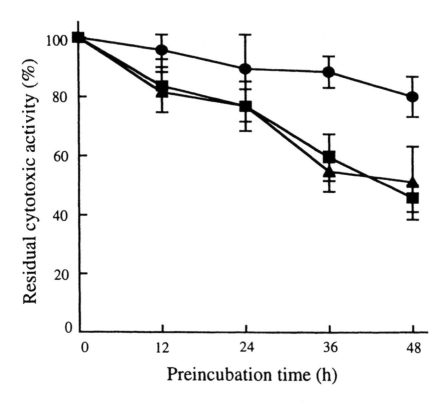

**Figure 2** Residual cytotoxic activity of OX-Dex/CDDP conjugate, DCM-Dex/CDDP conjugate, and free CDDP after preincubation in culture medium containing FCS against colon 26 cells for 48h *in vitro*. ■: free CDDP. ▲: OX-Dex/CDDP conjugate. ●: DCM-Dex/CDDP conjugate.

TABLE 1. Serum Pt Concentration after I.V. Injection of Free CDDP and DCM-Dex/CDDP Conjugate in Rats.

| Min. | 30 | 60 | 120 | 240 | 360 |
|---|---|---|---|---|---|
| Free-CDDP | | | | | |
| Free-Pt[a] | 0.86 ± 0.17 | 0.21 ± 0.00 | 0.02 ± 0.05 | 0.01 ± 0.00 | 0.01 ± 0.00 |
| Total-Pt[b] | 1.22 ± 0.23 | 0.64 ± 0.14 | 0.39 ± 0.07 | 0.29 ± 0.05 | 0.14 ± 0.08 |
| OX-Dex/CDDP | | | | | |
| Free-Pt[a] | 2.00 ± 0.65 | 1.20 ± 0.28 | 0.55 ± 0.05 | 0.00 ± 0.05 | 0.00 ± 0.30 |
| Total-Pt[b] | 8.10 ± 3.50 | 4.90 ± 1.30 | 2.45 ± 0.60 | 1.20 ± 0.60 | 0.85 ± 0.70 |
| DCM-Dex/CDDP | | | | | |
| Free-Pt[a] | 0.70 ± 0.05 | 0.75 ± 0.05 | 0.60 ± 0.05 | 0.50 ± 0.05 | 0.55 ± 0.05 |
| Total-Pt[b] | 13.65 ± 3.25 | 11.85 ± 4.15 | 11.00 ± 4.40 | 9.70 ± 3.50 | 10.55 ± 3.00 |

[a] Free Pt means serum Pt could pass through ultrafiltration filter (cut off: $M_w = 1.2 \times 10^4$). Pt not bound to protein or polymer.
[b] Total Pt means all of serum Pt (Pt bound to protein or polymer + free Pt).

TABLE 2.  Half-Life of Pt in Blood Stream after I.V. Injection Estimated by Compartmental Model Analysis.[a]

|  |  | CDDP | OX-Dex/CDDP | DCM-Dex/CDDP |
|---|---|---|---|---|
| $t_{1/2}$ (min) | Total Pt[b] | 50.0 ± 10.4 | 35.0 ± 7.9 | 787.5 ± 136.4 |
|  | Free Pt[b] | 15.6 ± 5.0 | 32.6 ± 3.9 | 316.6 ± 162.3 |

[a]Data are shown as the mean ± S.D.
[b]Free Pt means serum Pt passed through ultrafiltration filter ($M_w$ = 12,000): Pt not bound to protein or polymer. Total Pt means all of serum Pt: Pt bound to protein or polymer + free Pt.

Such high maintenance of cytotoxic activity in the *in vitro* serum medium and long *in vivo* circulation of Pt in the blood for DCM-Dex/CDDP conjugate were due to the strength of six-membered chelate-type coordination of the platinum atom with carrier polymer and steric hindrance of carrier polymer. Summarized in Table 2 are the results of the compartment model analysis. The show excretion rate of the Pt complex into urine by the DCM-Dex/CDDP conjugate compared with rapid excretion by the free CDDP in rats supported the longer circulation time for the Pt polymer complex in the blood (Figure 3).

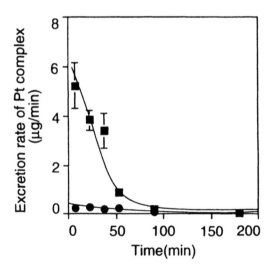

**Figure 3** Excretion rate of Pt complex into urine for free CDDP (■) and DCM-Dex/CDDP conjugate (●) in rats.

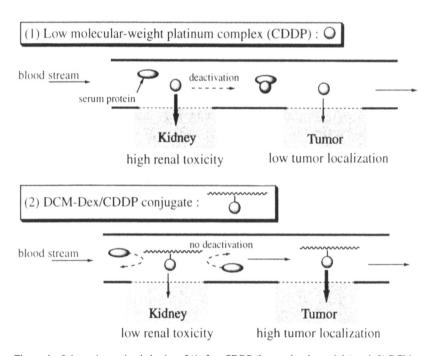

**Figure 4** Schematic moving behavior of (1) free CDDP (low molecular weight) and (2) DCM-Dex/CDDP conjugate from bloodstream (CDDP courtesy of Nihon Kayaku Co. Ltd.).

TABLE 3. Growth Inhibitory Effects of DCM-Dex/CDDP Conjugate and Free
CDDP against Colon 26 Tumor Cells Inoculated in Balb/c mice.[a]

| Day | 14 | 17 | 20 | 24 | 28 |
|---|---|---|---|---|---|
| Saline | 1.00 ± 0 | 9.33 ± 2.78 | 65.99 ± 18.93 | 138.95 ± 28.68 | 309.08 ± 52.63 |
| Free-CDDP | 1.00 ± 0 | 5.10 ± 2.21 | 29.63 ± 8.51 | 63.84 ± 14.27 | 112.00 ± 8.05 |
| DCM-Dex/CDDP | 1.00 ± 0 | 8.93 ± 2.70 | 22.67 ± 6.37 | 36.54 ± 15.61[b] | 57.17 ± 18.25[b] |

[a] The samples were injected intravenously to Balb/c mice on days 14, 17, and 20.
[b] $p < 0.01$ vs. CDDP by student's $t$ test.

Thus, the DCM-Dex/CDDP conjugate appeared to show a large accumulation at the inflammatory tumor site by the "enhanced perameability and retention" (EPR) effect (Figure 4). So, we evaluated the growth inhibitory effect of DCM-Dex/CDDP conjugate against colon 26 tumor cells in mice. The results shown in Table 3 supported an EPR effect by the DCM-Dex/CDDP conjugate.

Therefore, DCM-Dex/CDDP conjugates with passive targeting ability to tumor can be expected to alleviate severe renal toxicity CDDP in the clinical cancer therapy.

## REFERENCES

1. Rosenberg, R., Van Camp, L., Trosko, J., and Mansour, V. H. *Nature,* 1969, 222, 385.
2. Maeda, M., Oda, T., Matsumura, Y., and Kimura, M. *J. Bioact. Compat. Polym.,* 1988, 3, 27.
3. Ouchi, T. and Ohya, Y. *Progress in Polymer Science,* 1995, 20, 257.
4. Ohya, Y., Masunaga, T., Baba, T., and Ouchi, T. *J. Biomat. Sci. Polym. Ed.,* 1996, 7, 1085.
5. Ohya, Y., Masunaga, T., Baba, T., and Ouchi, T. *J. Macromol. Sci.-Pure Appl. Chem.,* 1996, A33, 1005.

# CHAPTER 18

# Synthesis of Linear and Hyperbranched Stereoregular Aminopolysaccharides by Oxazoline Glycosylation

JUN-ICHI KADOKAWA[1]
HIDEYUKI TAGAYA[1]
KOJI CHIBA[1]

## INTRODUCTION

R ECENT significant progress on synthetic reactions and technologies has made it possible to develop naturally occurring biopolymers such as nucleic acids, such as, DNA and RNA, and polypeptides. Polysaccharides, the rest of the most important biopolymers in nature, are high molecular weight carbohydrates formed as the result of polycondensation reactions through elimination of water between the hydroxy group at the C1 carbon atom of a monosaccharide and a hydroxy group of another sugar unit. Although the recent developments of new synthetic methodologies have enhanced the production of various polysaccharides [1], chemical approaches for the synthesis of well-defined polysaccharides require complicated procedures with many problems to be solved, control of the stereochemistry of anomeric carbons, control of regioselectivity of various hydroxy groups having a similar reactivity, and selective protection and deprotection of the hydroxy groups [2].

Natural and synthetic aminopolysaccharides have recently attracted much attention because of their unique structures and properties that are generally different from those of normal polysaccharides such as cellulose. For example, chitin is the most abundant aminopolysaccharide among the naturally occurring polysaccharides and has been of great interest in numerous scientific and application field (Scheme 1) [3].

[1]Department of Materials Science & Engineering, Faculty of Engineering, Yamagata University, Yonezawa 992-8510, Japan.

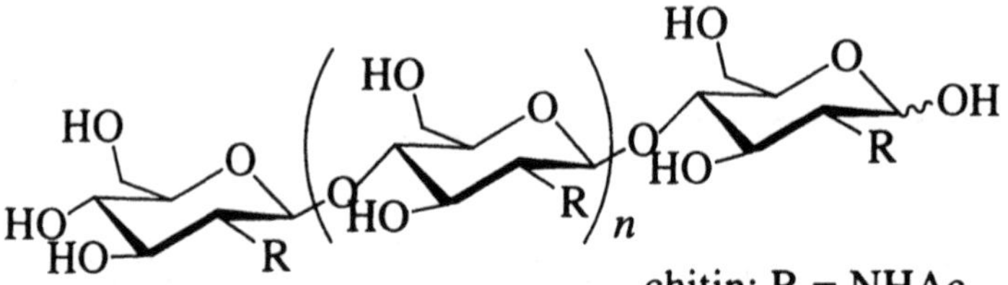

chitin; R = NHAc
cellulose; R = OH

**Scheme 1.**

Synthesis of aminopolysaccharides, therefore, is one of the important research areas in the field of functional materials, examples of biorelated polymers, antibacterial substance, and biodegradable polymers as well as materials for drugs and matrices of drug delivery systems. Only a few methods, however, such as ring-opening polymerization and enzymatic polymerization have been available for the precision synthesis of aminopolysaccharides [4,5].

Glycosylation, a reaction for the synthesis of glycosides, usually involves the condensation of a protected glycose derivative (glycosyl donor) with an appropriate aglycon derivative (glycosyl acceptor). The development of stereoselective glycosylation made it possible to synthesize virtually any desired sugar compound [6]. However, synthesis of polysaccharides by glycosylations as the polymerization reaction has not been well established yet. Only a few methods such as orthoester and Koenigs-Knorr glycosylations have been used for the synthesis of polysaccharides (Scheme 2) [2b, 7].

The oxazoline method is one of the excellent glycosylations for high stereoselectivity, in which the stereospecific glycosylation of alcohol with

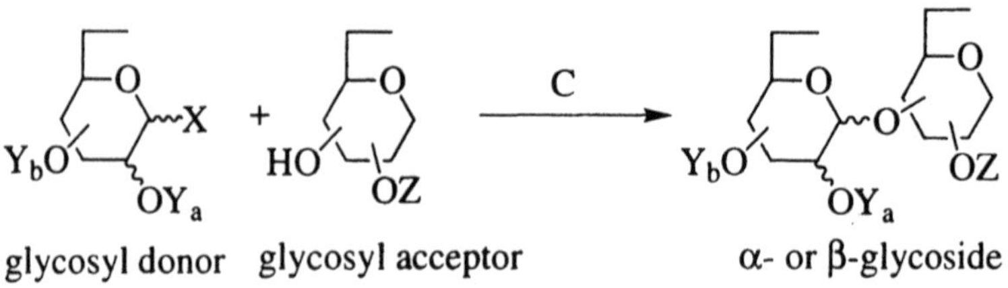

glycosyl donor    glycosyl acceptor               α- or β-glycoside

X = leaving group, C = promoter, $Y_a$, $Y_b$, and Z = protecting groups

**Scheme 2.**

an oxazoline of an amino sugar was achieved in the presence of an acid catalyst to give β-glycoside (Scheme 3) [8].

This glycosylation method has inspired us to employ sugar oxazoline compounds having hydroxy groups as a monomer for the synthesis of aminopolysaccharides with stereoregular β-glycosidic linkages. In this chapter we describe the synthesis of linear and hyperbranched stereoregular aminopolysaccharides by the oxazoline glycosylation method of sugar oxazoline monomers having hydroxy groups.

**EXPERIMENTAL**

POLYMERIZATION

A typical example for the polymerization of **1, 4,** and **8** was as follows. Under argon, monomer and acid catalyst (10 mol% for monomer) were dissolved in 1,2-dichloroethane at room temperature. Then the mixture was refluxed for an appropriate amount of time and cooled to room temperature. The products were isolated by general work-up procedures to give aminopolysaccharides **2, 5,** and **9.**

DEBENZYLATION

Debenzylation of **2** and **5** was carried out in a mixed solvent of DMF and water (3:1 v/v) in the presence of 10% Pd-C and a small amount of

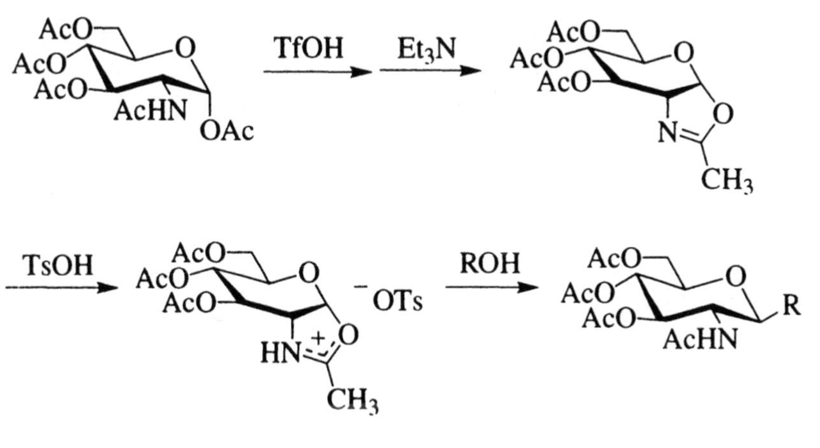

**Scheme 3.**

hydrochloric acid at 40°C for 24 h under a hydrogen atmosphere. After the residue was filtered off, the product was isolated as an acetone insoluble fraction.

### REACTION OF 9 WITH TIPDSCL$_2$

Polysaccharide **9** was treated with TIPDSCl$_2$ overnight in pyridine at room temperature. The reaction mixture was extracted with 2-butanone, and the product was isolated as a hexane-insoluble fraction.

### DETOSYLATION

Detosylation of **9** was carried out as follows. Sodium hydroxide and **9** were dissolved in 50% aqueous ethanol. The solution was heated under reflux for 5 h, cooled, neutralized with acetic acid, and concentrated under the reduced pressure. Methanol was added to the residue and the insoluble material was isolated by filtration and dried in vacuo to give the detosylated product.

## RESULTS AND DISCUSSION

### SYNTHESIS OF NATURAL- AND NON-NATURAL-TYPE LINEAR AMINOPOLYSACCHARIDES

Synthesis of natural-type aminopolysaccharide having dibenzylchitin structure was achieved by the polymerization of a sugar oxazoline monomer, **1** having one hydroxy group at position 4 (Scheme 4) [9]. The polymerization was carried out with an acid catalyst in 1,2-dichloroethane solvent at reflux temperature. All the $^1$H-NMR, $^{13}$C-NMR, and IR spectra as well as elemental analysis data of the isolated polysaccharide supported that the polymerization proceeded by the stereoregular glycosylation to give (1→4)-

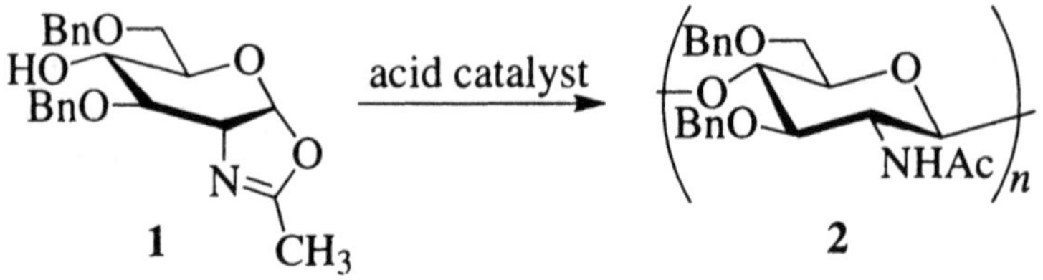

Scheme 4.

β-glucopyranane **2**, especially as there was no peak attributable to C1 of an α-glycoside (below δ 95) in the $^{13}$C NMR spectrum of the product, the β-configuration was tenable.

Shown in Table 1 are the polymerization results under various conditions. Three kinds of sulfonic acids, 10-camphorsulfonic acid (CSA), *p*-toluenesulfonic acid (TsOH), and trifluoromethanesulfonic acid (TfOH) were used as a catalyst for the polymerization of **1** to give aminopolysaccharide **2**. The highest molecular weight was 4,900 [degree of polymerization (DP) ca. 13] under the conditions of entry 3.

The debenzylation of aminopolysaccharide **2** by catalytic hydrogenation in the presence of Pd-C was carried out to give free aminopolysaccharide **3** (Scheme 5) [10]. Previously debenzylation of the synthetic polysaccharide derivatives had been carried out by Birch reduction (sodium in liquid ammonia) [11]. In the case of **2**, the Birch reduction was not effective. Consequently hydrogenation was used for the debenzylation of **2**. The $^1$H-NMR spectrum (acetic acid-$d_4$) of the debenzylated product obtained by the hydrogenation of **2** is very similar to that of natural chitin measured in formic acid-$d_2$, supporting the oligochitin structure of the debenzylated polysaccharide. However, absorption of aromatic protons was still present at δ 7.42, indicating the incomplete debenzylation. From the integrated ratio between this aromatic peak and the methyl peak of acetamido, it was calculated that ca. 90% of the benzyl groups were cleaved by the catalytic hydrogenation.

The molecular weight of the debenzylated polysaccharide was estimated to be 2,800 by GPC measurements with water eluent. This value is in very good agreement with the calculated value (2,600) based on the molecular weight of polysaccharide **2** before the debenzylation. The debenzylated

TABLE 1. Polymerization of **1** and **4** under Various Conditions.[a]

| Entry | Monomer | Catalyst | Time (h) | Yield (%) | $M_n$[b] |
|-------|---------|----------|----------|-----------|----------|
| 1 | **1** | CSA | 5 | 46 | 2,500 |
| 2 | **1** | CSA | 20 | 45 | 3,600 |
| 3 | **1** | CSA | 100 | 44 | 4,900 |
| 4 | **1** | TsOH | 5 | 47 | 3,400 |
| 5 | **1** | TfOH | 120 | 49 | 3,900 |
| 6 | **4** | CSA | 3 | 17 | 12,600 |
| 7 | **4** | CSA | 5 | 17 | 11,800 |
| 8 | **4** | CSA | 8 | 21 | 13,100 |
| 9[c] | **4** | CSA | 3 | 32 | 12,600 |

[a]Polymerization was carried out with 10 mol% catalyst at in 1,2-dichloroethane solvent at reflux.
[b]Determined by GPC.
[c]Polymerization was carried out in higher concentration solution.

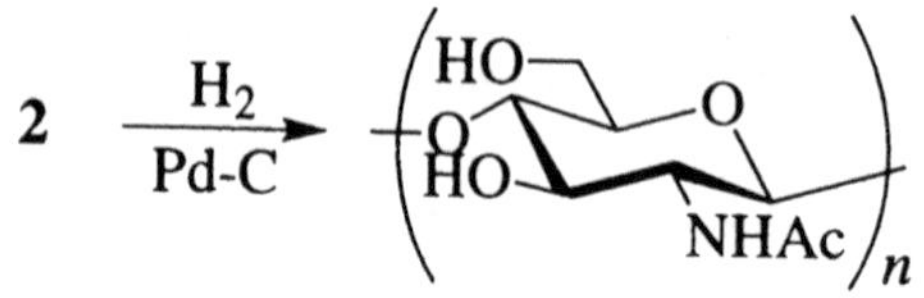

$$2 \quad \xrightarrow[\text{Pd-C}]{H_2} \quad \left[ \begin{array}{c} HO \\ O \\ HO \end{array} \hspace{-0.5em} \begin{array}{c} O \\ NHAc \end{array} \right]_n$$

**3**

Scheme 5.

polysaccharide is relatively soluble in water, acetic acid, DMSO, and DMF in spite of its oligochitin structure while natural chitin is insoluble in such solvents. The X-ray diffraction (XRD) measurement of the debenzylated polysaccharide did not show any peaks, indicating no crystallinity of the polysaccharide. This XRD result is completely different from that of natural chitin, which has shown the characteristic peaks due to the crystallinity of natural chitin [5]. This is probably due to the existence of benzyl groups in the synthetic polysaccharide. The lack of crystallinity may cause the higher solubility compared with natural chitin.

The polymerization of **1** mentioned above should be compared with the enzymatic synthesis of chitin reported by Kobayashi and coworkers, in which an oxazoline derivative of $N,N'$-diacetylchitobiose, the repeating unit of chitin was polymerized in the presence of chitinase enzyme via ring-opening addition process to give an artificial chitin (Scheme 6) [5]. The method using an enzyme, however, may not enable synthesis of non-natural-type aminopolysaccharide because the reaction catalyzed by chitinase enzyme is limited to the formation of $(1\rightarrow4)$-$\beta$-glycosidic linkage.

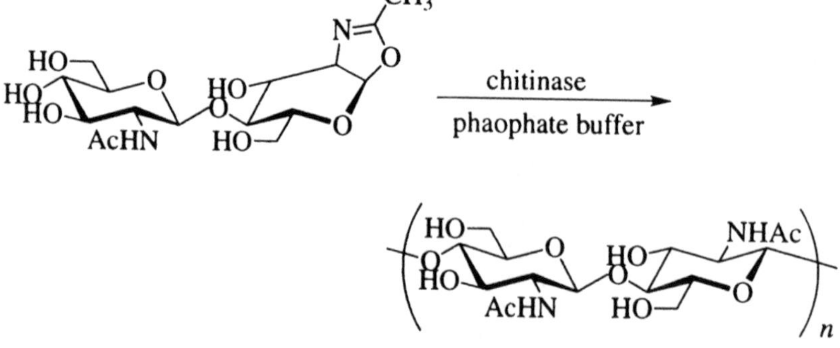

Scheme 6.

The following attempt at the polymerization of another sugar oxazoline monomer was made to synthesize a non-natural-type aminopolysaccharide. The polymerization of a sugar oxazoline monomer, **4,** with one hydroxy group at position 6 was carried out with CSA catalyst in 1,2-dichloroethane at reflux temperature [10]. The GPC chart of the reaction mixture, however, showed a bimodal profile consisting of a peak in the polymer region and a peak in the low molecular weight region; the former corresponded to the molecular weight of ca. 10,000 and the latter was located in the same region as the peak of monomer **4.** These GPC data indicated that the reaction mixture contained two products. To isolate the former polymeric material from the reaction product, the mixture was poured into a large amount of 1,2-dimethoxyethane (DME). The GPC profile of the DME-insoluble part showed one peak in the polymer region, indicating that the product polymer could be isolated. On the other hand, the GPC profile of the DME soluble part showed a major peak in the low molecular weight region accompanied by a small peak in the polymer region.

The structure of the DME insoluble polymer was determined, by $^1$H-NMR, $^{13}$C-NMR, and IR spectra as well as elemental analysis, to be $(1{\rightarrow}6)$-$\beta$-glucopyranan **5** obtained by stereoregular glycosylation. On the other hand, the low molecular weight product, isolated from the reaction mixture by using silica gel column chromatography, was determined to be an anhydro sugar **6.** These data indicated that the intramolecular cyclization of monomer **4** occurred during the reaction (Scheme 7). Furthermore, it was confirmed that the polymerization and the intramolecular cyclization of **4** that took place as a parallel reaction to each other and formed **6,** was stable under the reaction conditions.

The polymerization results of **4** catalyzed by CSA are shown in Table 1. The molecular weights of the polymers, obtained under the reflux tempera-

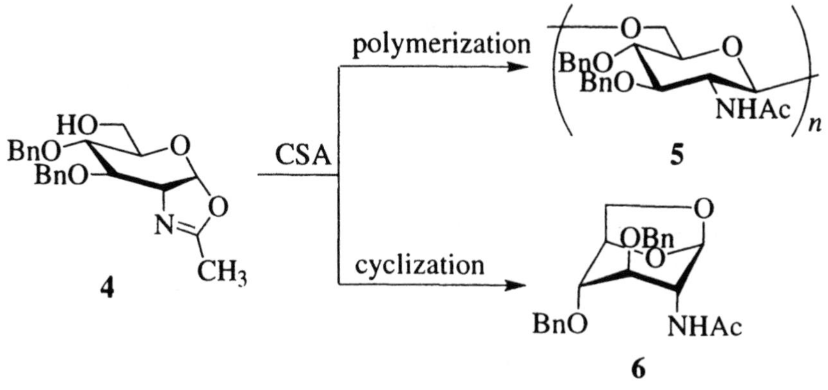

**Scheme 7.**

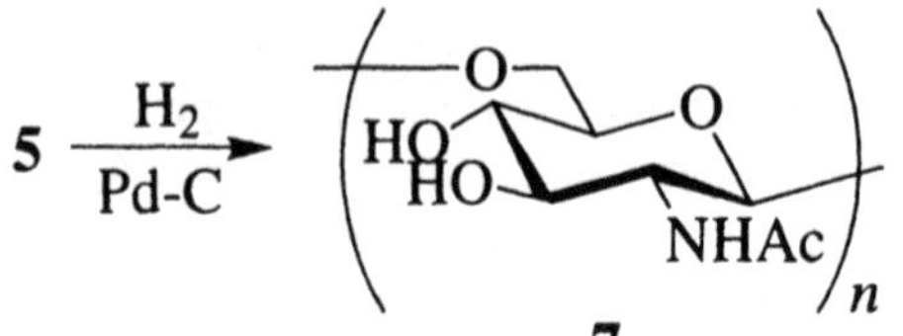

Scheme 8.

ture of 1,2-dichloroethane using CSA with 10 mol% monomer, were around 12,000. Under higher concentration condition, the yield of the polymer was higher. This is attributed to the prevention of the intramolecular cyclization under the higher concentration condition. The molecular weight of **5** is higher than that of **2,** probably due to the higher reactivity of a primary alcohol at position 6 of **4** than that of a secondary alcohol at position 4 of **1.**

The hydrogenation of **5** was carried out in an experimental manner similar to that of **2** (Scheme 8). No observation of the aromatic peak in the NMR spectra of the product indicated the occurrence of the perfect debenzylation, giving rise to polysaccharide **7.** The molecular weight determined by GPC with water eluent was 5,600, which is in good agreement with the calculated value (6,200). The solubility of **7** in water, DMSO, and DMF is higher than that of **3.**

## SYNTHESIS OF HYPERBRANCHED AMINOPOLYSACCHARIDE

The polymerization using sugar oxazoline monomers is not limited to the synthesis of linear aminopolysaccharides as described above and can be extended to formation of a hyperbranched material. Synthesis of hyperbranched aminopolysaccharide was achieved by acid-catalyzed polymerization of a sugar oxazoline monomer, **8,** having two hydroxy groups at position 3 and 4, which can be considered as an $AB_2$ monomer (Scheme 9) [12]. This is the first example of the synthesis of a hyper-

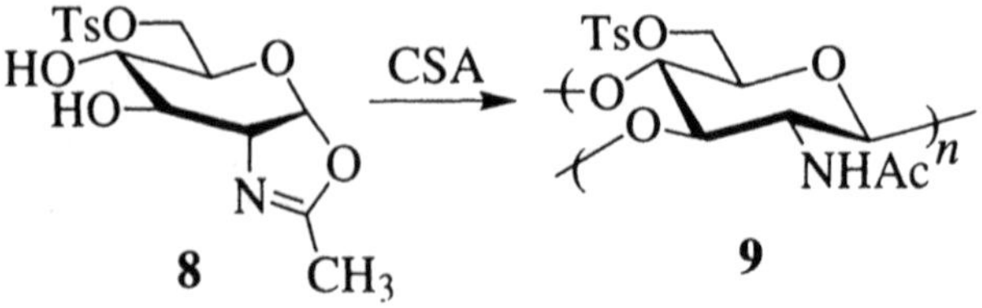

Scheme 9.

branched polysaccharide. The polymerization of **8** was carried out with CSA as an acid catalyst in 1,2-dichloroethane solvent. All the NMR spectroscopic data of the product supported that the each unit has β-glucopyranan structure **9**.

The number-average molecular weight ($M_n$) obtained after 3 h was determined to be 6,300 (DP ca. 17.6) by GPC with polystyrene standards with DMF eluent. On the other hand, the weight-average molecular weight ($M_w$) of the same product determined by the light scattering method was $4.5 \times 10^5$. Branched polymers are known to assume a spherical conformation in solution [13]. Therefore, it may be considered that the actual $M_n$ value of the present polysaccharide is larger than that estimated by GPC.

The degree of branching (DB) can be defined as the number of branched units relative to the number of total units. For high molecular weight polymers, the number of branching points is nearly equal to the number of terminal units. To determine the DB value by the molar ratio of terminal units to total units, the product polysaccharide **9** was treated with 1,3-dichloro-1,1,3,3-tetraisopropyldisiloxane (TIPDSCl$_2$), a reagent for the protection of two adjacent hydroxy groups in a saccharide [14]. Each terminal unit in **9** has two hydroxy groups, whereas only one hydroxy group exists in each linear unit. Therefore, the terminal units in **9** should be able to react quantitatively with TIPDSCl$_2$, whereas the linear and branching units should not react with TIPDSCl$_2$. The reaction of **9** with TIPDSCl$_2$ was carried out in pyridine (Scheme 10).

From the integrated ratio between the peak due to the tetraisopropyldisiloxanyl group and the other peaks of acetamido and tosyl groups, the

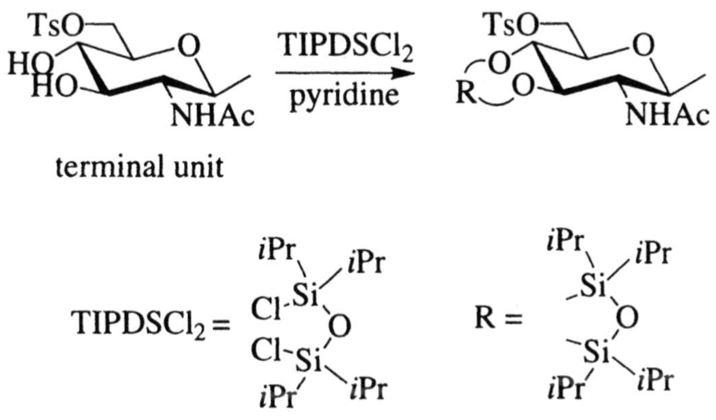

Scheme 10.

TABLE 2. Polymerization of **8** with Various Reaction Times[a]

| Entry | Time (h) | Yield (%) | $M_w$[b] | $M_n$[c] | Terminal Units[d] Total Units |
|-------|----------|-----------|----------|----------|-------------------------------|
| 1 | 1 | 57.3 | 230,000 | 5,500 | 0.45 |
| 2 | 2 | 48.5 | 290,000 | 4,800 | 0.44 |
| 3 | 3 | 49.4 | 450,000 | 6,300 | 0.51 |
| 4 | 5 | 57.5 | 760,000 | 6,600 | 0.49 |

[a]Polymerization was carried out with 10 mol% CSA in 1,2-dichloroethane at reflux.
[b]Determined by the light-scattering method.
[c]Determined by GPC.
[d]Determined by 1H-NMR spectroscopy. *Source:* adapted from Reference [12].

number of the terminal units in the polysaccharide was calculated as 0.51. In the complete hyperbranched polymer, the number of the terminal units is nearly a half of the total units when the molecular weight of the polymer is high. Therefore, the DB of **9** obtained with the reaction time of 3 h can be considered to be almost perfect. The reaction time of 5 h also gave **9** with almost perfect branching, as the ratio of the terminal units to the total units was about 0.5. The ratio of the terminal units in the polysaccharides obtained with reaction times of 1–2 h was 0.44–0.45, which indicates incomplete DB in these polysaccharides (Table 2).

Detosylation of **9** was carried out under alkaline conditions in aqueous ethanol at reflux temperature to yield a free hyperbranched aminopolysaccharide, **10a,** (Scheme 11). The product was isolated as a methanol insoluble fraction and is insoluble in common organic solvents and water but soluble in formic acid. The IR spectrum of the product (Figure 1) showed the disappearance of the absorptions due to the tosyl S $=$ O group, supporting occurrence of complete detosylation. Partial deacetylation of the acetamido groups of **9** is possible to give unit **10b,** although the absorption at 1,647 cm$^{-1}$, attributable to the C $=$ O moiety of the acetamido group, was present in the IR spectrum of the product.

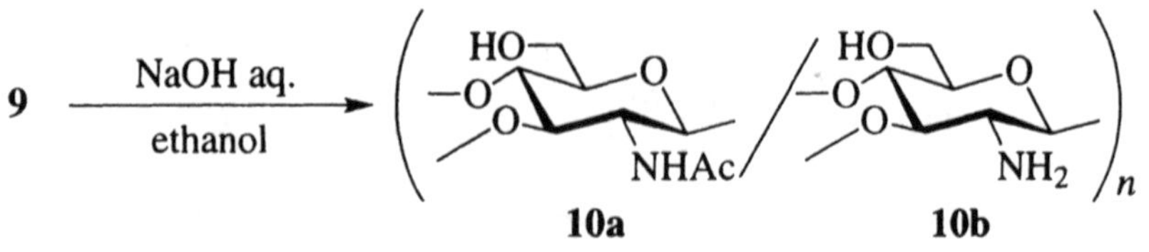

Scheme 11.

Aminopolysaccharides **9** and **10** can be expected to have unusual properties because of the hyperbranched structure. The thermal properties of **9** and
**10** were examined (Figure 2) by thermogravimetric analysis (TGA). The
TGA curve of **9** shows that decomposition of the tosyl group started at
166°C and showed a 45% weight loss up to 337°C. The residual material exhibited thermal resistance above 337°C, but a second weight loss occurred at

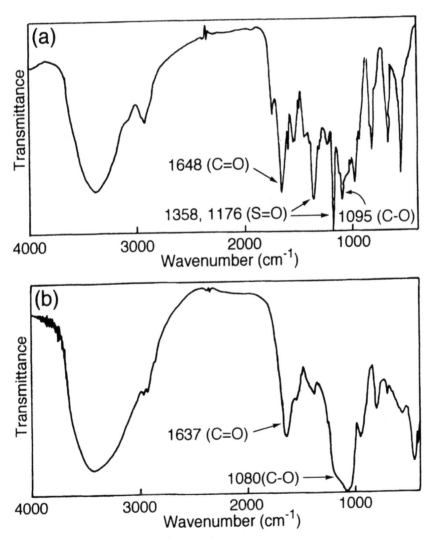

**Figure 1** IR spectra of tosylated (a) and detosylated (b) hyperbranched aminopolysaccharides **9**
and **10**.

510°C. Decomposition was complete at 600°C. As shown in Figure 2, the thermal decomposition of **10** occurred gradually with 65% residual weight still observable at 600°C. The TGA curves of **9** and **10** indicated that the thermal stabilities of the present hyperbranched aminopolysaccharides were much higher than those of normal linear polysaccharides. For example, chitin showed a significant weight loss at 275–428°C.

## CONCLUSIONS

Acid-catalyzed polymerization of sugar oxazoline monomers **1** and **4** proceeded by stereoregular glycosylation to give natural- and non-natural-type aminopolysaccharides **2** and **5**, respectively. The debenzylation of the product aminopolysaccharides was carried out to give free aminopolysac-charides. The debenzylation of **2** was incomplete, whereas the perfect debenzylation of **5** took place to produce polysaccharide **7**. Furthermore, hyperbranched aminopolysaccharide **9** was synthesized by acid-catalyzed

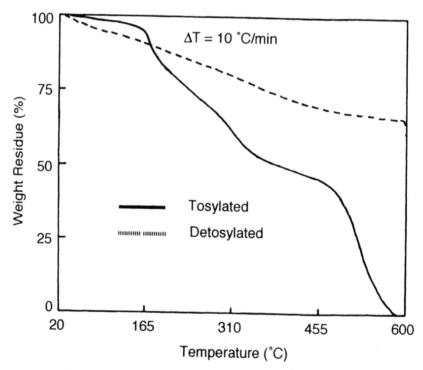

**Figure 2** TG curves of hyperbranched aminopolysaccharides in nitrogen.

polymerization of AB$_2$-type sugar oxazoline monomer **8**. The DB of the aminopolysaccharide obtained under the selected conditions was almost perfect. Detosylation of **9** proceeded under the alkaline conditions, giving rise to a free hyperbranched aminopolysaccharide. However, partial deacetylation of the acetamido groups of **9** was possible. Application of amino-polysaccharides synthesized here can be expected as materials for drugs and matrices of drug delivery systems. The hyperbranched aminopolysaccharide **9** is very interesting, for such polysaccharides could have biorelated and medical materials. Various oligosaccharides with highly branched chain structures have been found in glycoproteins at cell surface and intereceller systems [15]. The chemical syntheses achieved has allowed further study of their biological functions and unique structures [16].

## REFERENCES

**1a.** Lemieux, R. U. *Chem. Soc. Rev.* 1978, 7, 423.

**1b.** Paulsen, H. *Chem. Soc. Rev.* 1984, 13, 15.

**2a.** Schuerch, C. *Adv. Polym. Sci.* 1972, 10, 173.

**2b.** Schuerch, C. In *Advances in Carbohydrate Chemistry,* Tipson, R. S., and Horton, D., Eds., Academic Press, New York, 1981, Vol. 39, p. 157.

**2c.** Sumitomo, H., and Okada, M. In *Ring-Opening Polymerization,* Ivin, K. J., and Saegusa, T., Eds., Elsevier, London, 1984; Vol. 1, p. 299.

**3.** Muzzarelli, R. A. A., Jeuniaux, C., and Gooday, G. W. *Chitin in Nature and Technology,* Plenum Publishing, New York, 1986.

**4a.** Uryu, T., Hatanaka, K., Matsuzaki, K., and Kuzuhara, H. *Macromolecules* 1983, 16, 853.

**4b.** Kang, B. W., Hattori, K., Yoshida, T., Hirai, M., Choi, Y. S., and Uryu, T. *Macromol. Chem. Phys.* 1997, 198, 1331.

**5.** Kobayashi, S., Kiyosada, T., and Shoda, S. *J. Am. Chem. Soc.* 1996, 118, 13113.

**6a.** Paulsen, H. *Angew. Chem. Int. Ed. Engl.* 1982, 21, 155.

**6b.** Schmidt, R. R. Angew. *Chem. Int. Ed. Engl.* 1986, 25, 212.

**6c.** Kunz, H. *Angew. Chem. Int. Ed. Engl.* 1987, 26, 294.

**7a.** Kochetkov, N. K., Bochkov, A. F., and Yazlovetsky, I. G. *Carbohydr. Res.* 1967, 5, 243.

**7b.** Kochetkov, N. K. *Tetrahedron* 1987, 43, 2389.

**8.** Zurabyan, S. E., Volsyuk, T. P., and Khorlin, A. *J. Carbohydr. Res.* 1969, 9, 215.

**9.** Kadokawa, J., Watanabe, Y., Karasu, M., Tagaya, H., and Chiba, K. *Macromol. Rapid. Commun.* 1996, 17, 367.

**10.** Kadokawa, J., Kasai, S., Watanabe, Y., Karasu, M., Tagaya, H., and Chiba, K. *Macromolecules* 1997, 30, 8212.

**11.** For example: Ichikawa, H., Kobayashi, K., and Sumitomo, H. *Carbohydr. Res.* 1988, 179, 315.

12. Kadokawa, J., Sato, M., Kasai, S., Karasu, M., Tagaya, H., and Chiba, K. *Angew. Chem. Int. Ed. Engl.* 1998, 37, 2373.

13a. Bywater, S. *Adv. Polym. Sci.* 1979, 30, 89.

13b. Rempp, P., Franta, E., and Herz, J. E. *Adv. Polym. Sci.* 1988, 86, 145.

14. Verdegaal, C. H. M., Jansse, J. F. M., de Rooij, J. F. M., and van Boom, J. H. *Tetrahedron Lett.* 1980, 21, 1571.

15. Kornfeld, R., and Kornfeld, S. *Annu. Rev. Biochem.* 1976, 45, 317.

16. Ogawa, T., Katano, K., Sasajima, K., and Matsui, M. *Tetrahedron* 1981, 37, 2779.

# CHAPTER 19

# Diameter and Diameter Distributions of Poly(L-lactide) Microspheres by Ring-Opening Polymerization of L-Lactide and from Earlier Synthesized Polymers

STANLEY SLOMKOWSKI[1]
STANLEY SOSNOWSKI[1]

## INTRODUCTION

MICROSPHERES made of various polymers biodegradable to nonharmful compounds that could be metabolized and/or removed from organisms found many applications in medicine as carriers of drugs or other bioactive compounds [1–4]. Besides chemical composition there are also other properties of microspheres that are of primary importance to their medical applications. In particular, average diameters and diameter distributions illustrated in Table 1 based on data published in [5] are the relationship between the diameters of microspheres and their localization in various cells and tissues of the human body.

Biodegradable polymeric microspheres are usually obtained from earlier polymers synthesized by one of the following methods:

- polymer emulsification followed by solvent evaporation
- polymer emulsification followed by solvent extraction
- spray-drying of polymer solution

Over the years we have developed methods for the direct synthesis of hydrolytically and/or enzymatically degradable microspheres by ring-opening polymerization of ε-caprolactone and lactides [6–12]. The diameters of these microspheres are usually less than 3 μm. In this chapter we discuss

[1]Center of Molecular and Macromolecular Studies, Polish Academy of Sciences, Sienkiewicza 112, 90-363 Lodz, Poland.

TABLE 1. Diameters of Polymeric Microspheres and Their Localization in Human Body after Subcutaneous and/or Intravenous Injection (Based on Data from Reference [5]).

| Diameter of Microspheres | Localization in the Body |
|---|---|
| $D > 10\ \mu m$ | To large to pass capillary blood vessels of many organs. Localized activity. |
| $5\ \mu m < D < 10\ \mu m$ | Captured in lung capillary blood vessels. |
| $D < 5\ \mu m$ | Particles captured by the mononuclear phagocyte system (MPS). |
| $1\ \mu m < D < 3\ \mu m$ | Particles captured in spleen. |
| $0.1\ \mu m < D < 1\ \mu m$ | Particles captured in liver. |
| $D < 0.1\ \mu m$ | Particles captured in bone marrow. |

the relation between diameters and diameter distributions of microspheres and the parameters characterizing processes of their preparations from the earlier synthesized polymers. We describe also a method of seeded dispersion polymerization of L-lactide (Lc) leading to poly(L-lactide) microspheres (poly(L-Lc)) with diameters $\langle D_n \rangle > 5\ \mu m$.

**EXPERIMENTAL PART**

PREPARATION OF MICROSPHERES FROM THE EARLIER SYNTHESIZED POLY(L-LACTIDE)

Poly(L-lactide) microspheres were obtained from polymers synthesized as described by Duda and Penczek [13] in polymerization of Lc initiated with tin(II) 2-ethylhexanoate and carried out in 1,4-dioxane. Microspheres were prepared from a polymer with $M_n = 9,300$ and $M_w/M_n = 1.06$.

Poly(L-Lc) Microspheres by Solvent Evaporation Method

Microspheres were prepared by dissolving poly(L-Lc) (6.6 g) in $CHl_2Cl_2$ (6.6 mL). Then the poly(L-Lc) solution was added slowly to 150 mL of water containing poly(vinyl alcohol) (1% wt/v) stirred at 170 rev/min at room temperature for 30 min. During this time the $CH_2Cl_2$ evaporated. Thereafter, poly(L-Lc) microspheres were isolated by sedimentation.

Poly(L-Lc) Microspheres by Solvent Extraction Method

Poly(L-Lc) (10 g) was dissolved in 10 mL of (-caprolactone. The solution obtained was dispersed by ultrasound or by mixing in 40 mL of heptane containing sorbitan trioleate (Span 85, 3% wt/v). This dispersion was rapidly introduced into isopropanol containing polyvinylpyrrolidone (PVP, 5% wt/v). The microspheres produced were washed with isopryopanol containing PVP (0.1% wt/v) and water with PVP (0.1% wt/v) and isolated by centrifugation.

### Microspheres by Spray-Drying

There microspheres were prepared from a solution of Resomer RG756 (copolymer of D,L-lactide and glycolide, 75%:25%, product of Boehringer, Germany) in $CH_2Cl_2$ using Mini Spray Dryer 190 (Büchi, Switzerland).

### POLY(L-Lc) MICROSPHERES BY RING-OPENING POLYMERIZATION OF Lc

Polymerizations of L-Lc were carried out in a 1,4-dioxane:heptane mixture (1:4 v/v) in the presence of poly(dodecyl acrylate)-g-poly(ε-caprolactone) [denoted later as poly(DA-CL)] as a surfactant. Number average molecular weight of poly(DA-CL) was $\langle M_n(\text{poly(DA-CL)})\rangle = 62,000$. Ratio of molecular weight of poly(ε-caprolactone) grafts to molecular weight of poly(DA-CL) was $\langle M_n(\text{CL})\rangle/\langle M_n(\text{poly(DA-CL)})\rangle = 0.18$. Polymerizations were initiated with tin(II) 2-ethylhexanoate. Synthesis of surfactant and details of dispersion polymerizations of L-Lc are described in our earlier papers [6,8]. Seeded polymerizations of L-Lc were carried out in the following way. First, the seed microspheres were synthesized in a mixture containing monomer, initiator, and surfactant with initial concentrations: $[\text{Lc}]_o = 3.50 \cdot 10^{-1}$ mol/L, $[\text{tin(II) 2-ethylhexanoate}]_o = 5.70 \cdot 10^{-3}$ mol/L, $[\text{poly(DA-CL)}] = 1.67$ g/L dissolved in 1,4-dioxane:heptane. Polymerization was carried out under argon, at 95°C, with stirring 60 rev/min. At predetermined time periods, after microspheres were already formed, samples were taken for analysis and new portions of monomer were added.

### MEASUREMENTS OF DIAMETERS OF MICROSPHERES AND OF MOLECULAR WEIGHT OF POLYMERS IN MICROSPHERES

Microspheres were monitored by scanning electron microscopy (SEM; JEOL 35C), and their diameters were determined from the corresponding SEM microphotographs. Typically, ca. 500 particles in randomly sampled areas of microsphere specimens were analyzed. Molecular weight of poly(L-Lc) was determined by GPC. A system consisting of a LKB 2150 pump, Ultrastyragel 1,000, 500, 100, 100 columns, and Wyatt Optilab 903 interferometric refractometer was used for the measurements. GPC traces were analyzed by using calibration with narrow polydispersity ($M_w/M_n < 1.15$) poly(ε-caprolactone) samples.

## RESULTS AND DISCUSSION

## DIAMETERS AND DIAMETER DISTRIBUTIONS OF MICROSPHERES

Typical diameter distributions of poy(L-Lc) microspheres obtained by solvent evaporation, solvent extraction, spray-drying, and ring-opening polymerization methods are illustrated in Figure 1. The diameters of the microspheres by solvent extraction and solvent evaporation methods are much larger than the diameters of microspheres obtained by direct synthesis by ring-opening dispersion polymerization of L-Lc and the diameters of the particles (usually irregular) obtained by spray drying. Moreover, the diameter distributions of particles obtained from the earlier synthesized polymers were significantly greater than in the case of microspheres by dispersion polymerization of Lc. It is important to stress that the molecular weight of poly(Lc) in different types of microspheres was similar ($M_n =$ 9,300 and $M_w/M_n = 1.06$ for microspheres by solvent evaporation and solvent extraction method and $M_n = 8,360$ and $M_w/M_n = 1.06$ for microspheres by ring opening polymerization).

Microspheres by solvent extraction method were obtained with rate of mixing equal 300 rev/s. Particles by spray drying were produced with spray dryer operated with an inlet temperature of 50°C and outlet temperature of 45°C. The air flow indicator was set at 700 and the aspirator at 5. The polymer solution (concentration 0.5% wt/v) was supplied at 10 mL/min. The concentrations of monomer, initiator, and surfactant in ring-opening dispersion polymerization leading to microspheres were as follows: $[Lc]_o =$ $2.77 \cdot 10^{-1}$ mol/L, $[tin(II) \text{ 2-ethylhexanoate}]_o = 4.9 \cdot 10^{-3}$ mol/L, $[poly(DA-CL)] = 1.6$ g/L.

The number average diameter and diameter polydispersity factor for microspheres obtained by solvent evaporation method were $D_n = 34.4$ μm and $D_w/D_n = 2.68$. Solvent extraction gave microspheres with $D_n = 20.0$ μm and $D_w/D_n = 2.06$. Spray-drying yielded particles with much smaller diameters ($D_n = 1.54$ μm); however, their diameter polydispersity was still high ($D_w/D_n = 1.86$). The number average diameter of microspheres obtained by ring-opening polymerization was equal $D_n = 3.6$ μm, i.e., was slightly larger than for particles by spray-drying but significantly smaller than for microspheres by solvent evaporation and solvent extraction methods. The polydispersity of diameters of microspheres by ring-opening polymerization was very low ($D_w/D_n = 1.07$).

Attempts to obtain regular spherical shaped particles by spray drying were unsuccessful. Apparently, collisions of not completely solidified particles in the jet stream supplying primary droplets into the drying chamber result in their coalescence and/or distortion of shape. The perspective of obtaining microspheres with well-controlled shape, diameter, and diameter

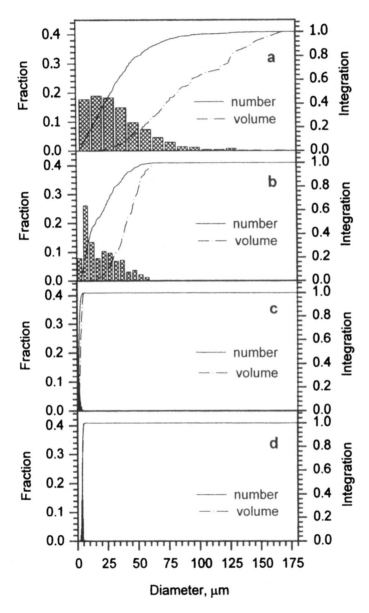

**Figure 1** Diameter distribution of microspheres obtained by (a) solvent evaporation, (b) solvent extraction, (c) spray drying, and (d) ring-opening polymerizations. Conditions at which particles were obtained are given in text.

distribution was rather vague. Therefore, we decided to investigate other methods, to obtain microspheres with controlled diameters and with low diameter polydispersity. Especially interesting to us was to obtain nearly monodisperse microspheres with diameters larger than 5 μm. Such microspheres, based on data in Table 1, could avoid fast *in vivo* removal by phagocytosis.

## DIAMETERS OF MICROSPHERES OBTAINED BY SOLVENT EXTRACTION METHOD: INFLUENCE OF RATE OF MIXING AND/OR OF POWER OF SONICATION

The rate of mixing was changed in these experiments from 100 to 300 rev/s. Examples of diameter distributions of microspheres obtained at the lowest and at the highest rates of mixing are shown in Figure 2. Plots in Figure 2 indicate that at both rates of mixing the size distribution of the microspheres is monomodal and that the difference between them is not very large.

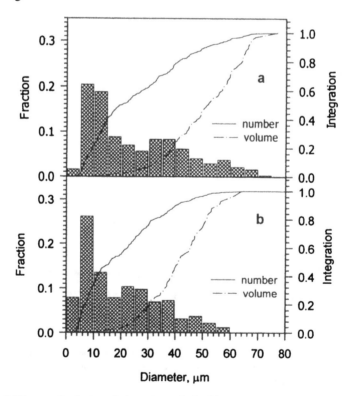

**Figure 2** Diameter distribution of microspheres obtained by solvent extraction method. Rate of mixing (a) 100 rev/s, (b) 300 rev/s.

Dependence of $D_n$ and $D_w/D_n$ on the rate of mixing is illustrated in Figure 3. From this plot it is evident that increasing the rate of mixing results in decreasing the diameters of the microspheres from 25.3 to ca. 20 μm. At rates of mixing lower than 100 rev/s it was difficult to obtain suspension of poly(Lc) microspheres due to the extensive coagulation of particles. However, at mixing rates from 100 to 300 rev/s the diameter polydispersity changed very little (from 2.00 to 2.07).

The results discussed above suggest that an adjustment in the rate of mixing in the formation of microspheres by solvent extraction method is not effective in achieving narrow dispersion of diameters of microspheres.

Plots illustrating the diameter distribution of poly(Lc) microspheres obtained by solvent extraction method with sonication equal 150 W and 300 W are shown in Figure 4.

Dependencies of $D_n$ and $D_w/D_n$ sonication power for poly(L,L-Lc) obtained by solvent extraction method are shown in Figure 5.

Based on the plots in Figure 5 it follows that sonication, in contrast to mixing, allows effective size regulation of microspheres. Microspheres with $D_n$ decreasing from 27.5 to 15.4 μm were obtained when the sonication power was increased from 150 to 300 W. However, variation of sonication power did not provide poly(Lc) microspheres with narrow diameter distribution. With sonication at 150 W, the diameter polydispersity parameter $D_w/D_n$ was equal to 1.94. With increasing power of sonication values,

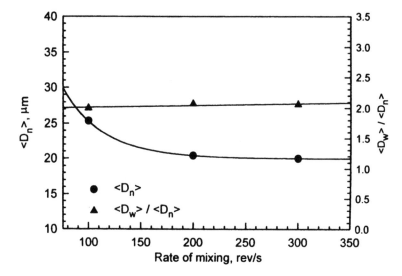

**Figure 3** Dependence of diameters and diameter distributions for microspheres obtained by solvent extraction method on the rate of mixing.

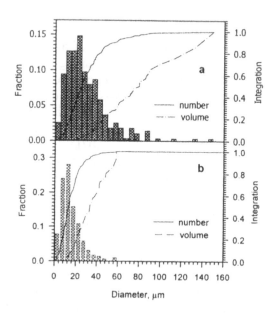

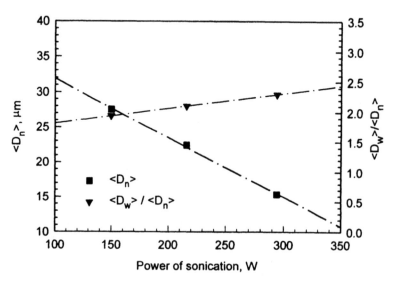

**Figure 4** Diameter distributions of microspheres obtained by solvent extraction method: (a) sonication 150 W, (b) sonication 300 W.

**Figure 5** Dependence of diameter and diameter distribution of microspheres obtained by solvent extraction method on sonication power.

$D_w/D_n$ slightly increased to 2.3. Possibly, sonication breaks polymer solution droplets into nonequal smaller ones. At higher power of ultrasound this process resulted in a higher degree of dispersion of primary droplets, producing smaller droplets with higher size dispersity.

## DIAMETERS AND DIAMETER DISTRIBUTION OF POLY(L-Lc) MICROSPHERES OBTAINED BY RING-OPENING POLYMERIZATION

After developing procedures for the polymerization of lactides and ε-caprolactone leading to polymers that form microspheres, we concentrated our attention on the mechanism of particle formation relative to polymerization conditions and size dispersity of microspheres. In these studies we found that nucleation of microspheres occurs during the initial period of polymerization. Aggregation of these primary microspheres is negligible, and the main fraction of monomer that has been added initially is converted into polymer with a constant number of particles in the polymerizing mixture [12,14]. Moreover, for ε-caprolactone polymerization, after microsphere nucleation all active centers are located inside of the growing particles and that polymerization occurs due to diffusion of monomer into the microspheres [9,12,14]. Narrow molecular weight distributions of polylactides obtained by dispersion polymerization suggests that during the initial polymerization period of this monomer not only are all the microspheres nucleated but also all macromolecules are initiated [8]. Thus, there are reasons to believe that in the dispersion polymerization of lactides, like in the dispersion polymerization of (-caprolactone, nucleation of microspheres results in confinement of all active centers inside of these particles. Apparently, size dispersity of microspheres depends mainly on the process of particle nucleation.

Nucleation of microspheres strongly depends on composition of poly(DA-CL) surface active agent. We found that microspheres with the narrowest size dispersity ($D_w/D_n < 1.15$) can be obtained when the ratio of molecular weight of poly(ε-caprolactone) grafts to the molecular weight of whole poly(DA-CL) is close to 0.25 [8,10]. When the concentration of poly(DA-CL) is close to the critical concentration at which macromolecular micelles are formed all the polylactide is in form of microspheres. The diameters of these microspheres are close to 3 (m. Decreasing of the concentration of poly(DA-CL) does not lead to the formation of microspheres with larger diameters but to the steadily increased fraction of polymer in form of shapeless aggregates. It became evident that by changing the concentration of surface active copolymer it is impossible to obtain uniform particles with diameters significantly larger 3 μm.

Rapid nucleation of uniform microspheres and their constant number during polymerization suggests that by using the appropriately high monomer concentration one could obtain larger microspheres with narrow diameter distribution. Unfortunately, the limited solubility of L-Lc in the reaction medium does not allow for polymerizations with the initial monomer concentrations higher than ca 0.4 mol/L. Therefore, we decided to check whether poly(L-Lc) micospheres with $\langle D_n \rangle > 5$ μm can be obtained by one pot seeded polymerization. In such synthesis, first, dispersion polymerization of L-Lc was carried out at typical conditions (e.g., at conditions described in experimental part, $[\text{L-Lc}]_0 = 3.50 \cdot 10^{-1}$ mol/L). Then, 1.5 h after initiation of the polymerization, a new portion of L-Lc was added, raising the overall concentration of monomer introduced to mixture ($[\text{L-Lc}] + [\text{L-Lc}$ units in polymer]) to $6.24 \cdot 10^{-1}$ mol/L. Subsequently, after the second 1.5 h another portion of monomer was introduced. At this moment, the total concentration of monomer and momomer units in poly(L-Lc) became equal to $9.10 \cdot 10^{-1}$ mol/L.

Before addition of the second and third portions of monomer and after final completion of the polymerization small samples of reaction mixture containing microspheres (ca. 10 μL) were taken for analysis. The size distributions of these microspheres are illustrated in Figure 6.

The number average diameter of microspheres obtained after the first step of polymerization was $D_n = 3.97$ μm, and parameter characterizing polydisperity of diameters was $D_w/D_n = 1.09$; after the second step $D_n = 5.44$ μm and $D_w/D_n = 1.13$; eventually, after completion of the polymerization after the third monomer addition $D_n = 6.36$ μm and $D_w/D_n = 1.20$. Thus, we noticed a substantial increase in the diameter of the microspheres without significant broadening diameter size dispersity.

Narrow size distribution of the microspheres obtained in the stepwise monomer addition polymerizations suggests that the number of microspheres is constant at each polymerization step, i.e., that polymerization after the addition of new portions of monomer does not lead to the formation of new particles but contributes to the growth of the already existing ones. This conclusion is supported by the plot in Figure 7 illustrating the dependence of the number average volume of the microspheres ($\langle V_n \rangle$), calculated according to Equation (1), on the total (combined after each addition) concentration of L-Lc introduced to the polymerizing mixture ($[\text{L-Lc}]_o$.

$$\langle V_n \rangle = \frac{(4/3)\pi \sum_{i=1}^{N} R_i^3}{N} \tag{1}$$

where $N$ denotes the number of analyzed microspheres and $R_i$ radii of the particles.

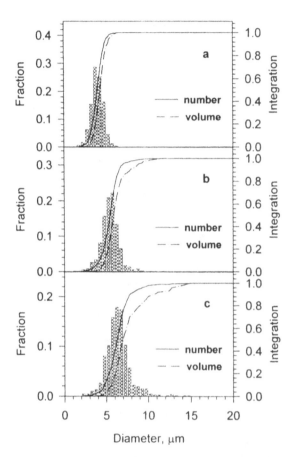

**Figure 6** Distribution of diameters of poly(ι-Lc) microspheres obtained in polymerization with stepwise addition of monomer after the (a) first, (b) second, and (c) third steps of polymerization. Conditions of the polymerization are given in the text.

It is important to stress that the relation between $\langle V_n \rangle$ and $[\text{L-Lc}]_o$ is described by the straight line (slope equal 220 $\mu m^3 \cdot L/mol$). Such dependence is possible only when the number of microspheres in the reaction mixture is constant. Concentration of microspheres ([microspheres] expressed as number of microspheres in 1 L of reaction mixture) can be calculated from equation:

$$[\text{microspheres}] = \frac{FW(\text{L-Lc}) \cdot 10^{12}}{\text{slope} \cdot d} \qquad (2)$$

in which $FW(\text{L-Lc})$ denotes the formula weight of L-Lc monomeric unit equal 114.13 and $d$ density of poly(L-Lc) for which we used value 1.25 $g/cm^3$

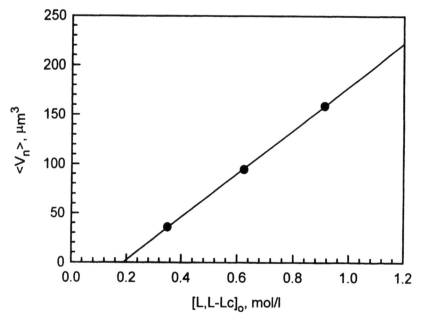

**Figure 7** Dependence of $\langle V_n \rangle$ on [L-Lc]$_o$ for dispersion polymerization of L-Lc with the stepwise addition of monomer.

($d$ equals 1.25 and 1.28 g/cm$^3$, for amorphous and crystalline poly(L-Lc), respectively [15,16]. Concentration of microspheres found from plot shown in Figure 6 was equal $5.24 \cdot 10^{-11}$ particles/L.

The line in the plot shown in Figure 7 intersects the abscissa for the concentration of L-Lc equal 0.18 mol/L. Therefore, under the conditions at which the polymerization was carried out only at [L-Lc]$_o$ higher than $1.8 \cdot 10^{-1}$ mol/L will particles with volume different from 0 can be formed. This limiting concentration is due to the reversibility of the polymerization of L-Lc and its value determined in the present work is close to the one ($1.4 \cdot 10^{-1}$ mol/L), which was found for similar systems by measuring the concentrations of unreacted monomer [8].

Seeded dispersion polymerization was extensively investigated for radical systems [17]. Much less is known about seeded dispersion polymerizations with propagation on ionic and/or pseudoionic active centers. Awan et al. reported seeded ionic polymerization of styrene, which at certain conditions produced particles with narrow diameter size dispersity [18,19]. We presented the first data on the seeded ring-opening polymerization with constant number of microspheres.

## CONCLUSIONS

The number average diameter of microspheres obtained from polymers synthesized, by emulsification of polymer solutions followed by solvent extraction and/or solvent evaporation methods, can be controlled by choosing the appropriate conditions at which particles are produced. However, by this method particles with $\langle D_n \rangle > 15$ $\mu$m and with $\langle D_w \rangle / \langle D_n \rangle > 1.9$ are produced. Spray drying did not provide poly(L-Lc) particles with regular spherical shape. Direct synthesis of poly(L-Lc) microspheres by ring-opening polymerization with stepwise monomer addition can be used as a method of choice for the production of microspheres with diameters controlled to ca. 6 $\mu$m and with diameter polydispersity parameter $\langle D_w \rangle / \langle D_n \rangle < 1.20$.

## REFERENCES

1. Rosoff, M., Ed. *Controlled Release of Drugs: Polymers and Aggregate Systems;* VCH: New York, 1989.
2. Whateley, T. L., Ed. *Microencapsulation of Drugs,* Harwood Academic, Reading, 1992.
3. Park, K., Shalaby, W. S. W., and Park, H., Eds. *Biodegradable Hydrogels for Drug Delivery,* Technomic Publishing Co., Inc., Lancaster, 1993.
4. Benita, S., Ed. *Microencapsulation, Methods and Industral Applications,* Marcel Deker, New York, 1996.
5. Domb, A. J., Ed. *Polymeric Site-Specific Pharmacotherapy,* John Wiley & Sons, New York, 1994.
6. Sosnowski, S., Gadzinowski, M., and Slomkowski, S. *J. Bioact. Compat. Polym.* 1994, 9, 345.
7. Slomkowski, S. *Macromol.Symp.* 1996, 103, 213.
8. Sosnowski, S., Gadzinowski, M., and Slomkowski, S. *Macromolecules* 1996, 29, 4554.
9. Gadzinowski, M., Sosnowski, S., and Slomkowski, S. *Macromolecules* 1996, 29, 6404.
10. Slomkowski, S., Gadzinowski, M., and Sosnowski, S. *Macromol.Symp.* 1997, 123, 45.
11. Slomkowski, S., Sosnowski, S., and Gadzinowski, M. *Polym.Degr.Stab.* 1998, 59, 153.
12. Slomkowski, S., Gadzinowski, M., and Sosnowski, S. *Macromol.Symp.* 1998, 132, 451.
13. Duda, A., and Penczek, S. *Macromolecules* 1990, 23, 1636.
14. Slomkowski, S., Sosnowski, S., and Gadzinowski, M. *Colloids and Surfaces. A: Physchemical and Engineering Aspects* 1999, 153, 111–118.
15. Miyata, T., and Masuko, T. *Polymer* 1997, 38, 4003.

16. Miyata, T., and Masuko, T. *Polymer* 1997, 39, 5515.
17. Lovel, P. A., and El-Aasser, M. S., (eds) *Emulsion Polymerization and Emulsion Polymers;* John Wiley & Sons, Chichester, 1977, chapters 4, 5, 7, 11, 12, and 16.
18. Awan, M. A., Dimonie, V. L., and El-Aasser, M. S. *J. Polym. Sci.: Part A: Polym.chem.* 1996, 34, 2633.
19. Awan, M. A., Dimonie, V. L., and El-Aasser, M. S. *J. Polym. Sci.: Part A: Polym.Chem.* 1996, 34, 2651.

# CHAPTER 20

# Examination of Fluorescent Molecules as in situ Probes of Polymerization Reactions

FRANCIS W. WANG[1]
DEBORAH G. SAUDER[2]

## INTRODUCTION

**P**RESENTED in this chapter are the preliminary results of a study designed to examine the feasibility of using a fluorescent dye as an in situ indicator of the physical condition of a bone cement sample. The fluorescence behavior of dyes is often affected by solvent–solute interactions. Variations in solvent dielectric constant, solvent polarity, pH, viscosity, or the presence of hydrogen bonding or other strong intermolecular interactions can all produce substantial changes in fluorescence behavior. Previous studies have illustrated the use of both exciplex [1] and charge transfer, (CT) [2,3,4], probes in monitoring the degree of polymerization in a variety of systems. In the experiments described here we have examined the fluorescence behavior of anthracene and $Re(CO)_3ClL$, (where L = 4,7-diphenyl-1,10-phenanthroline) in commercial bone cements under different concentration and laboratory temperature conditions in which the degree of polymerization is known as a function of time from previous work [5].

The ultimate success of methyl methacrylate bone cements in surgical arenas depends on its application at an appropriate viscosity. Recent studies have raised concerns that the long-term stability of bone cements may be compromised by the empirical way in which the setting of samples is determined [6]. The literature from one manufacturer states that, in addition to the concentration effects one would expect in a biphasic free-radical

[1]Dental and Medical Materials Group, Polymers Division, National Institute of Standards and Technology, Gaithersburg, MD 20899, USA.
[2]Department of Chemistry and Physics, Hood College, Frederick, MD 21701, USA.

system, ambient temperature and humidity can substantially affect the setting time of a sample. It suggests that ". . . the working time . . . is best determined by the experience of the surgeon . . . "[7]. Farrar and Rose [8] have shown the substantial effects that small ambient temperature variations can have on the dynamic viscosity of a sample of bone cement over time. This study represents an initial effort to understand the behavior of fluorescent probes in methacrylate cements. Eventually, fluorescence may prove useful in providing an in situ quantitative measure of the extent of polymerization of a cement sample by providing a measure of its viscosity.

The bone cement used in these studies was a two-component system. The liquid component [9.75 mL methyl methacrylate (MMA); 0.25 mL N,N-dimethyl-p-toluidine (DMPT); 75 mg/kg hydroquinone] was mixed with a solid component [3.0 g poly(methyl methacrylate) (PMMA); 15.0 g MMA-styrene copolymer; benzoyl peroxide, mass fraction 2%; 2.0 g $BaSO_4$] to form the cement. Dissolution of the solid component proceeded simultaneously with polymerization once the cement was mixed.

The samples used in this study polymerized via a free-radical, addition mechanism. Although most free radical polymerizations require initiation by the addition of certain labile compounds and/or exposure to heat or light, methyl methacrylate will spontaneously polymerize at room temperature. Hydroquinone is therefore added to the liquid component of the cement to act as an inhibitor — it scavenges the radicals that spontaneously form in the system, limiting polymerization processes during storage. The benzoyl peroxide in the solid component of the cement is present at a sufficiently high concentration that it overwhelms the trace amount of hydroquinone present and acts as a free radical initiator, which is accelerated by DMPT once the solid and liquid cement components are mixed. The $BaSO_4$ serves to make the cement visible via X-ray examination once it is set.

## EXPERIMENTAL

The fluorescent probes used in this study were dissolved in the liquid component of the cement before mixing. The probe concentrations were adjusted until a maximum fluorescence emission from the probe dissolved in the MMA liquid component of the cement at room temperature was observed. Anthracene was recrystallized from alcohol before use. $Re(CO)_3ClL$ was synthesized in the manner described by Salman and Drickmar [9]. In this study, samples were prepared for fluorescence measurement by placing 1 to 3 g of the powder component directly in a 1-cm glass fluorescence cell. A known mass of the liquid component (plus probe) was injected into the cell using a syringe and mixed thoroughly with the powder component.

The dissolution of the PMMA polymer and MMA-styrene copolymer in the MMA liquid component occurred simultaneously with polymerization of the cement after mixing. Observations were made on samples maintained at ambient conditions. The presence of $BaSO_4$ in the solid phase rendered the samples opaque, so fluorescence measurements were made over time in front-face reflectance mode. Steady-state fluorescence spectra were obtained using a commercially available spectrophotometer. The bandwidth of the spectrophotometer under experimental conditions was 10 nm. All spectra were taken in ratio mode, so that fluctuations in the incident intensity at the excitation wavelength did not affect the results. No attempts were made to monitor or control the temperature of the system. However, because the volume of sample observed was small, the polymerization was relatively slow, and the sample was in direct contact with the room temperature cell, we did not anticipate that the exothermicity of the polymerization would cause temperature fluctuations large enough to affect the measurements. All intensity values used in the analysis of these results are taken from single-scan, uncorrected fluorescence data. Based on previous experience, we estimate a relative standard uncertainty of $\pm 5\%$ in the peak intensities from single-scan, uncorrected spectra.

## RESULTS

As in previous studies [1–4], both substantial spectral shifts and enhancements in fluorescence intensity were observed from the anthracene and Re compound probe molecules as nonradiative energy disposal paths were restricted by the increasing local viscosity accompanying the polymerization processes. New in these studies was the observation that intermolecular quenching, impeded by increasing polymer concentration, allowed the recovery of normal fluorescence to be correlated with a sample's increasing viscosity. The data collected at short times (<5 min) following mixing show no evidence of a red shift, which would be expected if the system dynamics were dominated by local temperature increases due to the exothermicity of the polymerization process.

### ANTHRACENE

The intensity of anthracene fluorescence from liquid methyl methacrylate was examined over a wide range of concentrations. The fluorescence intensity measured following excitation at 350 nm was observed to be linear with anthracene concentration up to a mass fraction of $3 \times 10^{-5}\%$ anthracene in MMA. Experimental measurements were made with this concentration of anthracene in the liquid component of the cement. Fluorescence spectra taken as single scans at three different times after mixing a 2:1 powder:liquid (mass ratio) sample are shown in Figure 1. These uncorrected fluorescence spectra have a ±5% relative standard uncertainty in the maximum intensity of the individual fluorescence features. The excitation wavelength was 350 nm. In addition to the sharp features between 370 and 470 nm, which were attributed to fluorescence from isolated anthracene molecules in a polar solvent, there was a broad band centered at ≈540 nm that shifted to the blue as the cure proceeded. In considering previous studies [10] we attributed this band to an exciplex interaction between anthracene probe molecules and the dimethyltoluidine initiator in the cement. In the dilution studies of fluorescence intensity from the liquid component of the cement, the excimer intensity showed a linear dependence on anthracene concentration over the range where the intensity of anthracene monomer feature depended linearly on anthracene concentration.

As expected [2–4], when the solid and liquid cement components were mixed, the anthracene–toluidine complex fluorescence increased in intensity over time as the cure proceeded and nonfluorescence pathways for energy disposal were blocked. Although the change in peak shape made it difficult to comment on the relative fluorescence intensity from the exciplex compared to that from independent molecules, it was clear that the exciplex

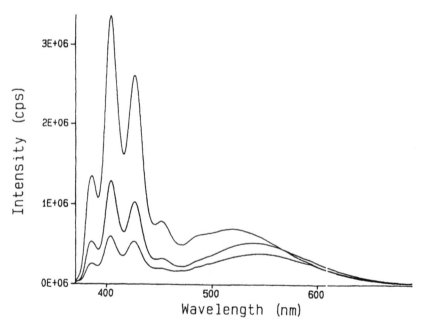

**Figure 1** Fluorescence spectra taken as single scans at 3 min (lower curve), 24 min (middle curve), and 121 min (upper curve) after mixing a 2:1 (by mass) powder:liquid sample.

intensity did not increase as substantially with the dissolution/polymerization process as the isolated anthracene fluorescence signal. These results are consistent with a picture that the diffusion of the probe and amine molecules were restricted fairly early in the cure process, thereby limiting the extent of anthracene–toluidine exciplex interaction. In contrast, the nonradiative pathways for exciplex fluorescence continued to decrease as long as the microviscosity continued to increase.

Previous studies [2] emphasized the difficulties in using absolute fluorescence intensity to determine the degree of polymerization. We avoided this problem by using the ratio of intensities of different features that were determined over time. The results from a single sample, followed over time after mixing, are shown in Figure 2. The ratio of fluorescence intensity between two monomer features at 405 and 427 nm remained almost constant over the 2 h time period during which the fluorescence was monitored. The slight decrease in the I(427)/I(405) ratio over the first 20 min of the polymerization was probably due to the wavelength and polarization-dependent transmission efficiency of the emission monochromator because the fluorescence polarization of the sample changed with the increase in viscosity accompanying the polymerization [11].

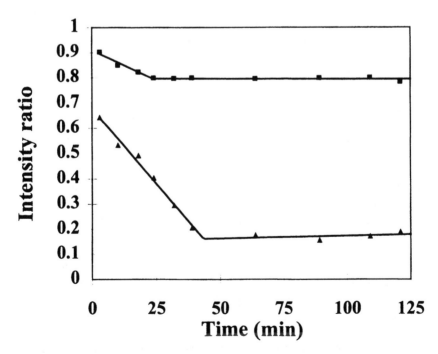

**Figure 2** Anthracene monomer, ■ I(427)/I(405) and anthracene–toluidine exciplex to monomer, ▲ I(540)/I(405) peak height ratios from a single sample, scanned repeatedly over the first 90 min of cure.

The exciplex intensity showed quite different behavior as the setting proceeded. A comparison of the (monomer peak/monomer peak) ratio to the (exciplex peak/monomer peak) ratio was quite illuminating. We considered the initial maximum wavelength of the exciplex emission at 540 nm, and compared its intensity to the monomer intensity at 405 nm as the dissolution/polymerization proceeded. A substantial decrease in exciplex intensity, compared to monomer intensity, was observed over the first 40 min of the cure. The ratio then leveled off, indicating that the local viscosity had reached a maximum after 40 min and that the dissolution/polymerization was considered to have reached completion at the ambient temperature of the laboratory. Since the working time for the cement was considerably less than the 40-min time period over which the exciplex/monomer intensity ratio was steadily decreasing, the intensity ratios served *as in situ* monitors of the cure.

## RE-COMPLEX

Compounds with the general formula $Re(CO)_3ClL$ are unusual among organometallic compounds in that they photoemit faster than they dissociate under illumination in most solvents over a broad range of temperatures [1]. These compounds also demonstrate a substantial shift in both the wavelength and the intensity of spectral emission in response to microviscosity changes in the solvent [1]. These characteristics made $Re(CO)_3ClL$ complexes particularly attractive as probes to monitor the microviscosity changes that accompany cement cure.

In this study, the complex where L = 4,7-diphenyl-1,10-phenanthroline was used. A solution in which the mass fraction Re-complex in methyl methacrylate was 0.12 % gave two absorption bands at 350 and 475 nm. Both absorption bands produced an emission feature at 612 nm. No other emission features were observed in the visible part of the spectrum.

According to Wrighton and Morse [12], in $CH_2Cl_2$ the Re-L $\pi^*$ charge transfer (CT) band is reported at 26,530 cm$^{-1}$ (377 nm) and the intraligand (IL) band is reported at 34,970 cm$^{-1}$ (286 nm). The free ligand has a maximum absorption at 36,760 cm$^{-1}$ (272 nm). In general, the maximum of the lower energy absorption band in these complexes shifts to the blue in more polar solvents. Wrighton and Morse also observed that the emission maximum in $ReCl(CO)_3L$ complexes shifts to the blue upon cooling a sample in EPA to 77 K. Similar studies by Hanna et al. [13], determined that the emission maxima of $ReCl(CO)_3$-2,2'-bipyridine shifted to the blue in both MMA and PMMA solvents when the temperature was reduced from 298 to 20 K.

Our study included only observations of the behavior of the CT band and did not consider the IL band, which should appear farther to the UV than the wavelength range in which these experiments were conducted. In our fluid MMA samples, the CT fluorescence was efficiently quenched, most likely by the amine accelerator used in the bone cement. However, as dissolution and polymerization proceeded following mixing of the cement components, the viscosity increased, and the Re-complex fluorescence reappeared, shifted to the blue, and increased in intensity. Approximately 15 min after mixing, the Re-complex CT band intensity reached a maximum. Measurements up to 2 weeks after the mixing of the cement showed a constant fluorescence intensity when the cured samples were stored in the dark under ambient laboratory temperature and humidity conditions.

To verify the quenching interaction between the Re-complex and the di-methyl-*p*-toluidine, a Stern-Volmer plot of the results of a concentration dependent study of Re-complex fluorescence intensity as a function of amine concentration in fluid MMA was prepared (Figure 3). The samples contained $1.6 \times 10^{-7}$ mol Re-complex, and up to a maximum of $2.6 \times 10^{-5}$ mol of amine, in $\approx 2.5$ g of MMA. Re-complex CT band peak heights at 612 nm were measured from uncorrected fluorescence spectra taken in single scans following excitation at 350 nm. The Stern-Volmer plot is linear over the range of amine concentrations studied. A linear Stern-Volmer plot,

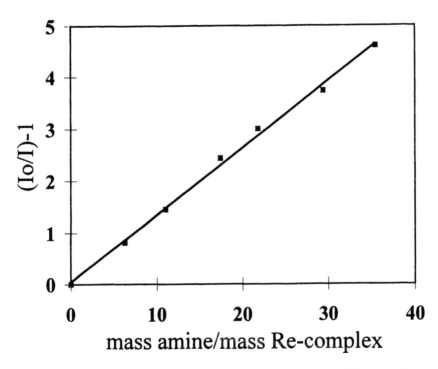

**Figure 3** Stern-Volmer plot of the results of a concentration dependent study of Re-complex fluorescence intensity as a function of amine concentration in fluid MMA.

together with the viscosity dependence of fluorescence shown below, indicates that a bimolecular quenching process is occurring. It also supports our assumption that the Re-complex probe is not being degraded by the DMPT on the time scale of interest in these experiments.

In the PMMA environment, as has been seen in previous studies [11], the Re-complex emission was considerably blue shifted. For all the cement samples, the complex was excited at 350 nm, and fluorescence intensity was monitored near 566 nm as a function of time.

The Re-complex fluorescence measured over time from two separate polymerizing samples were analyzed as $I(t)/I$(initial) at 566 nm. $I(t)$ is the intensity at 566 nm at some time $t$ after mixing of the solid and liquid components of the cement. $I$(initial) is the CT peak intensity at 566 nm, measured 2 or 3 min after mixing the components, depending on the run. Because the Re-complex intensity did not change in the first 5 min following mixing, the time of the initial reading was not closely controlled. The results from two separate runs showing the evolution of different samples over time are graphed in Figure 4. The fluorescence intensity at 566 nm showed a substantial increase over the time interval from 5 min to 11 min following mixing of the two cement components. A rapid increase in fluorescence intensity occurred during

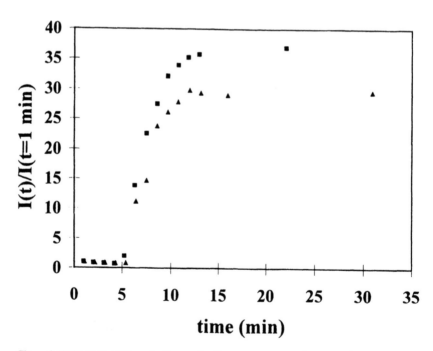

**Figure 4** $I(t)/I$(initial) at 566 nm for Re-complex fluorescence measured over time from two separate polymerizing samples.

the time interval in which the cement was expected to reach its working stage as a consequence of the impedance of fluorescence quenching due to the increasing local viscosity. This phenomenon should, therefore, provide a suitable method for in situ monitoring of the viscosity changes that occur during the dissolution/polymerization of PMMA-based cements.

## CONCLUSIONS

The results of this preliminary study have shown that anthracene and $Re(CO)_3ClL$ (L = 4,7-diphenyl-1,10-phenanthroline) could be used as in situ monitors of the microviscosity changes that occur as a bone cement sample cured. In addition, the results identified a novel technique—the impedance of quenching—for monitoring local viscosity.

### DISCLAIMER

Certain commercial materials and equipment are identified in this work for adequate definition of experimental procedures. In no instance does such identification imply recommendation or endorsement by the National Institute of Standards and Technology or that the material and the equipment identified is necessarily the best available for the purpose.

## REFERENCES

1. Kotch, T. G., Lees, A. J., Fuerniss, S. J., Papathomas, K. I., and Syder, R. *Polymer* 1992, 33, 657.
2. Wang, F. W., Lowry, R. E., and Fanconi, B. M. *Polymer* 1986, 27, 1529.
3. Loutfy, R. O. *Macromolecules* 1981, 14, 270; ibid., *J. Polym. Sci., Polym. Phys. Edn.* 1982, 20, 825.
4. Levy, R. L., and Ames. D. P. *Proc. Org. Coat. Appl. Polym. Sci.* 1991, 48, 116.
5. Wang, F. W., Lowry, R. E., and Cavanagh, R. R. *Polymer* 1985, 26, 1657.
6. Stone, J.J-S., Rand, J. A., Chiu, E. K., Grabowski, J. J., and An, K.-N. *J. Orthopaedic Res.* 1996, 14, 834.
7. Surgical Simplex P Bone Cement Monograph, Howmedica, Inc.
8. Farrar, D. F. and Rose, J. *Proceedings 24th Annual Meeting of the Soc for Biomaterials,* 1998, p. 287.
9. Salman, O. A. and Drickmar, N. G. *J. Chem. Phys.* 1982, 77, 3337.
10. Beens, H., Knibbe, H., and Weller, A. *J. Chem. Phys.* 1967, 47, 1183.
11. Lakowicz, J. R. *Principles of Fluorescence Spectroscopy,* Plenum Press, New York, 1988.
12. Wrighton, M. and Morse, D. L. *J. Am. Chem. Soc.* 1974, 96, 998.
13. Hanna, S. D., Dunn, B., and Zink, J. I. *J. Non-Cryst. Solids* 1994, 167, 239.

# CHAPTER 21

# Self-Etching, Polymerization-Initiating Primers for Dental Adhesion

CHETAN A. KHATRI[1]
JOSEPH M. ANTONUCCI[1]
GARY E. SCHUMACHER[2]

## INTRODUCTION

DENTIN is a complex heterogeneous substrate consisting mainly of water, collagen, and hydroxyapatite that varies in microstructure and composition between its enamel and pulpal boundaries [1]. Because of this complexity, the bonding of polymerizable resin-based materials such as dental composites to this substrate is not as straightforward as bonding to enamel that requires simple acid etching and, as a consequence, requires more sophisticated types of surface treatments. Effective adhesive bonds between dentin and dental restorative materials have been achieved by the sequential application of a series of solutions to the dentin surface. For example, application of a three-part adhesive system consists of (1) a conditioner or etchant such as aqueous nitric or phosphoric acid, (2) a primer such as N-phenylglycine (NPG) in acetone, and (3) an acetone solution of a multifunctional surface active monomer, such as 1,4-di[2'-(2'-methyl-2'-propenate)ethyl]phthalate-2,5-dicarboxylic acid (para-PMDM), the product from the reaction of pyromellitic dianhydride and 2-hydroxyethyl methacrylate [2]. The crystalline para product is separated by filtration from the meta isomer byproduct. Shown on Figure 1 is the synthetic scheme for the preparation of para-PMDM.

[1]Polymers Division, National Institute of Standards and Technology, Gaithersburg, MD 20899-8545, USA.
[2]American Dental Association Health Foundation, Paffenbarger Research Center, NIST, Gaithersburg, MD 20899-8545, USA.

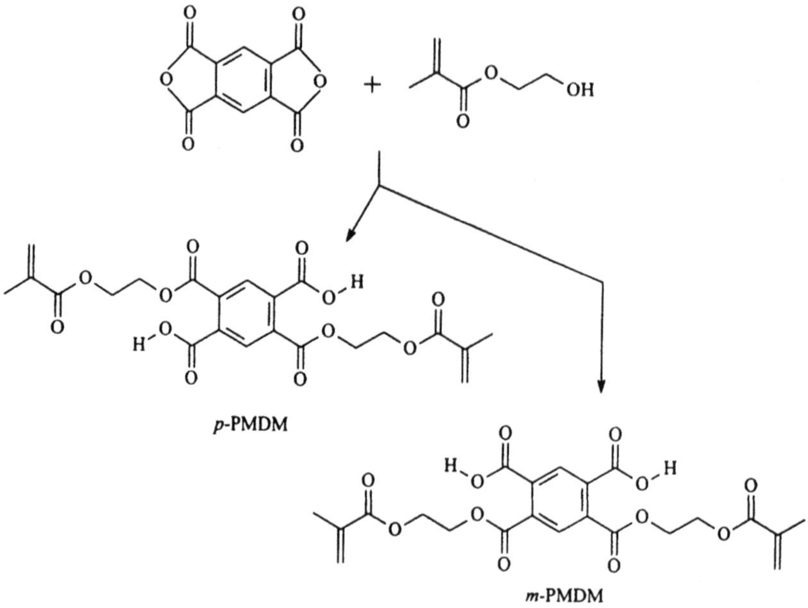

**Figure 1** Synthesis of 1,4-di[2′-(2′-methyl-2′-propenate)ethyl]phthalate-2,5-dicarboxylic acid (*p*-PMDM).

In the usual multistep adhesive protocol the role of nitric acid or similar strong acid is to act as an etchant or conditioning agent to cleanse the substrate by removing the smear layer from cut and ground dentin and also to create a microporous surface with open tubules. A surface-active primer such as NPG then diffuses into the conditioned dentinal surface where it stabilizes the demineralized collagen against collapse and facilitates diffusion of *para*-PMDM and by an acid-base reaction complexes with the carboxylic acid monomer. Previous studies indicate that such complexes are unstable and decompose into radicals, which are capable of initiating interfacial copolymerization of this carboxylic acid adhesive monomer with a bonding resin or with the resin phase of the composite as shown in Figure 2 [3,4]. NPG can act as an effective photoreductant for photosensitive oxidants such as camphorquinone to generate initiating radicals via an exciplex [3].

The use of strongly acidic conditioners such as nitric or phosphoric acid may excessively demineralize dentin and create a highly decalcified collagenous zone that is not optimal for bonding because it may not be completely infiltrated with the primer and adhesive resin. A less-aggressive decalcifying agent is the chelating acid ethylenediamine tetraacetic acid (EDTA), which is used in the form of a soluble salt because of its poor solubility. It would be desirable to have a soluble analog of EDTA that can

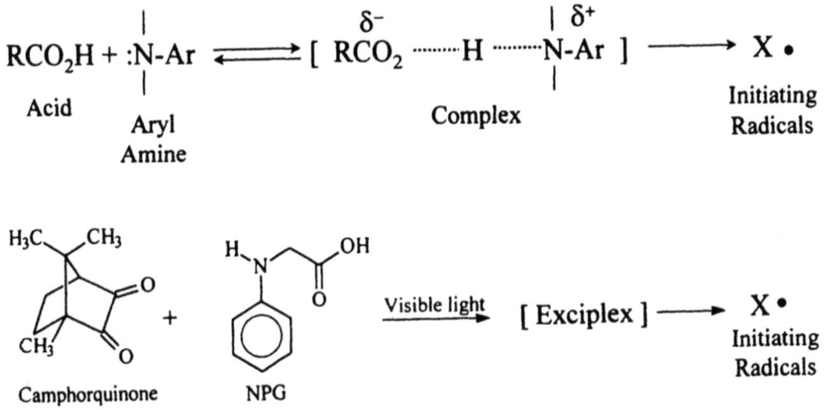

RCO₂H + :N-Ar ⇌ [ RCO₂ ········H········N-Ar ] ⟶ X •

Acid · Aryl Amine · Complex · Initiating Radicals

Camphorquinone + NPG →(Visible light)→ [ Exciplex ] ⟶ X • Initiating Radicals

**Figure 2** (a) Radical formation through complex formation between a carboxylic acid and an aryl amine such as NPG. (b) Photogeneration of initiating radicals via an exciplex formed by camphorquinone and NPG.

also act as a primer and, thereby, reduce the usual three-step bonding protocol to two steps by combining etchant and primer functions in one compound. In previous studies it was shown that N-phenyliminodiacetic acid (PIDAA) has these properties and also can self-initiate polymerization of several acrylic monomers [3,4]. PIDAA has a structure intermediate between NPG and EDTA as shown in Figure 3, and this unique structure enables it to function as a self-etching primer with the ability to initiate and co-initiate the polymerization of dental monomers.

Due to its aromatic iminodiacetic acid structure, PIDAA is not only more acidic than NPG but is also a more efficient chelator for $Ca^{+2}$. Thus it can effectively serve both as an etchant and as a primer like NPG because of its arylamine structure. To gain further insight into the mechanism of the spontaneous polymerization of acrylics caused by PIDAA, several derivatives of PIDAA with electron-withdrawing or -donating aromatic ring substituents were prepared. The rationale for this study was to ascertain, by

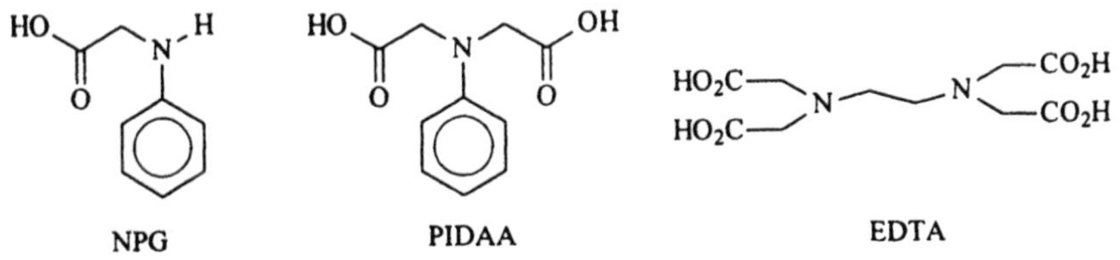

NPG · PIDAA · EDTA

**Figure 3** Chemical structures of N-phenylglycine, N-phenyliminodiacetic acid, and ethylenediaminetetraacetic acid.

substituting various groups on the aryl ring of the PIDAA molecule, the effect of changing the electron density of the "free" electrons of the nitrogen on the etchant and priming capacity of this arylimino diacid.

## EXPERIMENTAL

*Materials:* all reagents were used as received except tetrahydrofuran, which was dried over sodium/benzophenone under nitrogen and was disfilled directly into the reaction flask. The primary anilines, ortho and meta anisidine, were purified by distillation under vacuum.

*Instrumentation:* characterization of PIDAA derivatives was by NMR and FTIR spectroscop. All NMR spectra were measured on a JEOL GSX-270 instrument using DMSO-*d6* as the solvent and tetramethylsilane (TMS) as a reference at 0.00 ppm. The standard uncertainties are 0.02 ppm for $^1$H-NMR and 0.05 ppm for $^{13}$C-NMR, respectively. FTIR spectra of solids were recorded in KBr pellets on a Nicolet Magna 550 FTIR.

SYNTHETIC PROCEDURE I [5]

Scheme-1 was used for the synthesis of 3- and 4-methoxy-substituted PIDAA derivatives. Under an inert atmosphere to an oven dried flask was added 4.72 g (0.0383 mol) of the appropriate anisidine. About 50 mL of

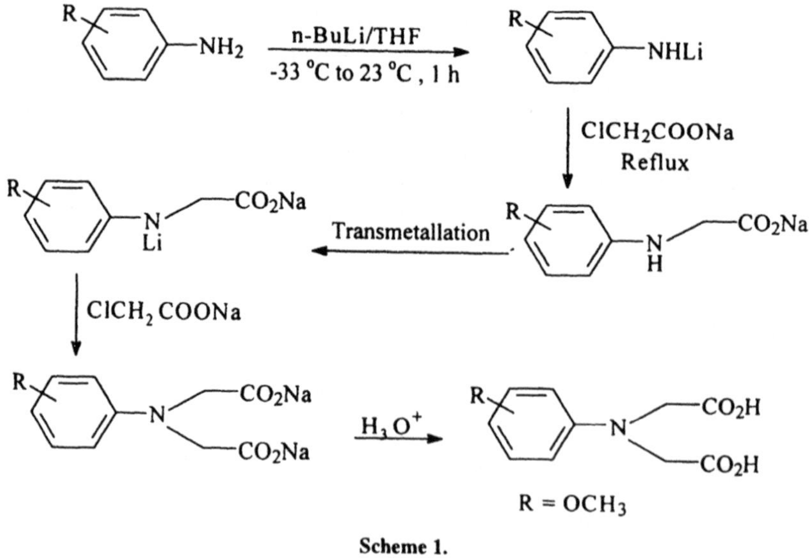

$R = OCH_3$

Scheme 1.

THF was then transferred by a vacuum transfer technique. After the solution was stirred at room temperature for 5 min and brought to − 30°C, 24.7 mL (0.0396 mol) of 1.6 mol/L *n*-butyl lithium in hexane was added with stirring. The solution was stirred at this temperature for 1 h and at 23°C for an additional 90 min; then 15.6 g (0.1339 mol) of sodium chloroacetate was added, and the mixture was refluxed for 24 h. The mixture was brought to room temperature and the solvent removed under *vacuo*. The residue was dissolved in 50 mL of water and extracted three times with 30 mL of dichloromethane. The aqueous layer was then acidified with 12 M. HCl until precipitation was observed. The flask was then warmed on water bath until a clear solution was obtained and then stored at 5°C overnight. Almost colorless crystals were obtained by filtration of the mixture. The product was dried under high vacuum and then stored in a tightly sealed vial in refrigerator. Yields were between 55% and 60% based on *n*-butyl lithium.

For 4-methoxy-PIDAA, $^1$H-NMR showed peaks at (12.44 (broad singlet), 6.77 (doublet), 6.45 (doublet), 4.05 (singlet), and 3.64 (singlet)) ppm. $^{13}$C-NMR showed peaks at 173.25, 151.88, 142.71, 115.15, 113.33, 55.86, and 53.97 ppm. For 3-methoxy-PIDAA, $^1$H-NMR showed peaks at 12.68 (broad singlet), 7.06 (triplet), 6.28 (doublet), 6.12 (doublet), 6.00 (doublet), 4.07 (singlet), and 3.67 (singlet) ppm. $^{13}$C NMR showed peaks at 172.92, 160.83, 149.73, 130.32, 105.18, 102.45, 98.73, 55.34, and 53.66 ppm.

## SYNTHETIC PROCEDURE II [6,7]

The following procedure [6] was used to synthesize 1,4-phenylenediiminotetracetic acid. 1,4-phenylenediamine (10.8 g, 0.1 mol), chloroacetic acid (37.8 g, 0.4 mol), sodium hydroxide (32.0 g, 0.8 mol), and potassium iodide (5.0 g, 0.03 mol) in 500 mL of water was refluxed for 2 h and then 40 mL of conc. HCl was added. The reaction mixture was cooled in an ice/water mixture. Slightly pink colored crystals separated out, which were vacuum filtered and dried in a vacuum oven; yield 21.5 g. For 1,4-phenylenediaminetetraacetic acid, $^1$H-NMR showed peaks at 11.99 (broad singlet), 6.42 (singlet), and 4.01 (singlet) ppm. $^{13}$C-NMR showed peaks at 173.56, 140.38, 113.49, and 54.06 ppm.

The 3- and 4-methoxy-substituted PIDAA derivatives were also synthesized by following the above procedure. The $^1$H- and $^{13}$C-NMR spectra were similar to those listed under synthetic procedure I. The 2-, 3-, and 4-carboxy-substituted PIDAA derivatives and 3-acetyl PIDAA derivatives also were synthesized by synthetic procedure II as described below [7].

To 10.3 g, 0.075 mol aminobenzoic acid neutralized with 5 mol/L sodium hydroxide (or to 3-acetyl aniline in 250 mL of water), was added sodium chloroacetate (26.2 g, 0.225 mol). The solution was refluxed, and the pH was maintained between 10 and 12 by the addition of 5 M aqueous sodium hydroxide solution. After the pH ceased to fall, the solution was refluxed for an additional 1 h and then cooled and acidified with 0.5 mol/L HCl. The crystals were vacuum filtered and dried under high vacuum. The product was recrystallized from an acetone/water mixture. The yield of these aryliminodiacetic acids reaction varied from 3.1 g to 5.3 g.

- For 4-carboxy-PIDAA, the $^1$H-NMR showed peaks at 12.55 (broad singlet), 7.71 (doublet), 6.53 (doublet), and 4.16 (singlet) ppm. $^{13}$C-NMR showed peaks at 172.13, 167.83, 151.93, 131.46, 118.98, 111.43, and 53.29 ppm.
- For 3-carboxy-PIDAA, the $^1$H-NMR showed peaks at 13.31 (broad singlet), 7.24 (singlet), 7.10 (multiplet), 6.77 (singlet), 6.69 (singlet), and 4.05 (singlet) ppm. $^{13}$C-NMR showed peaks at 173.76, 168.33, 147.94, 132.04, 129.69, 117.90, 115.82, 112.14, and 55.91 ppm.
- For 2-carboxy-PIDAA, the $^1$H-NMR showed peaks at 13.23 (broad singlet), 7.83 (doublet), 7.50 (multiplet), 7.39 (doublet), 7.17 (t), and 3.98 (singlet) ppm. $^{13}$C-NMR showed peaks at 171.52, 167.92, 149.69, 135.51, 131.71, 125.18, 124.61, 127.79, and 55.86 ppm.

When 2-anisidine, *p*-acetylaniline, 1,2-phenylenediamine, and 2-trifluro-methylaniline were used, the above procedure gave only monosubstituted or *N*-phenylglycine derivatives.

## SEM EVALUATION OF DENTIN TREATED WITH THE PIDAA DERIVATIVES

A scanning electron microscope, SEM (JEOL JSM-5300, JEOL USA, Inc., Peabody MA) was used to examine the morphology of dentin surfaces treated with solutions of various PIDAA derivatives at concentrations of 0.1 and 0.3 mol/L. The surfaces of dentin discs were treated for 60 s, placed in a vacuum desiccator (2.7 kPa) at room temperature for 24 h, gold sputter coated, and examined by SEM.

## POLYMERIZATION OF 2-HYDROXYETHYL METHACRYLATE (HEMA) AND METHYL METHACRYLATE (MMA)

To 2 g of HEMA or MMA in a vial was added a mass fraction of 0.37 PIDAA derivative. The vial is shaken to dissolve the PIDAA derivative and left standing at room temperature. Only the 3- and 4-methoxy-substituted PIDAA derivatives dissolved in HEMA and MMA. For other derivatives, a solution in acetone/water was used. Polymerization was noted visually by tilting the vial and observing the increase in viscosity until the solution gelled completely.

## RESULTS AND DISCUSSION

The ability of mineral acids such as dilute nitric and phosphoric acid to etch the dentinal surface and remove the smear layer is well known. *N*-Phenyliminodiacetic acid (PIDAA) is more acidic ($pK_1 = 2.5$) [8] than *N*-phenylglycine ($pK_a = 4.4$–$5.4$) [9] and also can chelate metal ions, including $Ca^{+2}$. Therefore, PIDAA has additional potential for modifying the smear layer created by cutting and grinding dentin in dental procedures. As expected from the structural similarity to NPG, PIDAA also has been shown to stabilize demineralized collagen and facilitate the diffusion of adhesive monomers into decalcified dentin. As in the case of NPG, PIDAA is capable of activating monomers by acid–amine complexation for radical polymerization. This self-initiation radical mechanism could either be due to intermolecular complex formation between the tertiary aryl amine and the carboxy group of the monomer or by the formation of an intramolecular zwitterionic dipolar species [4,10,11]. These potential pathways are shown in the Figure 4.

In either the intramolecular or intermolecular pathway, the ability to form such a complex could be enhanced by increasing the basicity or electron

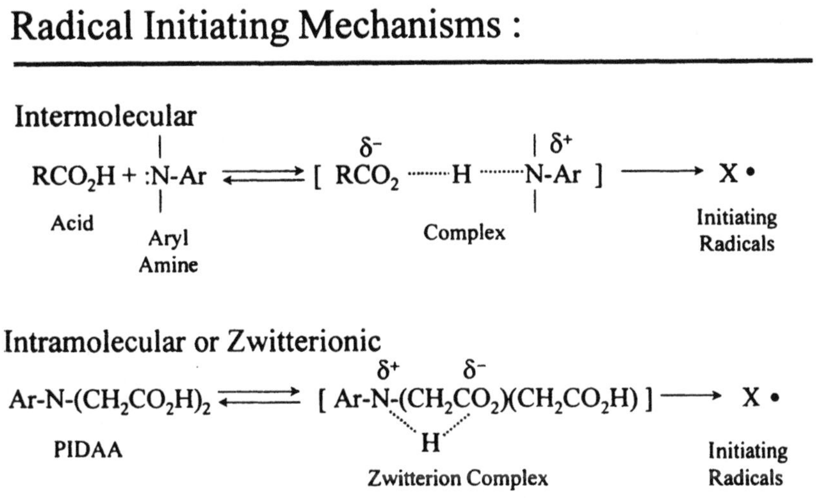

## Radical Initiating Mechanisms :

**Intermolecular**

$$RCO_2H + :N\text{-}Ar \rightleftharpoons [\ RCO_2 \overset{\delta^-}{\cdots\cdots} H \cdots\cdots \overset{\delta^+}{N}\text{-}Ar\ ] \longrightarrow X\bullet$$

Acid    Aryl Amine          Complex       Initiating Radicals

**Intramolecular or Zwitterionic**

$$Ar\text{-}N\text{-}(CH_2CO_2H)_2 \rightleftharpoons [\ Ar\text{-}N\text{-}(\overset{\delta^+}{CH_2CO_2})(\overset{\delta^-}{CH_2CO_2H})\ ] \longrightarrow X\bullet$$

PIDAA             H             Initiating Radicals

Zwitterion Complex

**Figure 4** Plausible self-initiation mechanism of acrylic monomers with PIDAA.

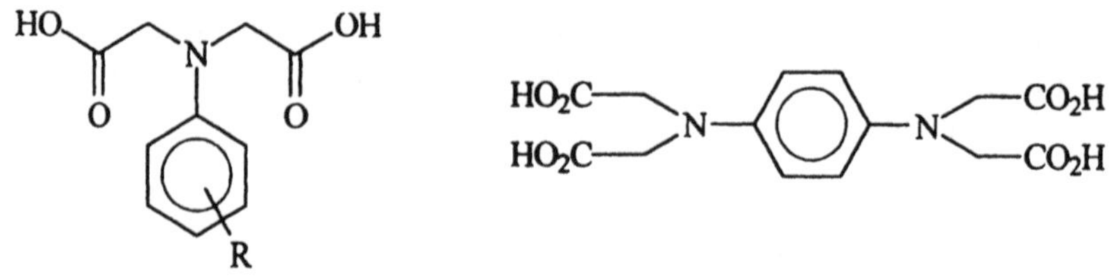

R = 2-CO$_2$H, 3-CO$_2$H, 4-CO$_2$H
3-OCH$_3$, 4-OCH$_3$, 3-COCH$_3$

1,4-phenylenediamine tetraacetic acid

**Figure 5** Chemical structures of PIDAA derivatives and 1,4-phenylenediaminetetracetic acid.

density of the nitrogen. One way to achieve this would be to substitute electron-donating groups on the aromatic ring. Accordingly, we synthesized several derivatives of PIDAA possessing either electron-withdrawing or electron-donating groups on the aromatic ring. Their structures are shown in Figure 5.

These derivatives are soluble in an acetone/water mixture with their $pK_1$ values similar to that of PIDAA. The phenylene analogs are similar to EDTA except that the two nitrogens are bridged by aromatic rings. These derivatives are soluble in acetone/water. They were characterized by measuring their $^1$H- and $^{13}$C-nuclear magnetic resonance (NMR) spectra and Fourier transform infrared (FTIR) spectra. All the PIDAA derivatives showed a peak near 53 ppm for the methylene carbons in carbon NMR spectra. The methylene carbon resonance appears around 44 ppm in the NPG derivative. Thus offers an easier way to characterize these materials. The FTNMR data are listed in Table 1 below.

TABLE 1. Proton and Carbon Chemical Shift Values
of the Methylene Groups.

| Compound | -NH-(-CH;l2CO;l2H) ppm ($^{13}$C) | -N(-CH;l2CO;l2H)$_2$ ppm ($^{13}$C) |
|---|---|---|
| PIDAA | | 52.91 |
| 2-HO$_2$C-PIDAA | | 55.86 |
| 3-HO$_2$C-PIDAA | | 53.41 |
| 4-HO$_2$C-PIDAA | | 53.30 |
| 3-H$_3$CO-PIDAA | | 53.66 |
| 4-H$_3$CO-PIDAA | | 53.97 |
| 3-H$_3$C(O)C-PIDAA | | 53.45 |
| NPG | 44.15 | |
| 2-H$_3$CO-NPG | 44.62 | |

FTIR studies on 2- and 4-carboxy-substituted NPG derivatives and for 2- and 4-carboxy-substituted PIDAA derivatives showed an absorbance at 3,460 cm$^{-1}$ for NPG derivative, which was absent in the PIDAA derivatives. Figure 6 and 7 shows these differences in 2- and 4-carboxy NPG and in 2- and 4-carboxy PIDAA.

The SEM micrographs of dentin surfaces treated with an aqueous acetone solution (mass ratio 1:1) containing 0.3 mol/L 3-methoxy PIDAA indicated significant removal of the smear layer by 3-methoxy PIDAA (3 MeOPID), similar to that achieved with PIDAA (Figure 8). Similar observations were made with the other PIDAA derivatives.

A preliminary study comparing the polymerization-initiating potential of the various PIDAA derivatives suggested the PIDAA derivatives with electron-donating groups were able to polymerize HEMA and MMA more rapidly compared with PIDAA. PIDAA alone was able to initiate the polymerization of these monomers faster compared with PIDAA derivatives with electron-withdrawing substituents. It is known that aliphatic amino acids

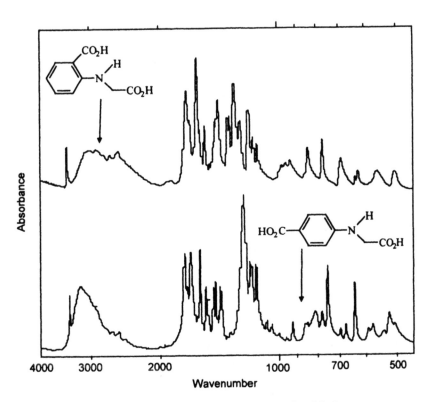

**Figure 6** FTIR spectra for 2- and 4-carboxy-*N*-phenylglycine.

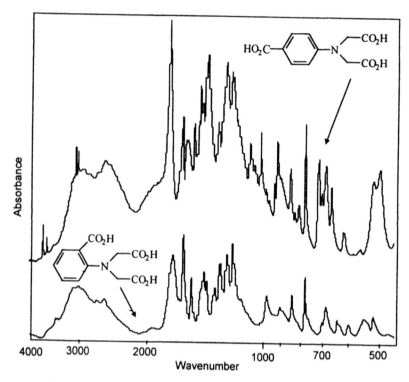

**Figure 7** FTIR spectra for 2- and 4-carboxy-$N$-phenyliminodiacetic acid.

exist in aqueous media as dipolar molecules (zwitterions). However, no conclusive evidence for the existence of dipolar species in aryliminodiacetic acids is available. Our observation that low concentrations of 4-methoxy PIDAA can polymerize MMA suggests the presence of zwitterionic species in this molecule may be the active species since the dielectric constant of MMA is not expected to promote ionization of PIDAA or its derivatives.

Preliminary bonding studies [11] to dentin treated with 3-methoxy-PIDAA or 2-carboxy-PIDAA (0.1 mol/L in aqueous acetone 1 : 1 mass) yielded composite to dentin mean shear bond strength values of 26.4 MPa ($\pm$4.5 MPa) and 21.0 MPa ($\pm$5.4 MPa), respectively, similar to the PIDAA control value of 24.2 MPa ($\pm$6.7 MPa), where $\pm$ represents standard uncertainty in these measurements.

*Disclaimer:* certain commercial materials and equipment are identified in this work for adequate definition of the experimental procedure. In no instance does such identification imply recommendation or endorsement by the National Institute of Standards and Technology or that the equipment identified is necessarily the best available for the purpose used.

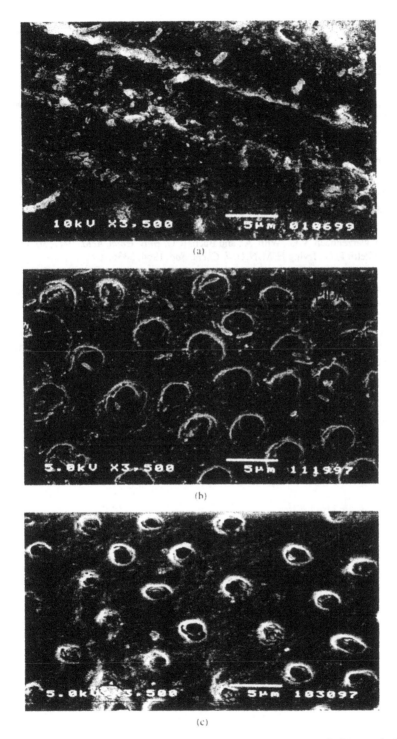

**Figure 8** SEM Photomicrographs of dentin with (a) smear layer untreated, (b) treated with PIDAA, and (c) treated with 3-methoxy-PIDAA (3 MeOPID).

299

# REFERENCES

1. Nakabayashi, N. and Pashley, D. H. *Hybridization of Dental Hard Tissues* 1998, Chapter 2, Quintessence Publishing Co., Ltd., Tokyo, Japan.
2. Bowen, R. L., Cobb, E. N., and Rapson, J. E. *Int. Dent. J.* 1987, 37, 158.
3. Code, J. E., Antonucci, J. M., Bennett, P. S., and Schumacher, G. E. *Dent. Mater.* 1997, 13, 252.
4. Schumacher, G. E., Antonucci, J. M., Bennett, P. S., and Code, J. E. *J. Dent. Res.* 1997, 76, 602.
5. Lin, S. T. and Huang, R. J. *Synthesis* 1989, 584.
6. Cox, J. R. and Smith, B. D *J. Org. Chem.* 1964, 29, 488.
7a. Schwarzenbach, V. G., Willi, A., and Bach, R. O. *Helv. Chim. Acta.* 1947, 30, 1303; **7b** Pettit L. D., Irving H. M. N. H. *J. Chem. Soc.* 1964, 5336.
8. Tichane, R. M. and Bennett, W. E. *J. Am. Chem. Soc.* 1975, 79, 1293.
9. Bryson, A., Davies, N. R., and Serjeant, E. P. *J. Am. Chem. Soc.* 1963, 85, 1933.
10. Farhani, M., Antonucci, J. M., Phinney, C. S., and Karam, L. R. *J. Appl. Polym. Sci.* 1997, 65, 561.
11. Antonucci, J. M., Khatri, C. A., Schumacher, G. E., Nikaido, T., and Code, J. E. *Proceedings of the 21st Annual Meeting of the Adhesion Society*, Savannah, GA., Feb. 22–25, 1998, 126–128.

# Bioactive Polymeric Composites Based on Hybrid Amorphous Calcium Phosphates

JOSEPH M. ANTONUCCI[1], DRAGO SKRTIC[2]
ARTHUR W. HAILER[3], EDWARD D. EANES[3]

## INTRODUCTION

$\mathbf{A}$ MORPHOUS calcium phosphate (ACP), an important precursor in the biological formation of hydroxyapatite (HAP), has recently been investigated for use as a bioactive filler in resin-based dental materials [1–3]. As its name suggests, ACP is a single phase calcium phosphate salt that lacks the long-range periodic atomic scale order of crystalline materials such as HAP [4]. This internal disorder is also reflected in particles with a spheroidal morphology whose surfaces lack flat, faceted crystalline-like features. On the other hand, ACP has a uniform composition very similar to that of a hydrated tricalcium phosphate $Ca_3(PO_4)_2 \times H_2O$ and a solution ion activity product that is constant over the pH range 7.4 to 9.2. These two features suggest that ACP has a well-defined solubility-controlling local unit. This unit, however, is much more soluble than HAP and, as a result, ACP readily converts into the latter at these pHs. The transformation can be slowed considerably by the inclusion of stabilizing ions such as pyrophosphate ($P_2O_7$). When used in its $P_2O_7$-stabilized form as a filler in polymeric composites, ACP can release supersaturating levels of $Ca^{2+}$ and $PO_4^{3-}$ ions over extended periods of time to form HAP external to the composite [1,2]. In addition, these composites can effectively remineralize *in vitro* caries-like

[1]Polymers Division, National Institute of Standards and Technology, Gaithersburg, MD 20899-8545, USA.
[2]American Dental Association Health Foundation, Paffenbarger Research Center, NIST, Gaithersburg, MD 20899-8545, USA.
[3]Craniofacial and Skeletal Diseases Branch, National Institute of Standards and Technology, Gaithersburg, MD 20899-8545, USA.

enamel lesions that were artificially induced in extracted bovine incisors [3]. ACP composites, however, lack the strength of conventional glass filled dental composites that are widely used in restorative and sealant applications.

The purpose of this study was to determine the feasibility of introducing glass-forming elements during the preparation of ACP so that the resulting hybrid fillers, e.g., silica- and zirconia-modified ACP, would have greater potential for strengthening the composite by improving interfacial interactions with the polymer phase. Specifically, the glass-forming agents tetraethoxysilane [$Si(OC_2H_5)_4$; TEOS], $Na_2SiO_3$ and zirconyl chloride [$ZrOCl_2$], were used to modify $P_2O_7$-stabilized ACP fillers. New composites containing these hybrid or modified ACP fillers were then evaluated to establish whether introduction of silica- or zirconia-ACP fillers improved their mechanical strength without compromising their remineralization potential.

**EXPERIMENTAL**

METHACRYLATE RESIN FORMULATIONS

The polymeric phases of the ACP comosites were derived from the materials shown in Table 1. The chemical structures of the matrix-forming chemicals are shown in Figure 1.

Two resins differing in degree of hydrophilicity were prepared from Bis-GMA, TEGDMA, HEMA, and ZrM. A photoinitiator system consisting of CQ and 4EDMAB was used to activate resin #1 and resin #2 (Table 2); hereafter designated as R #1 and R #2. R #1, consisting of equal mass fractions of Bis-GMA and TEGDMA, yielded the more hydrophobic composite matrix. Because of the substantial content of HEMA in R #2, the matrix derived from the polymerization of this resin was more hydrophilic than

TABLE 1. Monomers and Photoinitiator System Components Employed for Resin Formulations.

| Component | Acronym | Manufacturer |
|---|---|---|
| 2,2-bis[p-(2'-Hydroxy-3'-methacryloxypropoxy) phenyl]propane | Bis-GMA | Freeman Chemical |
| Triethylene glycol dimethacrylate | TEGDMA | Esstech |
| 2-Hydroxyethyl methacrylate | HEMA | Esstech |
| Zirconyl dimethacrylate | ZrM | Rohm Tech |
| Camphorquinone | CQ | Aldrich Chemical |
| Ethyl 4-N,N-dimethylaminobenzoate | 4EDMAB | Aldrich Chemical |

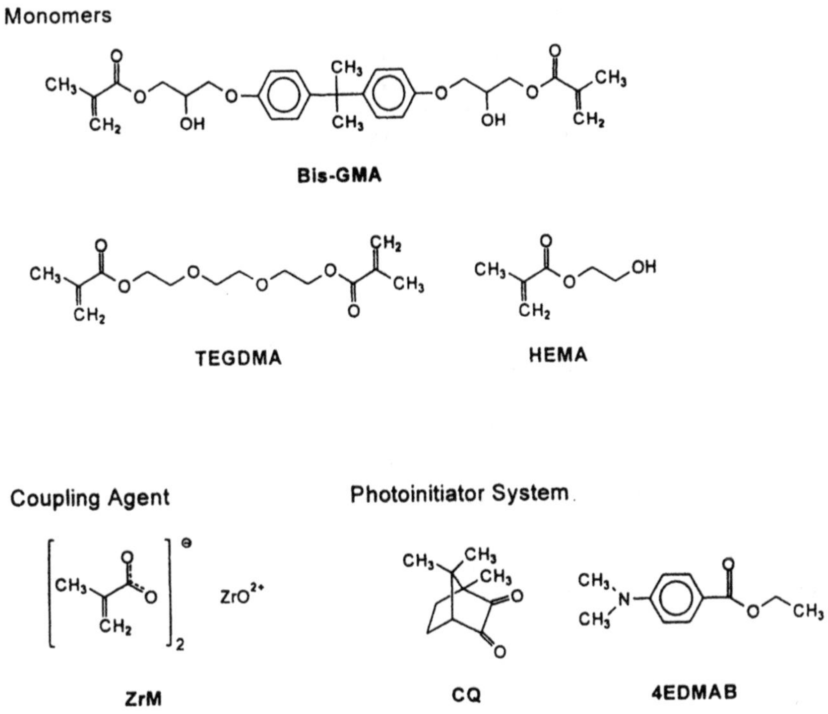

Monomers

**Bis-GMA**

TEGDMA          HEMA

Coupling Agent          Photoinitiator System

ZrM          CQ          4EDMAB

**Figure 1** Chemical structures of the matrix-forming chemicals used in the formulation of composites based on hybrid ACPs.

that derived from R #1. In addition, modest amounts of ZrM were included in R #2 as a dispersing agent for the ACP fillers.

PREPARATION AND CHARACTERIZATION OF THE FILLERS

The various types of fillers used to make the composite disk specimens are given in Table 3. ACP [5] was prepared by rapidly stirring an equal volume of an 800 mmol/L $Ca(NO_3)_2$ solution into a 536 mmol/L $Na_2HPO_4$ solution previously brought to pH 12.5 with 1 mol/L NaOH. The instantaneous precipitation was carried out in a closed system under $CO_2$-free $N_2$ at 22°C. In this way, $CO_2$ adsorption by the precipitate was minimized.

TABLE 2. Resin Composition (in Mass Fraction as Percent).

| Resin | Bis-GMA | TEGDMA | HEMA | ZrM | CQ | 4EDMAB |
|-------|---------|--------|------|-----|----|--------|
| R #1  | 49.5    | 49.5   | —    | —   | 0.2 | 0.8   |
| R #2  | 35.1    | 35.1   | 28.0 | 0.8 | 0.2 | 0.8   |

TABLE 3. Type of ACP Filler Used in the Preparation of Composite Disk Specimens.

| Filler | Stabilizer | Modifier |
|--------|------------|----------|
| Unstabilized ACP | None | None |
| $P_2O_7$-stabilized ACP | $P_2O_7$ | None |
| $ZrOCl_2$-ACP | $P_2O_7$ | $ZrOCl_2$ |
| TEOS-ACP | $P_2O_7$ | TEOS |
| $SiO_2$-ACP | $P_2O_7$ | $Na_2SiO_3$ |

After the pH stabilized at 10.5 to 11.0, which took less than 5 min, the suspension was centrifuged, the supernatant decanted, and the solid phase washed with ice-cold ammoniated water and then lyophilized. To prepare $P_2O_7$-stabilized ACP, 10.72 mmol/L of the $Na_2HPO_4$ was substituted with $Na_4P_2O_7$ before mixing with the $Ca^{2+}$ component. $ZrOCl_2$-, TEOS-, and SiO3-ACP fillers were prepared by adding, respectively, either a 0.25 mol/L $ZrOCl_2$ solution, a 5:1 volume mixture of TEOS reagent (10% TEOS, 10% ethanol, 10% tartaric acid, and 70% water; all mass fractions), or 120 mmol/L $Na_2SiO_3 9H_2O$ and 1 mol/L NaOH solution to the $P_2O_7$-containing $Na_2HPO_4$ solution simultaneously with the $Ca(NO_3)_2$ solution. $ZrOCl_2$, TEOS, or $Na_2SiO_3$ solutions were added to achieve molar $ZrOCl_2:Na_2HPO_4$, $TEOS:Na_2HPO_4$ and $Na_2SiO_3:Na_2HPO_4$ ratios of 0.1, 0.2, and 0.25, respectively.

The amorphous state of the lyophilized solids was verified by powder X-ray diffractometry (Rigaku Denki Co., Ltd., Danvers, MA), and their $Ca/PO_4$ ratios after dissolution in HCl were determined by atomic absorption (AAS, Perkin Elmer, Norwalk, CT) and UV spectrophotometric [6] (Varian Analytical Instruments, Palo Alto, CA) measurements of $Ca^{2+}$ and $PO_4$, respectively. Dissolution of the ACP fillers was studied by kinetically following the changes in $Ca^{2+}$ and $PO_4^{3-}$ concentrations in continuously stirred HEPES-buffered (pH = 7.4) solutions adjusted to 240 mOsm/kg with NaCl at 37°C. All solutions initially contained 0.8 mg/mL of the ACP filler.

## PREPARATION AND CHARACTERIZATION OF COMPOSITE DISK SPECIMENS

Composite pastes made of the different ACP fillers (mass fraction, 40%) and either R #1 or R #2 (mass fraction, 60%) were formulated by hand spatulation. In some preparations, a silanized BaO-containing glass (7724, Corning glass, mean particle size = 44 μm) was added to the $SiO_3$-ACP/R #2 formulation to produce pastes that consisted of a mass fraction of 24% $SiO_3$-ACP, 37% resin, and 38% BaO glass. The BaO glass was silanized with 0.5% 3-methacryloxypropyltrimethoxysilane (based on the mass of BaO glass according to a previous described method [7]) from cyclohexane using *n*-propylamine as a catalyst. The homogenized pastes were kept under vacuum (2.7 kPa) overnight to eliminate air entrained during mixing. The pastes were molded into disks (15.8 mm to 19.6 mm in diameter and 1.55 mm to 1.81 mm in thickness) by filling the circular openings of flat teflon molds, covering each end of the mold with a mylar film plus a glass slide, and then clamping the assembly together with a spring clip. The disks were photopolymerized by irradiating each face of the mold assembly for 2 min with visible light (Triad 2000, Dentsply International, York, PA). After postcuring at 37°C in air overnight, the intact disks were examined by XRD. Diffraction patterns of the flat surfaces of the disks were recorded at angles of $2\Theta$ between 20° and 45° on an X-ray powder diffractometer using graphite-monochromatized CuK( radiation at 40 kV and 40 mA.

## DISSOLUTION BEHAVIOR OF THE COMPOSITES

Each individual composite disk specimen was immersed in a 100 mL NaCl solution [HEPES-buffered (pH = 7.4), 240 mOsm/kg, 37°C, continuous magnetic stirring] for up to 264 h. Aliquots were taken at regular time intervals, filtered (Millex GS filter assemblies; Millipore, Bedford, MA), and the filtrates analyzed for $Ca^{2+}$ (AAS) and $PO_4$ (UV). Upon completion of the immersion tests, the disks were removed, dried, and again characterized by XRD. Variations in the total area of disk surface (A) exposed were taken into account and $Ca^{2+}$ and $PO_4^{3-}$ values normalized to an average surface area of 500 mm$^2$.

## MECHANICAL TESTING OF THE COMPOSITES

The mechanical strength of the composite disk specimens was tested, before and after immersion, under biaxial flexure conditionsn [8–10] with a universal testing machine (United Calibration Corp., Huntington Beach, CA). The biaxial flexure strength (BFS) of the specimens was calculated according to mathematical expression (1) [8–10]:

$$BFS = AL/t^2 \qquad (1)$$

where $A = -[3/4 \pi (X - Y)]$, $X = (1 + v) \ln (r_1/r_s)^2 + [(1 - v)/2] (r_1/r_s)^2$, $Y = (1 + v)1 + \ln (r_{sc}/r_s)^2]$, and where $v$ = Poisson's ratio, $r_1$ = radius of the piston applying the load at the surface of contact, $r_{sc}$ = radius of the support circle, $r_s$ = radius of disk specimen, $L$ = applied load at failure, and $t$ = thickness of disk specimen.

## RESULTS AND DISCUSSION

TEOS-modified $P_2O_7$-stabilized ACP-filled R #1 disks discharged into buffered saline solution more than three times the amount of both $Ca^{2+}$ and $PO_4^{3-}$ ions than did R #1 disks filled with unmodified $P_2O_7$-ACP (Table 4). On the other hand, ion release from $ZrOCl_2$-modified $P_2O_7$-stabilized ACP-filled R #2 disks was somewhat lower than ion release from unmodified

TABLE 4. Ion Release from the Composite Disks Specimens Made of Different Fillers and Resins.

| R #1 | | $P_2O_7$-ACP (3) | TEOS-ACP (7) | | |
|---|---|---|---|---|---|
| | Time (h) | $Ca^{2+}$ (mmol/L)[a] | | | |
| | 24 | 0.12 | 0.37 | | |
| | 72 | 0.23 | 0.68 | | |
| | 120 | 0.25 | 0.78 | | |
| | 264 | 0.27 | 0.91 | | |
| | Time (h) | $PO_4^{-3}$ (mmol/L)[a] | | | |
| | 24 | 0.08 | 0.22 | | |
| | 72 | 0.13 | 0.39 | | |
| | 120 | 0.14 | 0.42 | | |
| | 264 | 0.15 | 0.49 | | |
| R #2 | Unst. ACP (8) | $P_2O_7$-ACP (16) | $ZrOCl_2$-ACP (10) | $SiO_3$-ACP (4) | $SiO_3$-ACP + $BaOSiO_2$ (4) |
| Time (h) | | $Ca^{2+}$ (mmol/L)[a] | | | |
| 24 | 0.20 | 0.36 | 0.35 | 0.25 | 0.17 |
| 72 | 0.29 | 0.67 | 0.67 | 0.49 | 0.31 |
| 120 | 0.31 | 0.85 | 0.81 | | |
| 168 | | | | 0.66 | 0.46 |
| 264 | 0.32 | 1.02 | 0.86 | | |
| Time (h) | | $PO_4$ (mmol/L)[a] | | | |
| 24 | 0.10 | 0.21 | 0.20 | 0.14 | 0.10 |
| 72 | 0.13 | 0.40 | 0.39 | 0.28 | 0.19 |
| 120 | 0.14 | 0.50 | 0.47 | | |
| 168 | | | | 0.37 | 0.27 |
| 264 | 0.14 | 0.61 | 0.56 | | |

[a]Concentration is expressed as mean value for the number of individual runs (indicated in parenthesis next to each filler). Standard uncertainties were: <0.06 mmol/L ($Ca^{2+}$) and <0.03 mmol/L ($PO_4$) for $P_2O_7$-ACP/R #1; <0.08 mmol/L ($Ca^{2+}$) and <0.05 mmol/L ($PO_4$) for TEOS-ACP/R #1; <0.04 mmol/L ($Ca^{2+}$) and <0.03 mmol/L ($PO_4$) for unstabilized ACP/R #2; <0.11 mmol/L ($Ca^{2+}$) and <0.06 mmol/L ($PO_4$) for $P_2O_7$-ACP/R #2; < 0.08 mmol/L ($Ca^{2+}$) and <0.06 mmol/L ($PO_4$) for $ZrOCl_2$-ACP/R #2; <0.02 mmol/L ($Ca^{2+}$); <0.01 mmol/L ($PO_4$) for $SiO_3$-ACP/R #2; <0.03 mmol/L ($Ca^{2+}$) and <0.02 mmol/L ($PO_4$) for $SiO_3$-ACP/R #2 + BaO·$SiO_2$ glass.

$P_2O_7$-stabilized ACP-filled R #2 disks, particularly after time intervals >120 h (Table 4). However, these ion releases were still more than double the release from unmodified ACP composite disks specimens in which unstabilized ACP was used as the inorganic filler. Moreover, $ZrOCl_2$-modified $P_2O_7$-stabilized ACP-filled R #2 composite disks reimmersed in fresh saline after an initial 400 h of soaking released almost double the concentration of both mineral ions compared to identically treated unmodified $P_2O_7$-stabilized ACP-filled R #2 disks. The $Na_2SiO_3$-modified $P_2O_7$-stabilized ACP-filled R #2 disks released even lower amounts of $Ca^{2+}$ and $PO_4$ ions than did the $ZrOCl_2$-modified disks. The addition of silanized BaO glass suppressed the ion release even further, although the release was still better than that for unstabilized-ACP filled R #2 composite disks.

All ACP-filled composites, both cured and uncured, were stable when kept dry over $CaSO_4$ in a desiccator. Their XRD patterns (not shown) showed no signs of internal conversion of ACP into HAP. Upon immersion in saline buffer, conversion to HAP occurred more slowly when the modified ACPs were utilized in the polymerized composites. Rapid internal conversion into HAP is not desirable, as the latter phase is not sufficiently soluble to be effective as a remineralizing agent [3]. Additional evidence of the slower internal conversion of $ZrOCl_2$- and TEOS-modified $P_2O_7$-stabilized ACP fillers into HAP was obtained from a series of dissolution experiments with the ACP fillers alone in buffered NaCl solutions. It was found that solution $Ca/PO_4$ molar ratios for suspensions of both modified fillers were much higher (2.05 and 1.75 for $ZrOCl_2$ and TEOS, respectively) than for suspensions of $P_2O_7$-stabilized (0.91) or unstabilized (0.55) ACP filler and remained practically constant for 48 h. The drop in the solution $Ca/PO_4$ molar ratios of the unmodified ACP suspensions is ascribed to the formation of HAP with a $Ca/PO_4$ molar ratio greater than that of ACP [4]. It is, therefore, concluded that both $ZrOCl_2$ and TEOS extended the stability of the ACP filler reservoir in the polymer matrix by effectively inhibiting its crystallization into HAP.

However, preliminary studies indicate that solution dynamics may also have an important role in establishing the stability of the ACP component in composites exposed to buffered saline solutions. In these studies, $P_2O_7$-stabilized R #1 and #2 composites were molded and photopolymerized in situ in 20 mm long × 7 mm wide × 1 mm deep rectangular depressions in flat Teflon plates. When the plates containing the composites were immersed in 40 mL saline solutions for up to 330 h, we found that the ACP in the composite layer adjacent to the back surface of the mold converted more rapidly to HAP than did the ACP in the layer that was in direct contact with the bulk solution (Figure 2). This more rapid conversion suggests that microseepage created a stagnant solution layer between the backside of the composite and the mold. Because ions released from the ACP

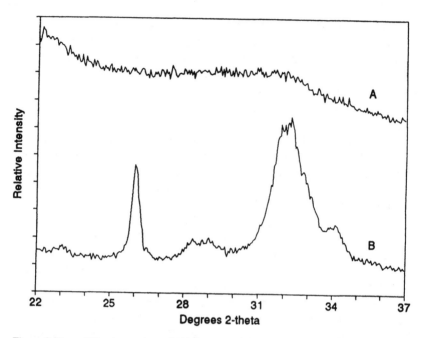

**Figure 2** X-ray diffraction patterns of (A) the exposed front surface and (B) the shielded back surface of a $P_2O_7$-ACP/R #2 composite slab molded in a Teflon holder and suspended in a HEPES-buffered saline solution for 167 h at 37°C. The relative intensity *(RI)* values for (A) ranged from 230 to 570 counts/sec and for (B) from 90 to 800 counts/sec. The standard uncertainty of these values is $1/(3RI)^{1/2}$ where 3 is the measurement time constant (in seconds).

could not readily diffuse into the bulk solution from the backside, the layer, once formed, rapidly equilibrated with the ACP. The resulting supersaturation with respect to HAP thus became much higher than in the solution layer adjacent to the front surface of the composite through which the ions could readily diffuse from the ACP into the bulk solution. The stronger thermodynamic driving force resulting from the higher supersaturation would, in turn, favor a more rapid conversion to HAP on the backside of the composite. This finding suggests that the fluid phase inside carious lesions coated with ACP composites may achieve rapid equilibrium with the filler phase. Consequently, a more efficacious remineralization of the lesion may occur than predicted from our ion-releasing solution model.

Results of BFS measurements are summarized in Table 5. In general, composites made of different ACPs had consistently lower mechanical strength than unfilled polymer disks regardless of the resin formulation used. Surprisingly, the mechanical strengths of the dry composites were

TABLE 5. BFS[a] of Unfilled and ACP-Filled Composite Disks Before (BFS$_{bi}$) and After (BFS$_{ai}$) 264 h Immersion in Buffered Saline Solutions.

| Resin | Filler | BFS$_{bi}$ (MPa)[a] | BFS$_{ai}$ (MPa)[a] |
|-------|--------|---------------------|---------------------|
| R #1 | None | 140 ± 16 (7) | 128 ± 25 (18) |
| | $P_2O_7$-ACP | 54 ± 9 (9) | 62 ± 12 (13) |
| | TEOS-ACP | 72 ± 17 (9) | 75 ± 15 (7) |
| R #2 | None | 107 ± 22 (21) | 100 ± 15 (7) |
| | $P_2O_7$-ACP | 55 ± 19 (11) | 51 ± 12 (16) |
| | $ZrOCl_2$-ACP | 69 ± 13 (12) | 65 ± 13 (10) |
| | $SiO_3$-ACP | 31 ± 14 (7) | 37 ± 8 (6) |
| | $SiO_3$-ACP + BaO•$SiO_2$ | 57 ± 23 (7) | 71 ± 15 (9) |

[a]BFS values are expressed as mean value ± standard uncertainty. ( ) Indicates number of runs in each experimental group.

not significantly affected after immersion in buffered saline for 264 h. Additionally, comparison of the BFS values of unmodified $P_2O_7$-ACP composite disks vs. $ZrOCl_2$ and TEOS modified $P_2O_7$-ACP composite disks revealed a uniform increase in the mechanical strength of the hybrid ACP composite disks. The relative increase was 25% (55 MPa to 69 MPa) and 27% (51 MPa to 65 MPa) for $ZrOCl_2$-modified ACP composites and 33% (54 MPa to 72 MPa) and 21% (62 MPa to 75 MPa) for TEOS-modified composites before and after immersion, respectively. The increases were statistically significant ($p < 0.05$, Student's t-test) except for TEOS after immersion ($p = 0.22$). The observed improvement in mechanical strength of composites based on $ZrOCl_2$ and TEOS modified ACP fillers probably resulted from better mechanical integration of such fillers with the polymerized resins. $ZrOCl_2$ and TEOS possibly changed the ACP particle morphology, intrinsic hardness, and/or surface activity in ways that permitted tighter spatial interlocking with the surrounding matrix, making the composites more resistant to crack propagation.

On the other hand, R #2 based composites containing $Na_2SiO_3$-modified, $P_2O_7$-stabilized ACP as a filler were significantly weaker mechanically than unmodified $P_2O_7$-ACP containing disks (Table 5). The relative decreases in BFS values were 44% and 27% relatively ($p < 0.05$). For reasons that are not clear, $Na_2SiO_3$, unlike $ZrOCl_2$ and TEOS appear to weaken the mechanical integration of filler with resin, making the composite less resistant to crack propagation. The observation that BaO glass reversed the negative effect of $Na_2SiO_3$ on mechanical strength suggests that this material could possibly be a useful co-filler in ACP resin composite applications. Unfortunately, the improvement in strength was offset by lower ion release (Table 4).

## CONCLUSION

The results demonstrate that it is possible to improve the mechanical properties of $P_2O_7$-ACP filled composites while retaining, if not enhancing, the high remineralization potential of these composites. These improvements were effected by introducing modifying or hybridizing agents, such as TEOS or $ZrOCl_2$, into the ACP component during synthesis. Such modified ACP fillers may be potentially useful for preparing bioactive composites suitable for more demanding restorative, sealant, and adhesive applications. Future studies will focus on the interactions of these modified ACPs with coupling agents and how these surface treatments affect the solubility, ion release, and the strength of these novel bioactive composites.

*Disclaimer:* certain commercial materials and equipment are identified in this work for adequate definition of the experimental procedures. In no instances does such identification imply recommendation or endorsement by the National Institute of Standards and Technology or that the material and the equipment identified is necessarily the best available for the purpose.

## REFERENCES

1. Antonucci, J. M., Skrtic, D., and Eanes, E. D. *Polymer Preprints* 1994, 35(2), 460.
2. Antonucci, J. M., Skrtic, D., and Eanes, E. D. *Polymer Preprints* 1995, 36(1), 779.
3. Skrtic, D., Eanes, E. D., Takagi, S., and Antonucci, J. M. *J. Dent. Res.* 1995, 74(SI), 185.
4. Eanes, E. D. In *Calcium Phosphates in Biological and Industrial Systems*, Amjad, Z., Ed. 1998, Kluwer Academic Publ., Boston, 21.
5. Eanes, E. D., Gillessen, I. H., and Posner, A. S. *Nature* 1965, 208, 365.
6. Murphy, J. and Riley, J. P. *Anal. Chim. Acta* 1962, 27, 31.
7. Antonucci, J. M., and Stansbury, J. W., and Venz, S. In *Synthesis and Properties of Polyfluorinated Prepolymer Multifunctional Urethane Methacrylate* Gebelein, C. G., and Dunn, R. L., Eds. 1990, Plenum Press, New York and London, 121.
8. Kirsten, A. F., and Woley, R. M. *J. Res. Natl. Bur. Stds.* 1967, 71C, 1.
9. Wachtman, Jr., J. B., Capps, W., and Mandel J. *J. Materials* 1972, 7,188.
10. Ban, S. and Anusavice, K. J. *J. Dent. Res.* 1990, 69, 1791.

# Index

**311**

Milton Keynes UK
Ingram Content Group UK Ltd.
UKHW020019071024
449327UK00032B/2856